AAPG REPRINT SERIES NO. 4

Carbonate Rocks I: Classifications— Dolomite—Dolomitization

Selected Papers Reprinted from

AAPG BULLETIN

Published by The American Association of Petroleum Geologists
Tulsa, Oklahoma, U.S.A., 1972

Carbonate Rocks I: Classifications—Dolomite—Dolomitization: Preface

Because of their worldwide distribution and economic importance, the diversity of environments and processes wherein and whereby they are formed, and their susceptibility to diagenetic alteration, carbonate sediments and rocks in general are the subject of more *Bulletin* pages than is any other rock type. Papers judged by the review committee to be significant enough to warrent reprinting total nearly 2,000 pages. A book of that magnitude would not conform with the concept of the Reprint Series, so the subject matter of the selected papers has been broken down into categories, and related papers have been grouped into volumes of appropriate size.

Carbonate Rocks I: Classifications—Dolomite—Dolomitization (this volume) and *Carbonate Rocks II: Porosity and Classification of Reservoir Rocks* (Reprint Series No. 5) are being published simultaneously. The remaining volumes—dealing with depositional environments (modern and ancient); reefs in general and modern reefs; and reefs in late Paleozoic, Devonian, and Silurian rocks—will follow as rapidly as practicable.

The papers included have been selected by Peggy Rice and A. A. Meyerhoff, with the aid of Jules Braunstein, L. L. Sloss, Frank E. Kottlowski, and the writer; they were assembled by Elizabeth Ross.

ROBERT H. DOTT, SR.
American Association of Petroleum Geologists
Tulsa, Oklahoma
June 5, 1972

GRAIN SIZE IN CARBONATE ROCK[1]

RONALD K. DeFORD[2]
Midland, Texas

INTRODUCTION

From June to November, 1945, W. E. Ham of Norman, Oklahoma, and the writer corresponded on the subject of grade scales for carbonate rocks. They did not reach agreement. The opinions and preferences expressed in this note are the writer's, but a large part of whatever clarity and merit they have attained is due to the discussion with Ham.

The purpose of this note is to raise the question of how best to describe the sizes of carbonate grains, not to solve the problem finally. A critical review of certain textural terms is attempted and some tentative suggestions are made.

Carbonate rocks are rocks, such as limestone and dolomite, composed largely of carbonate minerals. The grade scales for clastic rock are not suitable for carbonate rock, because the names of the scale units imply the clastic origin of the rock. Some carbonate rocks are clastic; that is, they are formed of fragments of older rocks. Even with regard to these it would be ambiguous to speak of clay grains or silt grains or sand grains. Carbonate grains could be periphrastically described, however, as "of clay size," "of silt size," "of sand dimension," and so forth.

An adequate scale for carbonate rocks would probably serve for the sulphate and chloride rocks as well, and perhaps also for chert and novaculite. Indeed the need appears to be for a scale of grain size without implications of origin.

DEFINITIONS AND CRITICISMS

Crystalline.—One aspect of crystallinity is internal molecular configuration, which differentiates crystalline solids from glasses. Payne[3] contrasts fragmental-textured and crystalline-textured rocks. The latter hyphenated adjective implies that the grains of the rock are more or less recognizable as crystals. It is hardly adequate to describe carbonate rocks composed of oölitic grains or fragments of older rocks. Some subsurface geologists in West Texas describe certain textural features of carbonate rocks other than grain size by contrasting the terms *granular* and *crystalline*. To describe a carbonate rock composed of grains of fine sand size as "finely crystalline" implies more than grain size. Therefore, *crystalline* is an unsuitable term for a carbonate grade scale.

Granular.—Because of past usages by geologists *granular* also implies more than that the rock is composed of grains. In the Wentworth scale granules range

[1] Manuscript received, August 26, 1946.

[2] Argo Oil Corporation.

[3] Thomas G. Payne, "Stratigraphical Analysis and Environmental Reconstruction," *Bull. Amer. Assoc. Petrol. Geol.*, Vol. 26, No. 11 (November, 1942), Table III, p. 1706.

in size between very coarse sand grains and pebbles. Payne used the phrase *granular crystalline* for this grade. Geologists should not use such terms as *microgranular* without defining them.

Grained.—The adjective *grained* may be used to describe any rock composed of grains. In many contexts it will be understood and can be omitted. Thus, a coarse-grained oölite can be described as a coarse oölite, and a cryptograined crystalline specimen can be called cryptocrystalline. *Grained* is an everyday English word, which descended from Latin through Old French, whence it gained its "i." The seeming miscegenation in such words as *micrograined* is not greater than is desirable in the melting pot of science.

Lithographic.—The term *lithographic* is very useful and will continue to be

TABLE I

Diameter Microns	Number of Grains	Cumulative Diameter	Percentage of Grains This Size	
11.90	8	95.20	1.68	1.9
10.20	1	10.20	0.21	
8.50	8	68.00	1.68	
7.65	2	15.30	0.42	9.8
6.80	23	156.40	4.82	
5.95	14	83.30	2.93	
5.10	34	173.40	7.13	34.4
4.25	39	165.75	8.18	
3.40	91	309.40	19.08	
2.55	78	198.90	16.35	39.2
1.70	109	185.30	22.85	
0.85	70	59.50	14.67	14.7
	477	1,520.65	100.00	100.0

1,520.65/477 = 3.19 microns = 0.0032 mm. mean diameter.

widely used. It implies, however, more of texture than grain size. It connotes smooth fracture and potential commercial use.[4] It would hardly be applied to a chalky limestone. It can be used only to describe the rock in aggregate and can not be applied to the individual grains. These considerations also suggest the occasional inadequacy of some of the terms in the standard grade scale for "fragmental-textured" rocks. Lithographic carbonate rocks are probably clastic, but who would call lithographic limestone a clay? Or a novaculite a claystone?

The mean grain diameter in typical lithographic limestone is approximately 0.003 millimeter. W. A. Waldschmidt made the following grain count (Table I) of a thin section of Solenhofen limestone; the specimen was obtained from Ward's Natural Science Establishment, Rochester, New York.

Two definitions of lithographic are proposed: (1) pertaining to the stone used

[4] P. E. Cloud, Jr., V. E. Barnes, and Josiah Bridge, "Stratigraphy of the Ellenburger Group in Central Texas—A Progress Report," *Univ. Texas Pub. 4301* (June, 1945), p. 136.

in lithography, "a compact fine-grained yellowish or grayish limestone, *lithographic limestone* or *stone*, the best variety of which is found at Solenhofen, Bavaria, in the Jurassic system";[5] (2) pertaining to compact carbonate rocks having approximately the same grain size and textural appearance as the stone used in lithography.

Sublithographic.—*Sub* and *super* go best with Latin roots; *hypo* and *hyper*, with Greek. *Subaqueous* and *hypodermic* are commonplace examples. *Sublithographic*, a Latin-Greek hybrid, is a recent coinage that has gained a certain currency in geologic literature. It belies its first impression; for it does not refer to rocks finer than and below lithographic in the grade scale. It means "somewhat coarser than lithographic." A grade scale for carbonate rocks would provide a grade name that would make this malformed pentasyllable unnecessary.

NATURAL TWO-FOLD GRADE SCALE: APHANIC AND PHANERIC

Ham pointed out that grains as small as 0.03 mm. in diameter were "clear to moderately clear" to the unaided eye with normal 20/20 vision in artificial or subdued natural light, and "clear" in bright sunlight or under a hand lens, and "very clear" under a 10-power binocular microscope. Grains between 0.03 and 0.004 mm. in diameter he described as follows.

Artificial or subdued light	Indiscernible
Bright sunlight	Faint to very faint
Hand lens	Clear to moderately clear
Binocular	Clear

Thus, Ham has established a natural division for field work. He illustrated this boundary by two specimens with grain diameters of 0.03 and 0.01 mm., respectively. The writer believes that if desirable the dividing line can be placed at either figure or in between.

The adjective *aphanitic* is used by a number of geologists to describe exceedingly fine-grained carbonate rocks. Since *aphanitic* also means "pertaining to aphanite," it is proposed that rocks above and below the natural boundary be described as phaneric and aphanic, respectively.

This proposal provides a simple two-fold grade scale that applies to grains either individually or in the aggregate. One may speak of "aphanic grains" or of "an aphanic rock."

GEOMETRIC INTERVALS

Krumbein and Pettijohn[6] "believe that a grade scale based on a fixed geometric interval and flexible enough to afford a number of relatively small sub-

---

[5] *Webster's New International Dictionary of the English Language*, 2d edition, unabridged, G. & C. Merriam Company (1934).

[6] W. C. Krumbein and F. J. Pettijohn, *Manual of Sedimentary Petrography*, D. Appleton-Century Company, New York. Chap. 4, pp. 76–90 (1938).

grades is to be preferred." The boundary values of geometric intervals form a geometric progression, such as: $a, ar, ar^2, ar^3, \ldots$. The symbols a and r stand for constants: a is any constant; r is the constant ratio or radix. If the number of terms in such a progression is n, the formula for the last term L is:
$$L = ar^{n-1}.$$
This may be written
$$\log L = \log a + (n-1) \log r.$$

In many grade scales (Wentworth's and Alling's for example) $a=1$ and the formulas become:
$$L = r^{n-1}$$
$$\log L = (n-1) \log r.$$

In Wentworth's well known grade scale for clastic rocks, the unit of measurement is one millimeter, the radix r is 2, and n varies by integral steps from minus 7 to plus 9 so as to create 17 boundary values. These define sixteen grades and fix a single boundary for two additional grades: a lower boundary for boulder size and an upper for clay size. Instead of sixteen names, however, Wentworth proposed only six for the range between 256 and 1/256 mm., but he increased these to eleven by differentiating five different grades of sand.

n	L mm.			
9	256	BOULDER		
8	128	COBBLE		
7	64			
6	32	PEBBLE		
5	16			
4	8			
3	4	GRANULE		
2	2			
		Very coarse		
1	1			
		Coarse		
0	1/2			
		Medium	SAND	
−1	1/4			
		Fine		
−2	1/8			
		Very fine		
−3	1/16			
−4	1/32			
−5	1/64	SILT		
−6	1/128			
−7	1/256			
		CLAY		

Alling[7] proposed a grade scale in which the unit of measurement is likewise one millimeter, but the radix r is 10, and n varies by integral steps from minus

[7] Harold L. Alling, "A Metric Grade Scale for Sedimentary Rocks," *Jour. Geol.*, Vol. 51, No. 4 (May–June, 1943), pp. 259–69.

3 to plus 4 so as to define seven grades (Fig. 2). Each of these major intervals he subdivided into four subgrades by allowing n to vary by quarter-integral steps: for example, $-3, -2\frac{3}{4}, -2\frac{1}{2}, -2\frac{1}{4}, -2 \cdots$.

Alling explicitly limited his proposed grade scale for use in thin-section and polished-section measurements of grain size. Obviously the average of measurements of grain diameters in a thin section is less than the average of the actual diameters of the same grains. The Wentworth scale has been used in recording

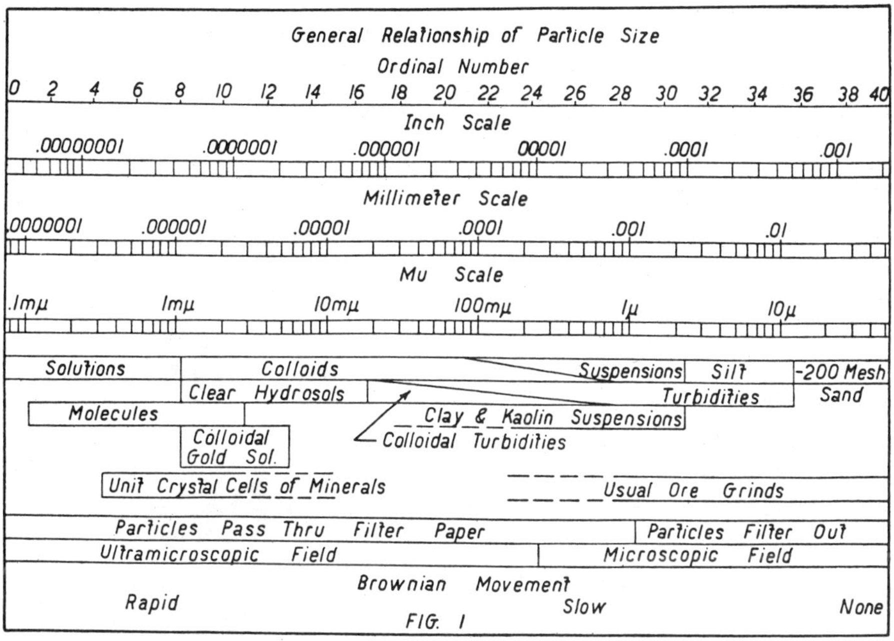

FIG. 1.—From Arthur J. Weinig, "A Functional Size-Analysis of Ore Grinds," *Colorado School Mines Quar.*, Vol. 28, No. 3 (July, 1933). The "Mu scale" is in microns. The "ordinal numbers" correspond with the exponent n in Weinig's scale: unit, one millimeter; radix, $r=\sqrt{2}$; constant, $a = 8.42937/10^8$. Reproduced by permission of author.

the diameters of three-dimensional grains. In Alling's view the use of the two different scales would help to avoid confusion. It is true that the two methods of measurement yield different means, modes, and medians of grain diameter. But there are still other methods, each yielding different means and medians. A common one is to classify the sample mechanically according to grain size, weigh the different fractions, and then to plot weight frequencies as ordinates against Wentworth grade units as abscissas. The resulting median grain size is larger than the median determined by a direct count of the frequencies of the diameters of three dimensional grains. Must a third grade scale therefore be invented? And a fourth and a fifth? The use of different scales sheds no light on the relation

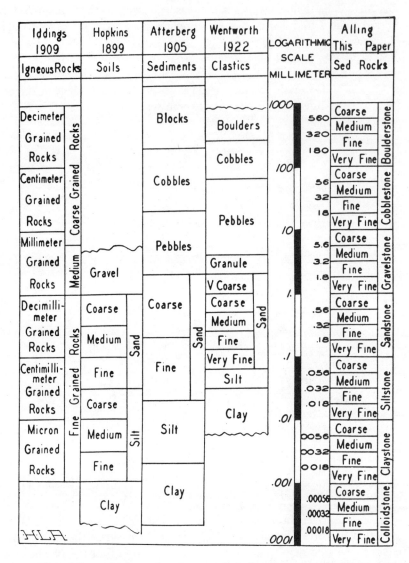

Fig. 2.—Plot of five scales for various purposes including the proposed scale for sedimentary rocks. Logarithmic scale, in millimeters
  Iddings, J. P. 1909. *Igneous Rocks*, Vol. I, p. 192. New York: John Wiley & Sons.
  Hopkins, C. G. 1899. *U. S. Dept. Agric., Dept. Chem. Bull. 56*, pp. 64–66.
  Grabau, A. W. 1913. *Principles of Stratigraphy*, p. 287. New York: A. G. Seiler & Co.
  Wentworth, C. K. 1922. *Jour. Geol.*, Vol. XXX, p. 384.
  Alling, Harold L. 1942. *Jour. Geol.*, Vol. LI, p. 266.
From *Jour. Geol.*, Vol. 51, No. 4 (May–June, 1943), p. 266.
Reproduced by permission of Harold L. Alling and the University of Chicago Press.

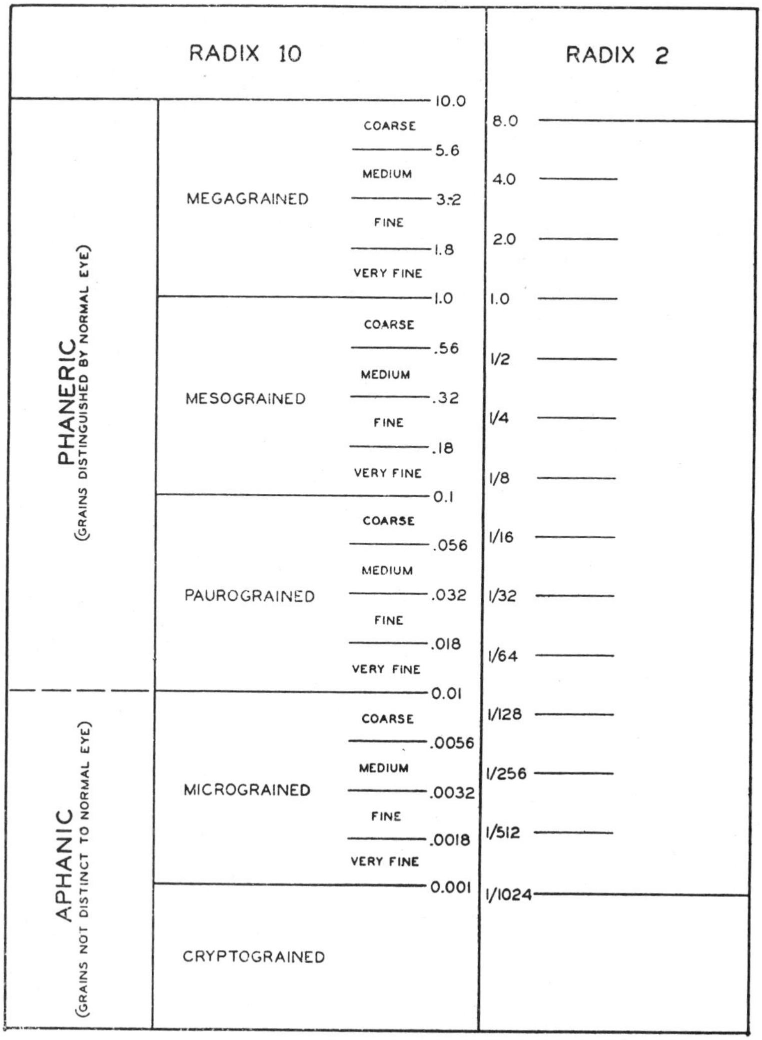

FIG. 3.—Tentative grade scale for carbonate rocks, grain diameters in millimeters, plotted logarithmically. Scale based on radix 2 (used by Wentworth) added for comparison.

between the results of the different methods. It appears that the simplest procedure is always to state precisely the kind of measurement made.

Alling's scale, however, has an orderly simplicity and appears to be mnemonically superior to Wentworth's. It is surprising that quantitative scales are still so little used in routine commercial work. They should be advocated. Alling's scale would be about as easy to introduce at this stage as Wentworth's, for clastic as well as for carbonate rocks.

GRADE SCALE FOR CARBONATE ROCKS

In establishing a scale of grain size for carbonate rocks the problem is threefold: (1) to choose a "number of relatively small subgrades"; (2) to name these; (3) to gain general acceptance and use of the proposed scale.

The problem can not be solved here and now. It is hoped, however, that the suggestions made will aid in the solution. Aspects *1* and *3* will at first prove somewhat contradictory. Many a geologist who now accepts the eleven divisions of the Wentworth scale as a matter of course (although he may not use it) will object to five major divisions in a scale of carbonate grain scale as "too complicated" for practical use. For his use the two-fold division into aphanic and phaneric is recommended, and he is requested to recognize the need that others may have for further subdivision.

The question of naming is even more delicate. Some of the names here suggested probably leave something to be desired. If the Alling scale (Fig. 2) is preferred, an attempt should be made at improving the names. If there is opposition to both scale and names, a new start should be made from the beginning. Figures 1 and 2 should aid in the endeavor.

The prefix *pauro-* means "small"; the *aur* is pronounced[8] like the *or* in orb. If no term were stated, *mesograined* would be understood: thus a "fine crystalline limestone" would be understood to have grain diameters between 0.18 and 0.32 mm. *Mega-* is preferred to *macro-*, because *macro-* may be too easily confused with *micro-*. One may speak either of the aggregate or of individual grains: of micrograins, of a micrograined limestone, or of a limestone with both micro- and paurograins. Oölites are mesograined; a pisolith is a megagrain.[9]

[8] *Webster's, op. cit.*

[9] Ronald K. DeFord and W. A. Waldschmidt, "Oölite and Oölith," *Bull. Amer. Assoc. Petrol. Geol.*, Vol. 30, No. 9 (September, 1946), pp. 1587–88.

PRACTICAL PETROGRAPHIC CLASSIFICATION OF LIMESTONES[1]

ROBERT L. FOLK[2]

Austin, Texas

ABSTRACT

Limestones are divisible into eleven basic types, which are relatively easy to recognize both in the laboratory and in the field. These rocks are made up of three constituents: (1) allochems, evidently transported or otherwise differentiated carbonate bodies; (2) 1–4-micron microcrystalline calcite ooze matrix, and (3) coarser and clearer sparry calcite, which in most rocks forms as a simple pore-filling cement (like the calcite cement in a quartz sandstone), and only uncommonly forms by recrystallization. Only four types of allochems are volumetrically important in limestones: (a) intraclasts (reworked fragments of penecontemporaneous carbonate sediment), (b) oölites, (c) fossils, and (d) pellets (rounded aggregates of microcrystalline calcite averaging .04–.10 mm.). Allochems provide the structural framework of limestones, just as sand grains provide the structural framework of sandstones; microcrystalline calcite and sparry calcite are analogous with the clay matrix and chemical cement of sandstones.

A triangular diagram showing the relative proportions of allochems, calcite ooze matrix, and sparry calcite cement is used to define three major limestone families. Family I consists of abundant allochems cemented by sparry calcite; these are the cleanly washed limestones, analogous with well sorted, clay-free sandstones and similarly formed in loci of vigorous currents. Family II consists of variable amounts of allochems embedded in a microcrystalline ooze matrix; these are the poorly washed limestones that are analogous with clayey, poorly sorted sandstones, and form in loci of ineffective currents. Family III limestones consist almost entirely of calcite ooze, hence are analogous with terrigenous claystones.

Just as clayey versus non-clayey sandstones can be divided mineralogically into orthoquartzites, arkoses, and graywackes, similarly the first two limestone families are subdivided by considering the nature of the allochems. Family I includes respectively intrasparite, oösparite, biosparite, and pelsparite; family II includes intramicrite and oömicrite (both rare), biomicrite, and pelmicrite. Family III includes homogeneous ooze (micrite), and disturbed ooze with irregular openings filled with spar (dismicrite). Rocks made up largely of organisms in growth position are considered as a separate family IV (biolithite). Properties and mode of formation of each of these types are discussed briefly.

Content of admixed terrigenous material or dolomite is shown by additional symbols; pure dolomites are classified on allochem content and crystal size. Recrystallization in limestone is believed to be locally abundant but of over-all minor importance. Among several types of recrystallization, that in which a former microcrystalline ooze matrix recrystallizes to 5–15-micron "microspar" is considered most common.

The term "calclithite" is suggested for the terrigenous carbonate rocks, e.g., limestone conglom-

[1] Read before the Society of Economic Paleontologists and Mineralogists at St. Louis, April 3, 1957. Manuscript received, November 16, 1957; revised, August 8, 1958.

[2] Department of Geology, University of Texas.

erates or sandstones made up of material eroded from outcrops of considerably older lithified-carbonate formations exposed in an uplifted source land.

INTRODUCTION

This classification was developed by the writer in essentially its present form in 1948, and first formalized in a Ph.D. dissertation on the Beekmantown (Lower Ordovician) carbonates of central Pennsylvania, submitted to the Pennsylvania State College in 1951, P. D. Krynine, supervisor (Folk, 1952). Modifications in terminology and the role of pellets were made in 1953, and the composite names were first coined in 1955. During this time, the classification has been used in description of several thousand carbonate thin sections from many areas. Hence it has undergone an extensive period of practical testing and revision, and is now in semi-final form. Imperfections will obviously arise as further samples are described because any classification is inevitably colored deeply by the limited experience of the investigator; however, the main foundation appears to be sound.

While first working on this classification, the writer was under the inspiring guidance of P. D. Krynine, and the stimulating mental climate engendered by this association contributed materially to the development of the scheme; further discussions have been carried on fruitfully with Krynine in later years. While using the classification during several sessions of a course in carbonate petrography at The University of Texas, the writer has also benefited by discussions with graduate students, in particular Thomas W. Todd, J. Stuart Pittman, and E. Hal Bogardus. The section of recrystallization has been largely developed through vigorous arguments with Robert J. Dunham of the Shell Research and and Development Company, who succeeded in proving to this stubborn writer that recrystallization was an important factor in the lithification of carbonate rocks. Constructive criticism by Dunham, J. L. Wilson, M. W. Leighton, and L. V. Illing has aided the writer in clarifying weak points before going to print.

This classification is intended for use with marine limestones. The writer has not examined enough fresh-water limestones to know if the same principles apply to them. Peculiar carbonate rocks such as caliche, travertine, cave deposits, vein carbonates, tufa, cone-in-cone beds, or spherulitic limestones are also excluded. The writer recognizes their existence and local importance but adding pigeonholes for them in this classification would serve no particularly valuable purpose at the present time.

For generations, geologists have been accustomed to using a dozen or so igneous rock terms as routine, and more than 2,000 types of igneous rock have been individually named. The utility of igneous classification is seen in the study of ore deposits, where certain metals are associated with monzonites, others with peridotites; in ordinary mapping, where different intrusions and extrusions are identified by differences in composition; and in geotectonics, where concepts such as petrographic provinces or the "andesite line" aid philosophical speculation. Similarly, mineralogists attacked with gusto the classification of metamorphic rocks, as an indispensable tool in field mapping and as an indicator of grade and type of metamorphism and of rock genesis. Half a century, however, elapsed before mineralogists turned their attention to classification of sandstones. Although many rival sandstone classifications have by now been proposed, the battle for acceptance is not yet won and most field geologists still go on describing stratigraphic sections as "sandstone" rather than using more exact terms such as fine sandstone:subarkose, or medium sandstone:orthoquartzite. Greater use of precise sandstone classification in routine field work would aid greatly in interpretation of environment and development of sedimentary petrographic provinces, just as nomenclatural precision has aided in igneous and metamorphic studies.

Limestones, however, have remained largely on the sidelines in the controversy over rock classification. The carbonates are scarcely touched on *as rock types* in college classrooms except to admire their fossil content. Thus when most geologists get out into their working life they have an inbred defeatist complex that limestones are much too complicated to bother studying closely, and if the rock fizzes in acid that is normally the terminus of the investigation. Occasionally a brief note is made that such-and-such a limestone is a calcarenite or calcilutite or contains crinoids, but beyond that point descriptions seldom go.

It is the purpose of this paper to show that limestones are not nearly so formidable as they might at first seem. There are only eleven basic types which are relatively easy to recognize both in the laboratory and the field. Four types of transported constituents may each occur with two types of interstitial material (ooze matrix or chemical cement) entirely analogous with sandstones wherein, again, four polar assemblages of sand grains may occur either with a clay matrix or with chemical cement. In addition to these eight types, there are three types of limestone that lack transported constituents.

Before classifying anything, it is necessary to determine what constituents occur; therefore, the six chief building-blocks of limestone are discussed first. Next, the principles of the rock classification scheme are introduced, and characteristics of the eleven rock types are briefly discussed. Finally, the effect of recrystallization in limestones is summarized.

CONSTITUENTS OF SEDIMENTARY ROCKS

As Krynine (1948) pointed out, all sedimentary rocks are composed of mixtures of end-members in various proportions (Fig. 1). Before classifying limestones, then, it is essential to determine what end-members are present. The main constituents are as follows.

I. Terrigenous constituents include all materials derived from erosion of source lands outside the basin of deposition and transported as solids to the sediment. Examples: quartz sand and silt, feldspar, clay minerals, zircon. This usage coincides with Krynine's (1948) definition of "detrital"; however, the word "detrital" is used by many others in an entirely different sense to include anything abraded or transported, even shell material or oölites in a limestone, and is thus rendered somewhat ambiguous. "Clastic" is also used differently by different persons, to some meaning land-derived material, to others including also broken shell material. Consequently to avoid confusion the writer recommends the use of the relatively unequivocal word "terrigenous."

II. Allochemical constituents, or "allochems," include all materials that have formed by chemical or biochemical precipitation *within* the basin of deposition, but which are organized into discrete aggregated bodies and for the most part have suffered some transportation ("allo" is from the Greek meaning "out of the ordinary," in the sense that these are not just simple, unmodified chemical precipitates, but have a higher order of organization). Allochems are by far the domi-

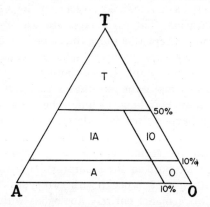

Fig. 1. Main divisions of sedimentary rocks, based on relative proportions of terrigenous (T), Allochemical (A), and Orthochemical (O) constituents. The Triangle is divided into fields corresponding with five types of sedimentary rocks: T, Terrigenous rocks (sandstones, mudrocks, conglomerates, etc.); IA, Impure Allochemical rocks (sandy oölitic limestones, silty pellet limestones, clayey fossiliferous limestones, etc.); IO, Impure Orthochemical rocks (clayey microcrystalline limestones, silty primary dolomites); A, Allochemical rocks (intraclastic, oölitic, biogenic, or pellet limestones, etc.); O, Orthochemical rocks (microcrystalline limestones, primary dolomite, halite, anhydrite, chert, etc.). Rocks in fields IA or IO may be collectively designated as Impure Chemical rocks; those in fields A or O can similarly be grouped as Pure Chemical rocks.

nant constituent of limestones, and only four types of allochems are of importance: intraclasts, oölites, fossils, and pellets.

1. *Intraclasts.*—This term is introduced to describe fragments of penecontemporaneous, usually weakly consolidated carbonate sediment that have been eroded from adjoining parts of the sea bottom and redeposited to form a new sediment (hence the term "intraclast," signifying that they have been reworked from *within* the area of deposition). Figures 5–8 illustrate typical intraclasts.

Intraclasts may be torn up from sedimentary layers almost immediately after they have been laid down, or under more severe conditions may be produced by erosion of layers that had become buried some feet below the sea floor.[3]

---

[3] The following discussion of intraclasts has been modified and considerably expanded in response to the very fine descriptive article of Beales (1958), which appeared after the present paper had been submitted.

Consequently, the sediment layers from which they are derived show a complete range of degrees of consolidation or lithification. Some intraclasts are reworked from surficial carbonate mud when that mud is still very plastic and barely cohesive; these on redeposition are usually plastically deformed and commonly have vague or mashed boundaries. Other early formed intraclasts are the Bahama "grapestone" aggregates of Illing (1954), which are clusters of pellets that have become stuck together by incipient cementation shortly after deposition; these later undergo erosion and various degrees of abrasion. In this writer's opinion, however, the most common mode of formation of intraclasts is by erosion of fragments of a widespread layer of semi-consolidated carbonate sediment, with erosion reaching to depths of a few inches up to a few feet in the bottom sediment. These fragments (which commonly show bedding) are then abraded to rounded or somewhat irregular shapes, and the abraded margin of the intraclast cuts indiscriminately across fossils, earlier intraclasts, oölites, or pellets that were contained inside the intraclast. This indicates abrasion of intraclasts that had become consolidated enough so that these included objects would wear equally with the matrix. These intraclasts could be formed either by submarine erosion (such as might be caused by storm waves or underwater slides), by mild tectonic upwarps of the sea floor, or by low tides allowing wave attack on exposed, mudcracked flats. Specifically excluded are fragments of consolidated limestone eroded from ancient limestone outcrops on an emergent land area (see later under "calclithites").

Intraclasts commonly range from very fine sand size to pebble or boulder size, as in the familiar "edgewise" limestone conglomerates. Usually they are well rounded, and the form varies from equant to highly discoidal. Less commonly they may be subangular to subround, and some may possess irregular protuberances like the grapestone of Illing (1954). Intraclasts may be composed of any type of limestone or dolomite, thus many have complex internal structure and contain fossils, oölites, quartz silt, pellets, and previously formed intraclasts; in fact, these are their most important diagnostic features. However, some are composed of homogeneous microcrystalline calcite (i.e., "lithographic" limestone) and these are difficult to differentiate from pellets if they are smaller than about 0.2 mm.

The term "intraclast" is thus used to embrace the entire spectrum of sedimented, aggregated, and then reworked particles, regardless of degree of cohesion or time gap between deposition of the original layer of sediment and later reworking of parts of it. After this paper was submitted, Beales (1958) showed that these objects were abundant in certain formations of Canada, and maintained that most of them had formed like the Bahama material, i.e., by *in situ* aggregation of pellets, somewhat analogous with the formation of lumps in a bowl of sugar. Indeed, he advocated using the term "bahamite" for the rock made of these particles. In this writer's opinion, use of the term "bahamite" implies that one knows that the aggregates formed like the grapestone of Illing (1954), hence has a very restricted genetic meaning. Usually, however, it would be almost impossible to prove that these objects formed by this exact mechanism rather than resulting from more ordinary modes of erosion, which this writer thinks of as being more common. Especially would this be difficult after abrasion has smoothed off their characteristic "grapestone" outer surface. Inasmuch as the descriptive term "intraclast" covers the whole range of particles regardless of the precise method of origin, it is certainly preferable. One can call an object an intraclast, then if detailed study shows that they have in fact formed like grapestone, it could be designated as the bahamite variety of intraclast—although, following priority, it would be both better and more specific to call it the grapestone variety of intraclast and drop the term "bahamite."

2. *Oölites.*—These particles must show either radial and/or concentric structure (Figs. 9, 10). Superficial oölites (Illing, 1954)—in which a large nucleus (which may be an intraclast, pellet, or fossil) is surrounded by a relatively thin oölite coat—are for point-count purposes considered as oölites. Spherical masses of microcrystalline calcite devoid of internal structure are usually either intraclasts or pellets. Pisolites might also be included as a form of oölite, although they are more probably to be considered as algal accretions and hence are genetically different.

3. *Fossils.*—The petrography of fossils is an enormous subject in itself and the main features of the different phyla are covered in Johnson (1951). For purposes of this classification, both sedentary and transported fossils are grouped together as allochemical constituents, except for coral or algal structures growing *in situ* and forming relatively immobile resistant masses (those are considered as a separate rock group later). It seems quite logical to class transported fossils as allochems, because they have been current-sorted and are often broken and abraded just like oölites or intraclasts. But there is some question as to the wisdom of including sedentary fossils within the same allochemical category, when there is no evidence that they have been transported or abraded. Yet it is very difficult and probably of very little significance to determine whether an entire brachiopod embedded in carbonate mud was actually living and died in that position or whether the animal was rolled over several times after death. Consequently, for the practical reason that the distinction is difficult to make and further of dubious petrologic significance, all fossils and fossil fragments are considered together as allochems. As shown later, one can, if desired, specify the sedentary nature of the fossils by using an adjectival modifier of the main rock name. Some rocks contain fossils with gobbets of carbonate sediment attached; since it is unfortunately necessary to draw lines, the writer considers these as intraclasts for counting purposes inasmuch as they at one time were part of a cohesive, deposited sediment.

4. *Pellets.*—These bodies are rounded, spherical to elliptical or ovoid aggregates of microcrystalline calcite ooze, devoid of any internal structure. In any one rock they show a surprising uniformity of shape and size (Figs. 20–22), ranging in

different specimens between .03 mm. and about .15 mm., although the most common size is .04-.08 mm. This writer follows Hatch, Rastall, and Black (1938) and considers them as probably invertebrate fecal pellets. They are distinguished from oölites by lack of radial or concentric structure, and from intraclasts by lack of internal structure, uniformity of shape, extremely good sorting, and small size. Usually one has no difficulty in identifying them with a slight amount of practice; however, the writer has seen some rocks in which he found it impossible to tell whether certain small rounded homogeneous objects were tiny intraclasts or large pellets.

It is possible that some pellet-looking objects may form by recrystallization processes, a sort of auto-agglutination of once-homogeneous calcareous mud; of such nature may be the "grumeleuse" structure of Cayeux (1935, p. 271). However, nearly all the pellets studied by the writer have been obviously current-laid grains because they are interbedded with quartz silt and usually are delicately laminated and cross-bedded. Pellets with vague boundaries are sometimes encountered; the seeming vagueness of the borders is partly an optical effect due to the small size of the near-spherical pellets and the thickness of the thin section, but in other rocks it is caused by recrystallization of pellets, matrix or both to produce microspar, which blurs the boundaries. Pellets are usually richer in organic matter than the surrounding material in the slides, thus showing as brownish objects when convergent light is used; in fact that is very helpful in recognizing them when they are embedded in a microcrystalline calcite matrix.

5. *Pseudo-allochems.*—This classification assumes that allochems, except for certain sedentary fossils, are transported constituents. This is true for the great majority of carbonate rocks; however, some limestones may contain pseudo-allochems which are objects that simulate the appearance of intraclasts, oölites, or pellets but which have formed in place by recrystallization processes. Some of the vague-looking pellets may be examples of this. Further, the writer has seen some thin sections in which it is logical to infer that oölites have grown *in situ* while remaining stationary, completely embedded in carbonate mud. Dunham (personal communication, 1955) has postulated that some intraclast-looking objects may actually be negatives from recrystallization, i.e., a once-homogeneous rock recrystallized in patches to sparry calcite, and the remnants of unaltered rock may be cut off and thus mimic intraclasts. If one were to erect a classification to include all these possibilities in separate pigeonholes, it would be far more complex than it is now. The writer feels very strongly that these pseudo-allochems are rare exceptions to the normal rule. Certainly one must be alert to catch such unusual lithologic features and they should be adequately described, but the basic classification need not be greatly expanded or distorted for the sake of such rare characteristics.

III. Orthochemical constituents or "orthochems." This term includes all essentially normal precipitates, formed within the basin of deposition or within the rock itself, and showing little or no evidence of significant transportation.

Only three orthochemical constituents require discussion at this point, microcrystalline calcite ooze, sparry calcite, and other replacement or recrystallization minerals.

1. *Microcrystalline calcite ooze.*—This type of calcite forms grains 1–4 microns in diameter, generally subtranslucent with a faint brownish cast in thin section (Figs. 17, 23, 24). In hand specimen, this is the dull and opaque ultra-fine-grained material that forms the bulk of "lithographic" limestones and the matrix of chalk, and may vary in color from white through gray, bluish and brownish gray, to nearly black. Single grains under the polarizing microscope appear to be equant and irregularly round, although electron microscope study by E. Hal Bogardus and J. Stuart Pittman at The University of Texas has shown that some microcrystalline calcite forms polyhedral blocks bounded by sub-planar (crystal?) faces much like the surfaces of novaculite-type chert (Folk and Weaver, 1952). Microcrystalline calcite ooze is considered as forming by rather rapid chemical or biochemical precipitation in sea water, settling to the bottom and at times suffering some later drifting by weak currents. This is analogous with the mode of deposition of snow which also is precipitated in a fluid medium (the atmosphere), then settles down and either lies where it falls, or may be swept into drifts. It is here considered as an orthochemical constituent because it is a normal chemical precipitate, despite the fact that it may undergo slight drifting; furthermore, some of it may form *in situ* as a diagenetic segregation or concretion. Consequently, lithographic limestone is considered an orthochemical rock (Fig. 1).

Conceivably, some 1–4-micron calcite may be "dust" produced by abrasion of shell debris, hence would not be a chemical precipitate; yet the writer thinks that this dust is quantitatively negligible and in any case it behaves hydraulically as ordinary ooze. As yet no criteria are known whereby it might be identified in thin section; therefore it is included with ordinary, chemically precipitated ooze in this classification. Microcrystalline ooze, in addition to being the chief constituent of lithographic limestone, also forms the matrix of poorly washed limestones and forms pellets, intraclasts, and some oölites.

2. *Sparry calcite cement.*—This type of calcite generally forms grains or crystals 10 microns or more in diameter, and is distinguished from microcrystalline calcite by its clarity as well as coarser crystal size (Figs. 7, 9, 11, 13). The name *spar* alludes to its relative clarity both in thin section and hand specimens, parallelling the term as used by Sander (1951, pp. 1, 3). It is difficult to draw a sharp boundary between these two types of calcite that are genetically different; the writer has vacillated at different times between grain-size boundaries of 10, 5, and finally 4 microns, but drawing the boundary strictly on grain size is not very satifactory. Clarity is certainly a distinguishing feature between the two types, but clarity in itself is partially a function of grain size and is almost impossible to define quantitatively for practical work. Morphology helps—for example, if the calcite grains encrust allochems in radial fringes, the writer terms them sparry calcite regardless of their precise crystal size—but the differentiation remains

very subjective in borderline cases which, fortunately, are uncommon.

Sparry calcite usually forms as a simple pore-filling cement, precipitated in place within the sediment just as salt crystallizes on the walls of a beaker. Grain size of the crystals of spar depends upon size of the pore space and rate of crystallization; in most limestones, the spar averages from .02 to .10 mm. although crystals of 1 mm. or more are not uncommon in limestones with large pore spaces. In some rocks, sparry calcite is not an original precipitate but has formed by recrystallization of finer carbonate grains or microcrystalline calcite.

3. *Others.*—Orthochemical constituents include not only (1) sedimented ooze and (2) directly precipitated pore-fillings, such as the two varieties of calcite discussed above, but also include minerals formed by post-depositional replacement or recrystallization. Recrystallized calcite belongs to this latter group. The mineral dolomite forms a series parallel with calcite, inasmuch as it may also occur as directly deposited (?) ooze, and directly precipitated pore-fillings; however, by far the greatest amount of dolomite occurs as an orthochemical replacement. Likewise some types of quartz and chalcedony, evaporites, pyrite, etc. may occur as orthochemical pore-fillings or as replacement minerals in some limestones.

CLASSIFICATION OF CARBONATE ROCKS

Field vs. laboratory use of classification.—Carbonate rocks are no different from igneous rocks or sandstones in that only in thin section can one fully describe, accurately classify, and interpret them. However, good approximations can be made with a binocular microscope or even in the field if the specimen has been etched in weak (10 per cent) hydrochloric acid for a few minutes. This can be performed on an outcrop by placing a drop of acid on the horizontally held surface of the specimen, letting the acid spend itself, then adding another drop in the same spot and repeating this until about 5 drops have been added. With practice, one can name a rock in the field and be correct about two-thirds of the time. In the laboratory, of course, one can submerge the specimen in weak acid for about 5 minutes or less and examine the etched surface with a binocular microscope (sawed faces are desirable but not at all necessary). Microcrystalline calcite appears dull and opaque, whereas sparry calcite is clear with a vitreous luster. With this method, one can attain about 80 per cent accuracy in classification, subject to the difficulty that it is hard to differentiate heavily abraded fossils from intraclasts, and furthermore, rocks containing abundant pellets (pelmicrite and pelsparite) almost invariably look like pure microcrystalline ooze rocks (micrites). Etching is superior to the thin section in that it brings out superbly the content and distribution of silt, clay, pyrite, dolomite, and authigenic silica. All limestones should be studied by etched surfaces as well as thin sections.

The use of acetate peels (Buehler, 1948) is of about the same level of accuracy as etching; it is much better for determining grain size and texture of the calcite particles, but does not reveal distribution of insoluble constituents. Once a "pilot suite" of rocks has been examined by thin section and the results correlated with etched surfaces or peels, one soon gets the "feel" of the rock suite, and interpretation of new specimens from the same suite becomes much more rapid and the number of thin sections can be cut down greatly.

Three main limestone families.—Almost all carbonate rocks contain more than one type of material; one may be a mixture of oölites, fossils, and sparry calcite cement; another may consist of quartz silt, pellets, and microcrystalline ooze partly replaced by dolomite and chert. Thus the problem of classification becomes one of systematizing these variations of composition and drawing significant quantitative limits between types.

Disregarding for a moment the content of terrigenous material or of later replacement minerals, fracture or vug fillings, it is possible to base a practical limestone classification on the relative proportions of three end-members: (1) allochems, (2) microcrystalline ooze, and (3) sparry calcite cement.

Allochems represent the framework of the rock and include the shells, oölites, carbonate pebbles, or pellets that make up the bulk of most limestones. Thus they are analogous with the quartz sand of a sandstone or the pebbles of a conglomerate. Microcrystalline ooze represents a clay-size "matrix" whose presence signifies lack of vigorous currents, just as the presence of a clay mineral matrix in a sandstone indicates poor washing. Sparry calcite cement simply fills up pore spaces in the rock where microcrystalline ooze has been washed out, just as porous, non-clayey sandstones become cemented with chemical precipitates, such as calcite or quartz cement. Thus the relative proportions of microcrystalline ooze and sparry calcite cement are an important feature of the rock, inasmuch as they show the degree of "sorting" or current strength of the environment, analogous with textural maturity in sandstones. If we plot these two constituents and the allochemical "framework" as three poles of a triangular diagram (Fig. 2), the field in which normal limestones occur is shown by the shaded area; divisions between the three major textural families of limestone are also shown on this figure. A similar field appears if one plots terrigenous rocks on a triangle with the three analogous poles of sand plus silt, clay, and orthochemical cement.

This classification is predicated on the assumption that the sparry calcite and microcrystalline calcite now visible in the rock are the original interallochem constituents—i.e., the sparry calcite has not formed by aggrading recrystallization of a fine calcite ooze, and that microcrystalline calcite has not formed by degrading recrystallization of coarser calcite. In most limestones the writer has examined, this assumption is believed to be correct and it is discussed more fully in the final section of this paper. Nevertheless, the writer agrees that recrystallization is a very important process in some limestone formations, and the classification proposed here does not apply to recrystallized rocks. However, this classification provides a necessary foundation for the study of recrystallized rocks because on original deposition these rocks all must have belonged to one of the groups here proposed.

Type I limestones (designated as Sparry Allochemical rocks) consist chiefly

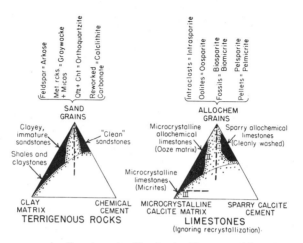

Fig. 2. Diagram comparing limestone classification in this paper with analogous classification of terrigenous rocks. Shaded areas are those parts of composition triangle which occur most commonly.

Terrigenous rocks could be classified by proportions of sand grains (structural framework fraction), clay matrix, and chemical cement, the proportions of the last two being an index to degree of sorting. Non-recrystallized limestone can be classified by the proportion of allochems (structural framework fraction), microcrystalline calcite matrix, and sparry calcite cement, the proportions of the last two also being an index of sorting.

Three basic limestone families are proposed: sparry allochemical limestone (type I), representing good sorting; microcrystalline allochemical limestone (type II), representing poorly winnowed sediments; and microcrystalline limestone (type III), analogous with claystone in terrigenous triangle. Just as one uses composition of sand grains for further classification of terrigenous rocks into arkose, graywacke, orthoquartzite, and calclithite, each ranging from clayey to non-clayey, so one uses composition of allochems for division of limestones into subvarieties such as intrasparite or biomicrite.

of allochemical constituents cemented by sparry calcite cement. These rocks are equivalent to the well sorted terrigenous conglomerates or sandstones in that solid particles (here intraclasts, oölites, fossils, or pellets) have been heaped together by currents powerful or persistent enough to winnow away any microcrystalline ooze that otherwise might have accumulated as a matrix, and the interstitial pores have been filled by directly precipitated sparry calcite cement. These sparry limestones have textures and structures similar to terrigenous rocks, e.g., cross-bedding and good grain orientation. The relative proportions of sparry calcite cement and allochems varies within rather restricted limits because of the limitations of packing.

1. There is a limit to the tightness with which allochems may be packed; thus there will always be some pore space available for cement to fill.

2. There must be a certain minimum amount of allochems present in order to support the structure—sparry calcite cement grows only in pore spaces and in general can not form a rock in its own right, unless recrystallization occurs. Similarly sandstones require a minimum amount of sand grains, on the order of 60 per cent, to support the rock structure. It may be noted that carbonate rocks on deposition may have porosity much greater than sandstones or conglomerates

of equivalent size because of the irregular shapes of fossils and some intraclasts. Coquinas like that shown in Figure 13 may have approached 80–90 per cent porosity prior to cementation with spar.

Type II limestones (designated as Microcrystalline Allochemical rocks) consist also of a considerable proportion of allochems, but here currents were not strong enough or persistent enough to winnow away the microcrystalline ooze, which remains as a matrix; sparry calcite is very subordinate or lacking simply because no pore space was available in which it could form. These rocks are equivalent texturally to the clayey sandstones or conglomerates, which also tend to have little chemical cement. In these rocks the restrictions of packing impose a certain maximum on the amount of allochems; yet there is no minimum, and Microcrystalline Allochemical rocks are found with percentages of allochems (intraclasts, oölites, fossils, or pellets) varying continuously from about 80 per cent down to almost nothing. The reason for this is that microcrystalline ooze can form a rock in its own right (comparable with claystone in the terrigenous series), and can accept any amount of allochem material that becomes mixed with it. Thus the boundary line between Microcrystalline Allochemical rocks and Microcrystalline rocks in entirely arbitrary, and has been set at 10 per cent allochems.

Type I limestones indicate strong or persistent currents and a high-energy environment, whereas type II limestones indicate weak, short-lived currents or a rapid rate of formation of microcrystalline ooze. Most limestones can be assigned readily to one or the other of these two classes because usually either sparry calcite or microcrystalline calcite is clearly dominant. In some rocks there are transitions inasmuch as washing may be incomplete and the ooze may be only partly removed. In normally calm environments with an abundance of ooze, momentary episodes of increased wave or current energy may sort laminae only a millimeter or so thick, whereas adjacent layers will be full of ooze; or a quick swash with rapid redeposition of allochems and small amounts of entrapped ooze may result in pores being partly and irregularly filled with carbonate mud (Fig. 19). Furthermore, in some pellet calcilutites the pore spaces between pellets are so tiny that the sparry calcite crystals are very minute, and can be distinguished from microcrystalline ooze only with great difficulty. All these transitional types can be designated by symbol I–II and given hybrid names (e.g., biomicrite-biosparite, or interlaminated intrasparite-intramicrite). Recrystallization of part of the ooze matrix to spar can mimic these "poorly washed" rocks and it is important to recognize these cases.

Type III limestones (the Microcrystalline rocks) represent the opposite extreme from type I, inasmuch as they consist almost entirely of microcrystalline ooze with little or no allochem material; "lithographic" limestone belongs to this class. These rocks imply both a rapid rate of precipitation of microcrystalline ooze together with lack of persistent strong currents. Texturally, they correspond with the claystones among the terrigenous rocks; they may form either in deep waters or in very shallow, sheltered areas.

Some microcrystalline rocks have been disturbed either by boring organisms or by soft-sediment deformation, and the resulting openings are filled with irregular "eyes" or stringers of sparry calcite. Other beds of microcrystalline ooze have been partially torn up by bottom currents and rapidly redeposited but without the production of distinct intraclasts. These are considered as Disturbed Microcrystalline rocks, and a special symbol and rock term ("dismicrite") is used for them (Table I).

Parts of some limestone are made up of organic structure growing *in situ* and forming a coherent, resistant mass during growth, as exemplified by parts of many bioherms (Cumings and Shrock, 1928). These rocks because of their unique mode of genesis are placed in a special class, Type IV. Formerly the writer called this type "biohermite" but that word was objectionable because it implied a mound-like form, which was a common but not universal attribute. As a substitute, Philip Braithwaite (student, University of Texas) suggested "biolithite." This is adapted from Grabau's (1913, pp. 280, 384) term "biolith," which he applied to rocks formed by organisms. This rock class is very complex and needs much subdivision itself, but no attempt to do so is made in this paper other than to suggest "algal biolithite" or "coral biolithite" as possibilites. If these organic structures are broken up and redeposited the resulting rock is considered to be made up of intraclasts or biogenic debris, and falls in type I or type II depending on the interstitial material. The name "biolithite" should be applied only to the rock made of organic structures in growth position, not to the debris broken from the bioherm and forming pocket-fillings or talus slopes associated with the reef. Study in the field is usually required to ascertain whether a specimen should be termed "biohermite."

Subdivisions of major limestone families.—After the main division of limestones into types I, II, and III—based chiefly on sorting—it is most essential to recognize whether the allochemical part consists of intraclasts, oölites, fossils, or pellets. In terrigenous sandstones, one desires to know not only whether the rock has a clay matrix or not, but what the composition of the sand is; hence geologists recognize arkoses, graywackes, and orthoquartzites, all of which types may or may not contain clay matrix (Fig. 2). It is just as important to recognize the radically different allochem types in limestones, and the scheme for classification is presented in Table I.

There would be few nomenclatural difficulties if all limestones were made up of only one allochemical constituent, such as all oölites or all intraclasts, for then there would be no need to encumber classifications with percentage boundaries. Although many limestones *are* almost pure end-members, most appear to be mixtures of several different types of allochems in varying proportions. Consequently it is not sufficient to define a pelsparite as a "rock consisting mostly of pellets" or an intrasparite as a "rock that contains abundant intraclasts." Classifications that sidestep the admittedly disagreeable problem of setting precise limits result in a triumph of vagueness and are entirely inadequate for quantitative work.

TABLE I. Classification of Carbonate Rocks
(see Notes 1 to 6)

Volumetric Allochem Composition	Volume Ratio of Fossils to Pellets		Limestones, Partly Dolomitized Limestones, and Primary Dolomites				Replacement Dolomites[7]
			>10% Allochems Allochemical Rocks (I and II)		<10% Allochems Microcrystalline Rocks (III)	Undisturbed Bioherm Rocks (IV)	
			Sparry Calcite Cement >Microcrystalline Ooze Matrix	Microcrystalline Ooze Matrix >Sparry Calcite Cement	1–10% Allochems	<1% Allochems	
			Sparry Allochemical Rocks (I)	Microcrystalline Allochemical Rocks (II)			Allochem Ghosts	No Allochem Ghosts
>25% Intraclasts		(i)	Intrasparrudite (Ii:Lr) Intrasparite (Ii:La)	Intramicrudite* (IIi:Lr) Intramicrite* (IIi:La)	Intraclasts: Intraclast-bearing Micrite* (IIIi:Lr or La)	Micrite (IIIm:L); if disturbed, Dismicrite (IIIm:X:L); if primary dolomite, Dolomicrite (IIIm:D)	Biolithite (IV:L)	Finely Crystalline Intraclastic Dolomite (Vi:D3) etc.	Medium Crystalline Dolomite (V:D4)
<25% Intraclasts	>25% Oölites	(o)	Oösparrudite (Io:Lr) Oösparite (Io:La)	Oömicrudite* (IIo:Lr) Oömicrite* (IIo:La)	Oölites: Oölite-bearing Micrite* (IIIo:Lr or La)			Coarsely Crystalline Oölitic Dolomite (Vo:D5) etc.	Finely Crystalline Dolomite (V:D3)
	<25% Oölites	>3:1 (b)	Biosparrudite (Ib:Lr) Biosparite (Ib:La)	Biomicrudite (IIb:Lr) Biomicrite (IIb:La)	Fossils: Fossiliferous Micrite (IIIb:Lr, La, or Ll)			Aphanocrystalline Biogenic Dolomite (Vb:Dl) etc.	
		3:1–1:3 (bp)	Biopelsparite (Ibp:La)	Biopelmicrite (IIbp:La)					
		<1:3 (p)	Pelsparite (Ip:La)	Pelmicrite (IIp:La)	Pellets: Pelletiferous Micrite (IIIp:La)			Very Finely Crystalline Pellet Dolomite (Vp:D2) etc.	etc.
			Sparry Allochem	Most Abundant Allochem				Evident Allochem	

NOTES TO TABLE I

* Designates rare rock types.

[1] Names and symbols in the body of the table refer to limestones. If the rock contains more than 10 per cent replacement dolomite, prefix the term "dolomitized" to the rock name, and use DIi or DILa for the symbol (e.g., dolomitized intrasparite, Li:DLa). If the rock contains more than 10 per cent dolomite of uncertain origin, prefix the term "dolomitic" to the rock name, and use DLr or DLa for the symbol (e.g., dolomitic pelsparite, Ip:dLa). If the rock consists of primary (directly deposited) dolomite, prefix the term "primary dolomite" to the rock name, and use Dr or Da for the symbol (e.g., primary dolomite intramicrite, IIi:Da). Instead of "primary dolomite micrite" (IIIm:D) the term "dolomicrite" may be used.

[2] Upper name in each box refers to calcirudites (median allochem size larger than 1.0 mm.); and lower name refers to all rocks with median allochem size smaller than 1.0 mm. Grain size and quantity of ooze matrix, cements or terrigenous grains are ignored.

[3] If the rock contains more than 10 per cent terrigenous material, prefix "sandy," "silty," or "clayey" to the rock name, and "Ts," "Tz," or "Tc" to the symbol depending on which is dominant (e.g., sandy biosparite, TsIb:La, or silty dolomitized pelmicrite, TzIIp:DLa). Glauconite, collophane, chert, pyrite, or other modifiers may also be prefixed.

[4] If the rock contains other allochems in significant quantities that are not mentioned in the main rock name, these should be prefixed as qualifiers preceding the main rock name (e.g., fossiliferous intrasparite, oölitic pelmicrite, pelletiferous oösparite, or intraclastic biomicrudite). This can be shown symbolically as Ii(b), Io(p), IIb(f), respectively.

[5] If the fossils are of rather uniform type or one type is dominant, this fact should be shown in the rock name (e.g., pelecypod biosparrudite, crinoid biomicrite).

[6] If the rock was originally microcrystalline and can be shown to have recrystallized to microspar (5–15 micron, clear calcite) the terms "microsparite," "biomicrosparite," etc. can be used instead of "micrite" or "biomicrite."

[7] Specify crystal size as shown in the examples.

Of all the allochemical particles, intraclasts are regarded as the most important because of their implication of lowered wave base or possible tectonic uplift. Therefore in this classification a rock is called an intraclastic rock if the *allochems* consist of more than 25 per cent intraclasts by volume (Fig. 3) even if it contains 70 per cent fossils, pellets, or oölites. If the rock has less than 25 per cent intraclasts, next determine the proportion of oölites; if the rock contains more than 25 per cent oölites, it is here called an oölitic rock. If the rock has less than 25 per cent intraclasts and less than 25 per cent oölites, then it consists largely of either fos-

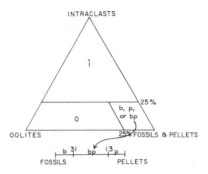

Fig. 3.—Triangular diagram to show method of classifying limestones based on volumetric allochem proportion. If allochems consist of more than 25 per cent by volume of intraclasts, the rock is Intraclastic (i) limestone. If there are less than 25 per cent intraclasts, then determine proportion of oölites; if allochems consist of more than 25 per cent oölites, rock is classified as Oölitic (o) limestone. If rock fits none of these categories, it consists either of pellets or of fossils, and linear scale below triangle is used to name them. If fossil:pellet ratio is more than 3:1, rock is Biogenic (b) limestone, and if this ratio is less than 1:3 it is Pellet (p) limestone. Intermediate specimens with subequal pellets and fossils can be termed Biogenic Pellet limestones (bp).

sils or pellets. If the volume ratio of fossils to pellets is greater than 3:1, it is a biogenic rock; if the ratio is less than 1:3, it is a pellet rock; and if the fossil:pellet ratio is between 3:1 and 1:3, it may be called a biogenic pellet rock (Fig. 3). Whether a rock is intraclastic, oölitic, biogenic, or pelletic is indicated by adding "i", "o", "b", or "p" to the symbol I or II or III, as in Sparry Intraclastic rocks (Ii), or Microcrystalline Biogenic rocks (IIb). The limits proposed here are unavoidably determined by the writer's experience and the rocks with which he has personally worked. There is no particularly valid reason for choosing a boundary at 25 per cent intraclasts as opposed to 20 per cent or 33.3 per cent. However, the writer feels that the limits proposed herein are close to the "natural" breaks, assuming there are such, and there is no particular advantage in delaying a classification for 20 years while waiting for enough data to fix divisions more precisely.

So far, gross texture (whether Sparry Allochemical, Microcrystalline Allochemical, or Microcrystalline) and composition of allochems (whether intraclasts, oölites, fossils, or pellets) have been included in the classification, but nothing has been said about grain size of the allochems. If the allochems average coarser than 1 mm., the rock is a calcirudite (or dolorudite); if they lie between

.0625 and 1 mm., the rock is a calcarenite or dolarenite; if under .0625 mm., calcilutite or dololutite. Although the traditional boundary between "rudite" and "arenite" has been 2.00 mm. (Grabau, 1913; Wentworth, 1922; Udden, 1914) the writer prefers to use 1.0 mm. because this appears to be a more meaningful bound-

	Transported Constituents	Authigenic Constituents	
64 mm	Very coarse calcirudite	Extremely coarsely crystalline	
16 mm	Coarse calcirudite		
4 mm	Medium calcirudite		4 mm
1 mm	Fine calcirudite	Very coarsely crystalline	1 mm
0.5 mm	Coarse calcarenite	Coarsely crystalline	
0.25 mm	Medium calcarenite		0.25 mm
0.125 mm	Fine calcarenite	Medium crystalline	
0.062 mm	Very fine calcarenite		0.062 mm
0.031 mm	Coarse calcilutite	Finely crystalline	
0.016 mm	Medium calcilutite		0.016 mm
0.008 mm	Fine calcilutite	Very finely crystalline	
0.004 mm	Very fine calcilutite		0.004 mm
		Aphanocrystalline	

TABLE II. GRAIN-SIZE SCALE FOR CARBONATE ROCKS

Carbonate rocks contain both physically transported particles (oölites, intraclasts, fossils, and pellets) and chemically precipitated minerals (either as pore-filling cement, primary ooze, or as products of recrystallization and replacement). Therefore, the size scale must be a double one, so that one can distinguish which constituent is being considered (e.g., coarse calcirudites may be cemented with very finely crystalline dolomite, and fine calcarenites may be cemented with coarsely crystalline calcite). The size scale for transported constituents uses the terms of Grabau but retains the finer divisions of Wentworth except in the calcirudite range; for dolomites of obviously allochemical origin, the terms "dolorudite," "dolarenite," and "dololutite" are substituted for those shown. The most common crystal size for dolomite appears to be between .062 and .25 mm. and for this reason that interval was chosen as the "medium crystalline" class.

ary in the size distribution not only of sandstones but also of limestones. For detailed grain size terminology in carbonates, Table II is suggested. In determining the grain-size name, only the size of the allochems is considered; percentage or crystal size of microcrystalline ooze or sparry calcite and grain size of terrigenous material is ignored. If this definition is strictly applied, a rock consisting of 20 per cent brachiopod shells embedded in microcrystalline ooze and quartz sand should be classified as a calcirudite, just as much as a rock consisting of 80 per cent limestone pebbles cemented by sparry calcite. Thus the word "calcirudite" is intended to be used strictly in a descriptive sense. Genetically there would be a vast difference between the two rocks just cited. For this reason the writer seldom in practice uses the term "rudite" to describe an assemblage of large fossils in an ooze matrix, because this usage imparts a rather misleading idea as to the formation of the rock.

In theory, the three-fold size classification of calcirudite, calcarenite, and calcilutite just given is valid; but in practice, rocks with allochems averaging in the calcilutite range are very rare. The only allochemical rock types with representatives in this size class are pellet rocks or some rare biogenic rocks (those made of tiny foraminifera, for example) and in both of these the pellets or fossil fragments average no smaller than .03–.06 mm., just barely under the limit of calcarenite; pellet rocks nearly always hover on the borderline between calcilutite and calcarenite (since the average pellet size is between .04 and .08 mm.). Setting a new rock class apart on such an artificial and insignificant boundary seems to be an unnecessary complication; hence the writer has lumped these rare allochem calcilutites together with the calcarenites in the classification scheme. The only common true calcilutites are the "pure" microcrystalline oozes, although many pellet rocks appear to be calcilutites in the field and under a binocular microscope.

Rock names.—All the rock characteristics discussed above can be combined in a single name, shown in Table I and diagrammatically in Figure 4. At first the writer used such cumbersome terms as "sparry intraclastic calcarenite" for intrasparite, "microcrystalline biogenic calcirudite" for biomicrudite, etc.; but these names, although self-explanatory, were soon found to be too awkward to use in descriptions. As an alternative the introduction of locality nomenclature was considered, but the localities would be difficult to choose and formidable to memorize. Finally, it was decided to use a composite word, the first part of which refers to the allochem composition: thus "intra-" for intraclastic rocks, "oö-" for oölitic ones, "bio-" for biogenic types, and "pel-" for pellet rocks. Whether the rock is type I or type II is shown by the second part of the name, "-sparite" for those with sparry calcite cement and "-micrite" (pronounced "mick-rite") for those with microcrystalline ooze matrix. The two word segments are combined to designate the eight types of allochemical limestone, for example, biomicrite, pelsparite, or oösparite. Type III limestones, almost entirely ooze, are designated simply as "micrite" without any allochem prefix. Inasmuch as most limestones are calcarenites, no further syllable is added if the size falls in that category, and as explained, the rare and somewhat artificial calcilutites are lumped together with the calcarenites in this table. However, if it is felt important to differentiate the calcirudites, the word segment "rudite" may be added if the size falls in that class, e.g., "intramicrudite" (pronounced "intra-mick'-rude-ite"). Examples, to-

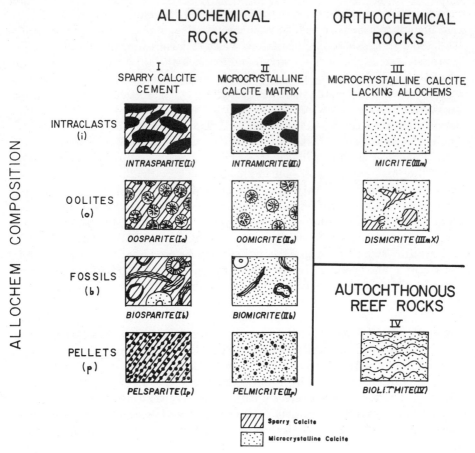

Fig. 4.—Graphic classification table of limestones. For determining allochem composition see Figure 3; for full details of classification, including method of denoting grain size and dolomite content, see Table I.

gether with the symbols used, are presented in Table I. On first acquaintance, such names as biosparite, intrasparite or pelmicrite sound odd, but so do gabbro, tourmaline, and brachiopod to the beginning geologist! Further, the names proposed here have the advantages of not requiring memorization because they can be deciphered by a simple syllabic code.

The principles of the classification are felt to be far more important than the use of the names, however, and some of the writer's colleagues prefer to use descriptive phrases like "sparry oölitic calcarenite" instead of "oösparite"; either method of nomenclature is completely functional and acceptable within the framework of the classification.

Some rocks classified as oösparite, intramicrudite, etc. may have significant amounts of other allochems which do not appear in the name. These may be specified at the discretion of the worker, such as fossiliferous oösparite Io(b), oölitic intramicrudite IIi(o), etc. Biogenic rocks, if composed largely of one type of organism, should always be described as brachiopod biomicrudite, gastropod biosparite, algal biomicrite, etc. If desirable, and if differentiation is possible, rocks containing fossils in growth position may be specifically designated as "autochthonous brachiopod biomicrite" etc.

Carbonate composition.—All of the rock types described, and listed in the table, can occur either as limestone or dolomitized limestone, and some may possibly occur as primary (directly deposited) dolomite. Over-all texture is combined with carbonate composition in a double symbol linked with a colon, as shown in the table. If the rock is a limestone, the rock name (e.g., oösparite or pelmicrite) is used unmodified and the symbol applied is Lr or La (for calcirudites and calcarenites, respectively). Intrasparrudite, for example, would be Ii:Lr, biomicrite, IIb:La. If the rock contains more than 10 per cent replacement dolomite, "dolomitized" is prefixed to the main rock name and the symbols DLa or DLr are used (e.g., dolomitized oösparite Io:DLa, or dolomitized pelmicrite IIp:DLa). If the dolomite is of uncertain origin, the term "dolomitic" and the symbols dLr and dLa are suggested, using a lower-case "d" to distinguish from known replacement dolomite. If the rock is a primary (early, non-replacement) dolomite, this term is prefixed to the rock name and Dr or Da are used for the symbol (e.g., primary dolomite intramicrudite IIi:Dr). Primary dolomite ooze may be called "dolomicrite," IIIm:D (name suggested by Thomas W. Todd).

Limestones that have been completely replaced by dolomite offer considerable difficulty since in many specimens the original structure is partly obliterated. Fine-grained clastic particles such as pellets or finely broken fossils are especially prone to vanish during dolomitization. Likewise, one does not know the original proportion of microcrystalline ooze versus sparry calcite cement. In such cases it is very difficult if not impossible to allot a dolomite to either classes I, II, or III, and it seems best to arbitrarily lump all such completely metasomatized rocks into a distinct class, type V; if ghost oölites, fossils, intraclasts, or pellets are present, that fact can be indicated by a symbol such as Vo, Vb, Vi, or Vp, respectively, and if no allochem ghosts are recognizable, it is simply listed as class V. The crystal size of these rocks is a very important characteristic and should be shown by the following terms and symbols, using divisions based on the Wentworth scale (Wentworth, 1922) but with a constant ratio of 4 between divisions.

Aphanocrystalline	D1	under .0039 mm.
Very finely crystalline	D2	.0039–.0156 mm.
Finely crystalline	D3	.0156–.0625 mm.
Medium crystalline	D4	.0625–.25 mm.
Coarsely crystalline	D5	.25–1.00 mm.
Very coarsely crystalline	D6	1.00–4.00 mm.
Extremely coarsely crystalline	D7	over 4.00 mm.

Examples of rock names in type V are medium crystalline intraclastic dolomite (Vi:D4), finely crystalline biogenic dolomite (Vb:D3), or for a rock with no visible allochems, coarsely crystalline dolomite (V:D5).

Terrigenous admixture.—So far in this paper, the content of terrigenous particles has been ignored. If the rock contains more than 50 per cent terrigenous material, it is a Terrigenous rock and not further discussed here. If it contains less than 10 per cent terrigenous material, it is a Pure Chemical rock and the terrigenous content is so low that it is not mentioned in the classification (Fig. 1).

However, if the rock contains between 10 and 50 per cent terrigenous material, that is regarded as important enough to be mentioned in the name and in the classification symbol. These rocks as a class are known as Impure Chemical rocks; a specimen of this type is classified just as previously described (i.e., as a biomicrite, oösparite, etc.), but to identify it as an Impure Chemical rock the following letters are prefixed to the symbol: Ts for rocks in which the terrigenous (T) material is dominantly and, Tz for those in which silt prevails, and Tc for rocks with clay as the most important terrigenous constituent.

The following list shows examples of this usage.

Clayey biopelmicrite, TcIIbp:La
Silty coarsely crystalline dolomite, TzV:D5
Sandy dolomicrite, TsIIIm:D2
Sandy dolomitized intrasparite, TsIi:DLa

The classification used here is determined necessarily by *relative* rates of formation of each constituent, not on absolute rates. Thus an abundance of terrigenous material in a limestone may mean (1) that uplift or proximity of the source area caused a more rapid influx of detritus; (2) a change of conditions in the depositional basin suppressed chemical activity, so that terrigenous minerals accumulated by default; or (3) current velocities were such as to concentrate terrigenous material of a certain size in preference to allochemical material of different size.

CHARACTERISTICS OF ROCK TYPES

It is not possible yet to give any quantitative estimate of the relative proportions of all the limestone types in the stratigraphic section as a whole. However, within each allochemical division it is possible to estimate, based on slides so far examined, whether intrasparite is more abundant than intramicrite, and to give some idea as to the usual petrography of these rock types. This summary, again, is based almost entirely on the writer's personal experience with some supplementation from published descriptions of limestones.

Intraclastic rocks.—The great majority of intraclastic rocks have a sparry calcite cement, inasmuch as currents that are strong enough to transport fairly large carbonate rock fragments are also usually capable of washing away any microcrystalline ooze matrix. Thus intrasparite (type Ii, Figs. 5–7) is much more common than intramicrite (type IIi, Fig. 8) which is relatively rare. Texturally, intraclastic rocks are about equally divided between calcirudites and calcarenites, and few occur beyond the size range of fine calcarenite through coarse calcirudite (Table II). Small amounts of fossils, oölites, or pellets with terrigenous sand or silt may occur between the intraclasts; thus sorting is variable. So-called "edge-

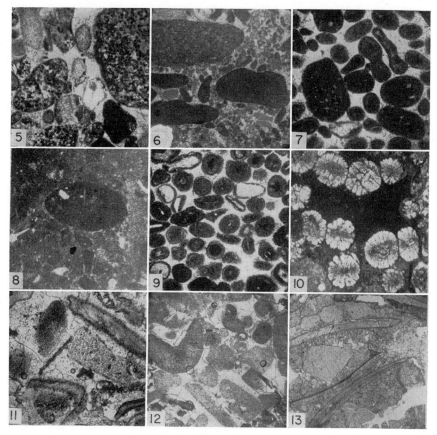

PLATE 1

FIG. 5. ×6. Intrasparrudite, Ii:Lr; larger intraclasts themselves are composed of pelsparite; a few fossils and smaller micrite intraclasts occur between large intraclasts. Packing of allochems is normal for well washed limestones, and sparry calcite cement is well developed. Lower Ordovician Axemann limestone, Centre County, Pennsylvania.

FIG. 6. ×6. Pelletiferous intrasparrudite Ii(p):Lr; intraclasts, elongate because of bedding, consist of micrite (dark) and pelmicrite. Between large intraclasts occur smaller intraclasts, common pellets, and a few fossil fragments, Lower Ordovician Ellenburger limestone, Blanco County, Texas. Collected by J. Stuart Pittman.

FIG. 7. ×10. Intrasparite, Ii:La; well sorted, sand-size intraclasts composed chiefly of micrite but with some pelmicrite. A few superficial oölites are present. Packing is normal for this rock type; sparry calcite is well shown. Permian Capitan reef, Eddy County, New Mexico. Collected by Robert J. Dunham.

FIG. 8. ×25. Intramicrite, IIi:La; intraclasts of various sizes, "floating" in an ooze matrix; packing irregular with large areas of "pure" ooze. Intraclasts are themselves composed of micrite. Lower Ordovician Ellenburger limestone, Blanco County, Texas.

FIG. 9. ×10. Oösparite, Io:La; well sorted oölites, closely packed in sparry calcite cement. A few fossil fragments, some coated with very thin oölitic rims. Lower Pennsylvanian Wapanucka limestone, Johnston County, Oklahoma.

FIG. 10. ×20. Oömicrite, IIo:La; oölites with well developed radial structure in an ooze matrix. Evidence from other parts of this slide shows that these "oölites" may actually have grown in place. Upper Cambrian Point Peak limestone, Burnet County, Texas.

FIG. 11. ×25. Brachiopod-crinoid biosparite, Ib:La. Fossils much abraded, well sorted, and rather tightly packed. Excellent sparry calcite. Middle Ordovician Trenton limestone, Harrison County, Kentucky.

FIG. 12. ×6. Crinoid biosparite, Ib:La. Crinoid fragments tightly packed because of their more equidimensional shape (compare pelecypod biosparite, Fig. 13). Sparry calcite forms overgrowths in optical continuity with crinoid pieces. Mississippian Mission Canyon limestone, Valley County, Montana. Collected by Daniel N. Miller, Jr.

FIG. 13. ×6. Pelecypod biosparrudite, Ib:Lr. Pelecypod fragments, originally aragonite, have inverted to calcite with disappearance of original delicate shell structure; all that is left is mosaic of sparry calcite. Rock originally had extremely high porosity because of way in which curved shells were cupped on top of each other during deposition. Loose packing produces illusion that spar has formed by recrystallization, but in three dimensions the shells are self-supporting and never had an ooze matrix. Pleistocene Miami limestone, Brevard County, Florida.

wise conglomerate" or "flat-pebble conglomerate," so common in lower Paleozoic limestones, almost without exception has a sparry calcite cement and may be classed as intrasparrudite. The discoidal pebbles (intraclasts) of these limestones are very often composed of pelsparite (Fig. 5) with thin laminae of quartz silt; this lamination is responsible for the extremely discoidal shape.

Intrasparite specimens imply a two-phase genesis: (1) a fine-grained calcareous sediment, usually microcrystalline calcite ooze, pellets, or fine-grained fossils, was deposited in a protected, calm-water environment, probably at shallow depths, and became weakly consolidated or cohesive; (2) a sudden change, such as caused by either a storm or relative lowering of sea-level, lowered wave base and tore up the previous sediment; the intraclasts were then ultimately deposited in the new high-energy environment. Such environments of prolonged calm-water conditions interrupted by sudden pulses of greatly strengthened wave or current energy would seem to require shallow-water deposition. The significance of intramicrite is not yet clear; it may form if the intraclasts are produced in a high-energy environment and then carried by environmental accidents and dumped into other (deeper and calmer) areas where there is an abundance of microcrystalline ooze.

Oölitic rocks.—These rocks with their high degree of sorting imply fairly vigorous current action; therefore as one would expect oösparite (type Io, Fig. 9) is much more abundant than oömicrite (type IIo, Fig. 10). Texturally both types are almost always fine to coarse calcarenites, although rare calcirudites may occur if the oölites are larger than 1.0 mm. Oösparite frequently contains no other allochem in significant amounts except for oölites, but in some strata intraclasts, fossils, and pellets are present. Oösparite forms in high-energy environments such as tidal channels (Illing, 1954), or the oölites may be drifted into submarine dunes and show cross-bedding near or in the loci of strong offshore currents. In oösparite, the sparry calcite cement often grows in radial fringes in continuity with the radial calcite crystals of the oölite itself. Oömicrite, like intramicrite, is an exceptional rock type implying formation of the oölites in a high-energy environment and their "accidental" transportation into a low-energy environment; perhaps these would be fairly common in zones of mixture where tidal channels with swift currents adjoined shallow protected marine flats. The writer has seen rare oömicrite specimens in which the "oölites" have grown in place while suspended throughout a microcrystalline matrix.

Just as some well sorted sandstones contain occasional clay stringers, similarly some oösparite specimens contain scattered stringers of oömicrite, representing a very brief slackening of current velocity. These may have been caused when wave agitation ceased for a short period (hours to days) and suspended ooze settled out to form a thin blanket on the oölitic sediment.

Pisolitic rocks, like oölitic ones, usually have sparry cement and if one wished to be consistent, could be called pisosparite.

Biogenic rocks.—These occur just as commonly with a microcrystalline ooze matrix (biomicrite, type IIb) as with a sparry calcite cement (biosparite, type

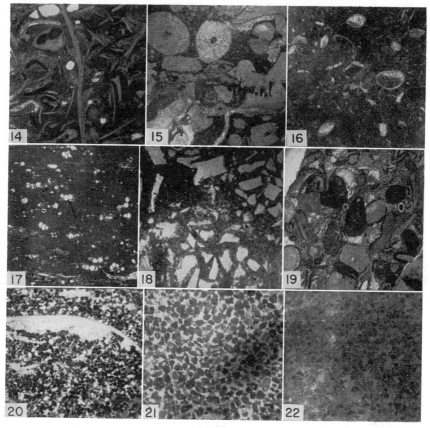

PLATE 2

Fig. 14. ×5. Brachiopod biomicrite, IIb:Lr. Brachiopods rather closely packed but randomly oriented; a few rounded quartz grains present. Limestone bed in Middle Silurian Rochester shale, Morgan County, West Virginia.

Fig. 15. ×10. Crinoid biomicrite, IIb:La. Crinoid fragments rather tightly packed in ooze matrix. Mississippian Madison limestone, Bonneville County, Idaho. Collected by Henry H. Gray.

Fig. 16. ×20. Ostracod biomicrite, IIb:La; fossils very loosely packed and randomly oriented in ooze matrix. Some fossils articulated, others greatly broken. Articulated fossils have geopetal accumulations of ooze, and upper part of shell (originally pore space) is now filled with spar. Devonian Hunton limestone, Murray County, Oklahoma.

Fig. 17. ×20. Foram biomicrite, IIb:La (almost a foraminiferal micrite, as rock contains only about 10 per cent fossils). Fossils very loosely packed in carbonate mud matrix. Middle Cretaceous limestone, San Luis Potosi, Mexico.

Fig. 18. ×10. Algal biomicrite, IIb:La. Angular chunks of spar represent fragmented algal plates. At first this rock might simulate dismicrite. Pennsylvanian (Allegheny) limestone, Butler County, Pennsylvania.

Fig. 19. ×10. Crinoid biosparite-biomicrite transition, I–IIb:La. This is interpreted as poorly washed limestone in which currents were unable to winnow out all the ooze. A few intraclasts, an oölite, and some other fossil fragments present; rock is tightly packed and rather well oriented. Lower Pennsylvanian Marble Falls limestone, Lampasas County, Texas. Collected by Daniel N. Miller, Jr.

Fig. 20. ×15. Silty pelsparite, TzIp:La. Rock contains large, curved brachiopod shell which fell concave-side down on bed of pellets, and resulting space underneath is now filled with spar. Rock also contains some small fossil fragments and quartz silt. Limestone stringer in upper Silurian Bloomsburg formation, Washington County, Maryland.

Fig. 21. ×30. Pelsparite, Ip:La. Pellets show typical excellent uniformity of size (.05 mm.) and shape, and have normally tight packing. Sparry calcite cement is very finely crystalline because of small size of available pores. Middle Ordovician Trenton limestone, Centre County, Pennsylvania.

Fig. 22. ×30. Pelmicrite, IIp:La. Pellets very loosely packed in ooze matrix that has partly recrystallized in places to microspar. Middle Ordovician Trenton limestone, Centre County, Pennsylvania.

Ib). Biomicrite (Figs. 14–18) signifies either that the fossils were sedentary or else that currents were calm in the depositional area and the microcrystalline ooze was not winnowed out from the shell material. Such a condition could prevail either in deep waters or in shallow, protected areas. Biosparite (Figs. 11–13) usually forms in environments with more vigorous current action where the microcrystalline ooze is washed away; thus fossils and fossil fragments generally show some abrasion. Biosparite may form also if no ooze is being produced in the environment. As both the intraclastic rocks and the oölitic rocks require vigorous current action in order to form, they are almost invariably -sparites; however, shelled animals may live and become deposited under a wide variety of current conditions and thus many have either ooze matrix or sparry cement with about equal frequency. Biogenic rocks range in grain size from coarse calcilutites, such as some foraminiferal limestones, to coarse calcirudites, but most appear to lie in the medium calcarenite to fine calcirudite range. The cement of biosparite shows diverse morphologies: around certain fossils (brachiopods, ostracods, and trilobites) the sparry calcite quite commonly forms radially oriented stubby fibers in continuity with the fibrous calcite of the fossil; in some trilobite biosparites, growth of the calcite fibers may actually spread the fossils apart perpendicular to the bedding, expanding the sediment to more than double its original bulk volume! This dilatant precipitation of calcite is similar to the mechanism of frost-heaving. Around echinoid fragments, large overgrowths of clear calcite develop as singly oriented clear crystals in optical continuity with the fragment, and in many the original boundary of the fossil is very hard to see (Fig. 12). These rocks (crinoidal biosparites) are often termed "recrystallized" by the field geologists though actually it is a simple matter of pore-filling overgrowths, exactly analogous with growth of quartz cement in continuity with detrital quartz grains in sandstones. Biomicrite and biosparite specimens in which algal fragments are abundant commonly produce puzzling rocks (Fig. 18) because many types of algae recrystallize readily to sparry calcite and the rocks look as if they had angular chunks or curving plates of spar embedded in a micrite matrix. Other types of algal structures may resemble intraclasts.

Biosparite and biomicrite show great variation in grain size, sorting, and orientation of fossil fragments. Sorting and orientation (Fig. 14) are normally poorer in biomicrites than biosparites because of the difference in energy of the environments under which they accumulate. There are numerous exceptions to this generalization; if the fossils are all of one type (e.g., all foraminifera or all crinoids, Figs. 15, 17), then they will be well sorted even in biomicrite. If many diverse types of fossils occur in one specimen (e.g., a mixture of bryozoans, foraminifera, brachiopods, and crinoids), then even the most winnowed biosparites will be poorly sorted considering the size distribution of the fossil fragments. The degree of damage to the fossils is another important characteristic to be noted in the description of biogenic rocks: whether the fossils are still articulated, or whether they are disarticulated or broken; and, if the fossils are fragmented, the degree of rounding of the particles. Normally, rounded and heavily abraded fossils occur in the high-energy biosparites (Fig. 11), whereas the best preservation of delicate structures is found in biomicrites (Fig. 16). Broken fossils occur commonly in both limestone types.

Some biomicrite has only a small proportion (10–40 per cent) of fossils (Figs. 16, 17), and there is a complete gradation between this rock type and Fossiliferous Micrite. The boundary line is entirely arbitrary, but with more thin-section data we may able to set a more "natural" boundary. However, it is possible that no natural boundary exists, because it seems logical that any amount of fossil material from 1 to 80 per cent could fall on and become incorporated in a calcareous mud bottom. Some biomicrite contains considerable clay, because of the hydraulic similarity of clay flakes and microcrystalline calcite particles. Most chalk is here termed foraminiferal biomicrite.

The reader may ask what is wrong with using such commonly-understood words as "coquina," "encrinite," "rudistid limestone," etc. In this writer's opinion, such words are definitely useful but only in a rather broad and vague sense, similar to the usage of "trap," "grit," or "puddingstone." They certainly convey meaning, but are not specific enough; "coquina" might be pelecypod biosparite, or brachiopod biomicrudite, and "encrinite" does not tell one the important fact whether the rock has an ooze matrix or is cemented with spar. Further, these are isolated and unquantified words, set off by themselves beyond the pale of a systematic scheme of nomenclature, and typical of the early nomenclatural history of the sciences.

Pellet rocks.—These are common, especially in lower Paleozoic limestones. However, they are so fine-grained that in the field they are almost without exception mistaken for micrite, and even with the binocular microscope under the most favorable observing conditions and highest power, this writer wrongly identifies most pellet rocks as micrite. To identify them with any confidence takes an acetate peel or a thin section, although one can make a shrewd guess at their identity by knocking a small chip off the corner of a hand specimen, placing the thin chip in index oil under a petrographic microscope and turning up the converger; if pellets are present they can be seen easily.

Usually pellet rocks have a very finely crystalline sparry calcite cement; thus most are pelsparite (type Ip, Figs. 20, 21), although some have a microcrystalline matrix (pelmicrite, type IIp, Fig. 22). In many instances, the sparry calcite is so fine that it is difficult to decide whether it should be called pelsparite or pelmicrite; furthermore, pelmicrite is sometimes difficult to distinguish from micrite even in thin section because it consists of small aggregates of microcrystalline ooze in a matrix of microcrystalline ooze. But the convergent lens helps bring out the pellets because they usually have more organic matter.

Texturally, pelmicrite and pelsparate are borderline between coarse calcilutite and very fine calcarenite. They may contain significant amounts of quartz silt, which is hydraulically equivalent to the pellets; furthermore, they usually show very fine lamination and sometimes delicate cross-bedding but they rarely contain any clay. The extremely discoidal pebbles of "edgewise conglomerate" are usually composed of silty pelsparite (Fig. 5).

The environment of deposition of pelsparite is not known. The lamination would seem to require an environment of gentle, persistent currents probably in water of moderate depth. If the pellets are truly fecal, and the writer believes they are, then mud-feeding organisms must have been common in the environment. Pelmicrite may form if occasional burrowing animals defecate sporadically within the calcareous mud, or if calcite ooze is subjected to a gentle rain of infalling pellets; in either case it indicates a very calm, current-free environment. Alternately, it may perhaps originate by partial recrystallization of the calcareous mud.

A fairly common rock type in the Silurian limestones of eastern West Virginia is composed of a subequal mixture of fossils and pellets. It is not known whether this is a valid rock class or not, but the writer has tentatively called these common hybrids "biopelsparite" (Ibp) and "biopelmicrite" (IIbp). It is quite logical that pellets and fossils should commonly be associated.

Microcrystalline rocks.—All rocks with less than 10 per cent allochems, that is, those that consist very largely of microcrystalline calcite, are classed as microcrystalline rocks (type III). This group itself should be split, however, to separate the rocks with less than 1 per cent allochems—the "lithographic limestone" of long usage, here called simply "micrite" (IIIm, Fig. 23)—from rocks with 1–10 per cent allochems. The writer feels that for the sake of precise usage it is better to introduce a quantitatively defined new word ("micrite") rather than trying to redefine the former words for this rock type like "lithographic limestone," "vaughanite," or "calcilutite" which are used chiefly in a loose megascopic sense. The term "micrite" should be reserved strictly for those rocks that, *under the petrographic microscope*, are seen to consist almost entirely of microcrystalline calcite. Many rocks that in the field appear like micrite actually turn out upon closer examination to be very fine-grained biomicrite, pelmicrite, or pelsparite. It might be better for preliminary field usage, during the measurement of stratigraphic sections, to continue to call these aphanitic rocks calcilutites until they can be properly named under the microscope or by acetate peeling.

Rocks consisting of microcrystalline ooze with 1–10 per cent scattered fossils (termed "fossiliferous micrite," IIIb, Fig. 24) are quite common, and as noted before, grade continuously into biomicrite (IIb); this rock type is common in chalk. If the fossils are dominantly of one type, this should be specified, e.g., foraminiferal micrite, crinoidal micrite. Pelletiferous micrite (IIIp, 1–10 per cent pellets; transitional to pelmicrite, IIp) is not uncommon, but micrite with 1–10 per cent intraclasts or oölites is rare.

All the microcrystalline rocks presumably indicate calm-water conditions because of the very fine grain size of the constituent particles. These rocks could accumulate either in shallow, protected shelves or lagoons as in the Bahamas, or

CLASSIFICATION OF LIMESTONES

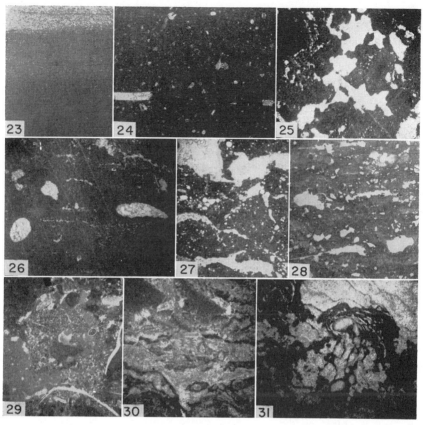

PLATE 3

Fig. 23. ×10. Micrite (IIIm:Ll). Homogeneous ooze with bed of silty pelsparite (clear) at top of slide. Upper Silurian Wills Creek formation, Washington County, Maryland.

Fig. 24. ×15. Fossiliferous micrite, IIIb:La. Contains about 5 per cent of diverse fine-grained fossil fragments loosely packed in carbonate mud. Cretaceous Buda limestone, Terrell County, Texas.

Fig. 25. ×20. Dismicrite (IIImX:Ll) with irregular patches of spar. Their origin is not known. Middle Ordovician limestone, Franklin County, Kentucky.

Fig. 26. ×30. Dismicrite, IIImX:Ll. Here "eyes" of sparry calcite are crudely cylindrical and obviously represent borings, probably of worms. Apparent compaction of ooze (darkening) in roughly concentric zone surrounding burrows. Smaller patches of spar of unknown origin; they may represent collapsed burrows or general stirring of soupy semi-congealed carbonate mud by small organisms. Middle Ordovician limestone, Franklin County, Kentucky.

Fig. 27. ×10. Sandy dismicrite, TsIIImX:Ll. Irregular areas of spar tend to have crudely horizontal orientation. Genetically, this is probably an algal reef (Ham, 1954), therefore a more proper name would be sandy algal biolithite. Appearance is practically identical with Recent laminated (algal mat?) sediment, Fig. 28. Upper Ordovician McLish limestone, Pontotoc County, Oklahoma.

Fig. 28. ×10. Thin section of Recent lithified, laminated (algal mat?) sediment from Florida keys. If the pores were filled with sparry calcite, this rock would simulate dismicrite (compare Fig. 27). Collected by Robert Ginsburg.

Fig. 29. ×12. Algal biolithite, IV:L. Very complex rock with algal ooze, entrapped small fossils, pellets, and irregular patches recrystallized to microspar. Cambrian Snowy Range formation, Park County, Montana. Collected by Richard E. Grant.

Fig. 30. ×10. Algal biolithite, IV:L. Algal structures with former pore space, entrapped pockets of pellets, and very irregular bedding. Permian Capitan reef, Eddy County, New Mexico. Collected by Robert J. Dunham.

Fig. 31. ×10. Algal biolithite, IV:L. Irregular algal structures, spar partly occupies pore spaces and partly is result of recrystallization. Pennsylvanian (Virgilian) reef, Otero County, New Mexico.

in calm, deep waters. In themselves, therefore, they do not seem to be a depth indicator; depth should be indicated by the rock types that are interbedded with the micrite, faunal assemblages, bedding, or other features.

Although many micrite specimens contain very little insoluble residue, others contain abundant clay. Silt and sand, however, are uncommon in the micrite specimens the writer has examined. This appears to be simply a matter of hydraulic equivalence. Silurian limestones in West Virginia commonly contain micrite and pelsparite interbedded on a scale of 1 cm. or less (Fig. 23) and the pelsparites contain abundant silt and no clay whereas the micrites contain abundant clay and very little or no silt.

Dismicrite (IIImX, disturbed micrite) is a sack term for a group of rock types of diverse and obscure origin (Figs. 25–28). These rocks all consist dominantly of microcrystalline calcite, but contain irregular patches, tubules, or lenses of sparry calcite, almost invariably with sharp boundaries. This type of limestone has been called "birdseye" by many geologists. It seems to form in at least six ways. (1) Animals (worms, mollusks, etc.) burrow in and "chew up" what was originally a soft, homogeneous carbonate mud (Figs. 25, 26); spar is later chemically precipitated in the resulting tubules and irregular openings. The resulting openings (later filled by chemically precipitated spar) range from a few distinct, cylindrical tubes (commonly with fecal pellets accumulated at intervals along the bottoms of the tubes) to sediments that are riddled with completely irregular openings showing no tubular aspect, possibly the result of collapse of soft sediment into larger burrows. (2) Some dismicrite specimens are apparently ancient algal mats, such as the birdseye McLish limestone described by Ham (1954); a thin section of the McLish looks almost identical with a thin section of recent lithified laminated sediment from Florida (Figs. 27, 28), thought by Robert N. Ginsburg (personal communication, 1957) to be bound by the action of blue-green algae in the intertidal zone. Here the spar lies in irregular patches parallel with the bedding, and apparently represents interspaces between irregularly shaped subhorizontal algal-bound layers. If it is certain that the rock originates in this way, it is preferable to call it by the genetic term "algal biolithite" rather than the purely descriptive "dismicrite." Vertically trending algal growths can also give rise to rocks simulating dismicrite, as can accumulations of recrystallized algal plates in calcareous mud. (3) Other dismicrites are the result of soft-sediment slumping or mudcracking, which creates openings having more the appearances of fractures, although the walls are commonly deformed by flowage in the plastic sediment. These should not be confused with tectonically induced fractures. (4) Some dismicrite specimens contain irregular vertical patches of spar, which may represent openings in semi-coherent calcareous mud caused by passage of gas bubbles. (5) If a soft, unconsolidated calcareous mud is partly torn up by an increase of current velocity, and then rapidly redeposited, the result is a rock with very vaguely defined proto-intraclasts, semi-coherent clouds of calcareous mud, and irregular patches of spar. This type of dismicrite is hence transitional with, and sometimes hard to distinguish from, intrasparite. (6) If a calcareous mud begins to recrystallize in patches, the resulting rock would simulate a dismicrite. Although an entirely possible mechanism, the writer knows of no such examples.

In many dismicrite specimens the origin of the spar-filled openings is not known.

Biolithite (type IV) is another sack term. Inasmuch as these rocks have been studied very little by this writer, he proposes no detailed classification other than to suggest that they be specified as algal biolithite, coral biolithite, etc. The name should be used only for these specimens that are in growth position and it does not include broken-off fragments of corals or algae. These are allochemical constituents and a rock composed of them would be algal or coral biosparite or biomicrite, depending on the matrix. Numerous specimens of algal biolithite (Figs. 29–31), especially in lower Paleozoic rocks, are very complex with crumpled banding, irregular pockets of winnowed pellets, scattered animalian fossils, and patches of sparry calcite, some of it formed by recrystallization and some representing fillings of hollows and burrows. Certain biolithites may also simulate dismicrite.

RECRYSTALLIZATION IN LIMESTONES

In order to avoid introducing unnecessary complications while discussing the classification, the writer has purposely evaded the role of recrystallization in limestones. But to ignore this important process entirely would give a misleadingly simple picture of limestone petrography; therefore, brief discussion of this phenomenon appears to be necessary. When a mineral undergoes "recrystallization," this term signifies that original crystal units of a particular size and morphology become converted into crystal units with different grain size or morphology, but the mineral species remains identical before and after recrystallization. Such changes would be the alteration of 3-micron, equant grains of microcrystalline calcite to 20–50 micron, equant mosaic crystals of sparry calcite; change of the same microcrystalline calcite to fibers of calcite 100 microns long; or (theoretically possible) conversion of large crystals of spar, such as are found in sparry calcite cement or echinoderm fragments, to very fine-grained calcite. It would not include conversion of calcite to dolomite, which is properly termed replacement; and would also not include solution of one type of calcite, leaving an open cavity existing for a significant time interval, and later or much later filling of that cavity with a different type of calcite. Recrystallization is really a special case of metasomatism or replacement in which the original and the "replacing" mineral are identical mineralogically, although different in grain size, morphology, and orientation. In common usage, the conversion of unstable aragonite to calcite is loosely called recrystallization, although this process is more properly termed inversion inasmuch as these two minerals are not the same, differing in ionic lattice, crystal system, density, and other properties.

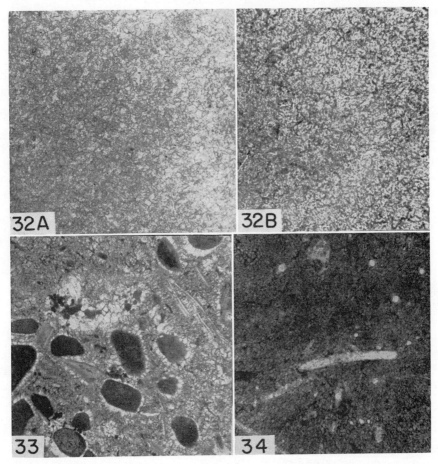

PLATE 4

FIG. 32. Microsparite, RIIIm:L. Rock was formerly homogeneous microcrystalline calcite ooze which recrystallized to 10–15 micron microspar. Photo A magnified 45×, photo B, 225×. Devonian Stribling limestone, Blanco County, Texas.

FIG. 33. ×45. Glauconitic biomicrosparite, RIIb:La. Small shell fragments and glauconite grains both have radiating fringes of microspar, and original ooze matrix has been completely recrystallized. Limestone in Mississippian Barnett formation, San Saba County, Texas. Collected by E. Hal Bogardus.

FIG. 34. ×30. Spiculiferous microsparite, RIIIb:Ll. Spicules (cross- and longitudinal-sections) originally were deposited in ooze matrix, which has recrystallized to 10-micron microspar. Pennsylvanian Marble Falls limestone, San Saba County, Texas.

Recrystallization and inversion manifest themselves in diverse ways in limestones. Some types of recrystallization leave obvious clues behind; others are extremely difficult to prove or disprove. The latter kind of recrystallization as a consequence gives rise to heated and often insoluble controversy. As in the granitization argument, the criteria for proving a recrystallization or replacement origin are usually simple, obvious, and incontrovertible if found; the defender of a primary (direct-precipitation), non-replacement, or non-recrystallization origin has no such firm ground to stand on, but is forced to fall back upon vague argu-

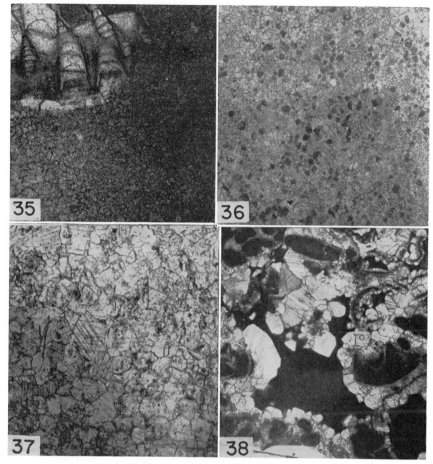

PLATE 5

Fig. 35. ×48. Coral biomicrosparite, RIIb:La. Microspar shows gradual coarsening in crystal size toward lower left part of picture. Upper Ordovician (Richmond) limestone, Green County, Ohio.

Fig. 36. ×45. Rock originally pelmicrite, but former ooze matrix has completely recrystallized to 15–20-micron microspar. Spar must have formed by recrystallization because pellets are so loosely packed. Illustrates completion of process which began in Fig. 22. Upper Ordovician Trenton limestone, Centre County, Pennsylvania.

Fig. 37. ×45. Rock formerly contained abundant (?) fossil fragments; not known whether it originally had ooze matrix or not. Now completely recrystallized to medium crystalline (.10–.15 mm.) sparry calcite, and fossils have been all but obliterated. Rare rock type tentatively designated "pseudosparite." Upper Cambrian Point Peak limestone, Burnet County, Texas.

Fig. 38. ×13.5. Intramicrite in process of recrystallizing. Rock originally consisted of intraclasts, fossils, and pellets loosely packed in microcrystalline calcite ooze matrix. Sparry calcite, starting from allochems as nuclei, has grown out in large crystals to replace most of original ooze matrix. Completion of process would yield rock with allochems floating loosely in sparry calcite. This rock type believed to be rare. Upper Cambrian San Saba limestone, Blanco County, Texas.

ments such as statistical associations, "common-sense," and other lines of reasoning that rarely give iron-clad proof. The four major types of recrystallization and inversion are depicted in Figure 39 and are ranked in the following paragraphs in the approximate order of increasing difficulty of proof.

1. *Inversion of originally aragonitic fossils to calcite.*—Aragonitic shell material

(most pelecypods, many gastropods) inverts to calcite with time, and the original delicately fibrous or prismatic structure of the shell is replaced by a structureless, interlocking, semi-equigranular mosaic of anhedral sparry calcite, apparently of the same appearance and crystal size as ordinary pore-filling sparry calcite (Fig. 13). Some algae also recrystallize or invert to sparry calcite easily (Fig. 18). Worn or rounded fragments of algal structures or pelecypod shells often are difficult to identify because after inversion they look like rounded allochems of mosaic sparry calcite and lack any internal structure as a clue to their true organic origin. Inversion is generally regarded as a function of time, but many aragonite shells of Pleistocene or Recent age have already inverted to calcite, while some aragonite shells have persisted un-inverted since the late Paleozoic. This type of inversion can be recognized simply by knowing which types of fossils have aragonite shells in life.

2. Recrystallization (or inversion) of an original microcrystalline calcite (or aragonite) ooze matrix to microspar.—This appears to form a common limestone type developed from rocks that would have been originally classified as micrite, type III, or biomicrite, type IIb (i.e., nearly pure microcrystalline ooze, or fossils in an ooze matrix, respectively) were it not for the fact that the microcrystalline calcite is coarser than normal—averaging 5–15 microns instead of 1–4 microns, although the grains are still equidimensional and uniform in size (Figs. 32–36). Because this relatively coarser material occupies large areas or makes up the entire specimen, the looseness of packing of the embedded allochems requires that it can not have formed as a cement, and probably represents aggrading recrystallization of a "normal" microcrystalline ooze matrix. These rocks the writer has designated as microsparite (corresponding with micrite) and biomicrosparite (corresponding with biomicrite), with symbols respectively RIIIm and RIIb. In these rocks, the allochems seem to remain unaffected by recrystallization, barring the previous existence of aragonitic shells. Unfortunately, microsparite looks exactly like micrite in hand specimen, but it is easily identifiable in thin section or by chipping off a sliver of limestone and examining in oil.

How do we know that the microspar did not accumulate as an originally deposited ooze, which because of unusual physico-chemical conditions (speed of crystallization, saturation, etc.) simply grew larger particles than "normal" ooze? In many microsparite specimens, the microspar occurs as irregular patches grading by continual decrease of grain size into areas of "normal" microcrystalline calcite (Fig. 35). Furthermore, the microspar may begin to crystallize about allochems (or even quartz grains) as an outwardly advancing aureole of recrystallization. In some samples the grains of microspar calcite in these recrystallization fringes have a vaguely rod-like to radial-fibrous form, oriented perpendicular to the allochem surface (Fig. 33). In most microsparites, the matrix has been converted completely; therefore in these there is no direct evidence of its origin (Figs. 32, 34). The reason why some micrite or biomicrite specimens are partly or entirely converted to microsparite is unknown; one can only advance the truism that it may be due to an original difference in the ratio of calcite ooze to aragonite ooze, to the influence of trace elements or clay minerals in the environment, or to other as yet unrevealed factors.

3. *Recrystallization transecting allochems.*—This uncommon type of recrystallization occurs when parts of allochems (e.g., intraclasts, non-aragonitic fossils) are recrystallized to sparry calcite, or when allochems and matrix are attacked indiscriminately. The areas recrystallized to spar may take the form of irregular patches, advancing massive fronts which may leave unrecrystallized relics behind, or vein-like areas of recrystallization, transecting both allochems and matrix. The criteria for recognition of this process are the same as those for recognizing any type of irregular replacement such as dolomitization or silicification. In extreme examples, the entire rock may be converted to spar so that no vestige of original structure remains (Fig. 37). These have been tentatively termed "pseudosparite."

4. *Recrystallization of original ooze matrix to coarse spar, leaving allochems essentially unaffected.*—The preceding examples of recrystallization are relatively non-controversial, inasmuch as they leave unimpeachable evidence behind. In this type of recrystallization—thought to be rather rare by this writer but considered very common by many others—most of the rocks that now have a sparry calcite cement are thought to have accumulated as a sediment consisting of allochems in a microcrystalline ooze matrix, and the spar later developed by recrystallization of the ooze (Fig. 38). This recrystallization spar supposedly imitates exactly the morphology of pore-filling spar. If this hypothesis were generally true, it would completely destroy the genetic principle of the classification proposed herein, by denying the role of hydraulic action in producing the cleanly washed (type I) versus the non-winnowed ooze-rich limestones (types II and III); instead, this difference would be ascribed to recrystallization in the former and lack of it in the latter limestone families.

This hypothesis is a very serious challenge. To meet it, the writer offers the following argument. In most limestones the presence or absence of spar is simply a function of packing (Fig. 40). Where allochems are closely packed (and especially if they are well sorted), there is sparry calcite cement; where allochems are loosely packed like fruits in a fruit cake, then a microcrystalline ooze matrix is present; and where there are no allochems, the rock consists of ooze. This is exactly the relation that one would expect if current action accounted for the presence or absence of ooze, just as is the case with terrigenous sandstones, clayey sandstones, and claystones. If sparry calcite were usually the result of recrystallization, then the presence or absence of spar should have no relation to packing; rocks with allochems spread far apart in a sparry calcite matrix (which was originally an ooze matrix) should be abundant, and many patches of recrystallization spar should be present in sections of dominant micrite. If the ooze recrystallizes in the closely packed, well sorted calcarenites, then it should also recrystallize in rocks made largely or entirely of ooze. Yet it is a monotonous rule that one can measure hundreds of feet of limestone sections and find that spar occurs only in closely packed calcarenites, and loosely packed or widely dispersed allochems occur only with ooze matrix—exactly analogous with sandstone-claystone sequences in terrigenous rocks. Those who claim a recrystallization origin for most spar are denying the evidence of recent carbonate sediments, most of which are porous calcarenites devoid of ooze matrix and with plenty of pore space waiting to receive their sparry calcite cement.

The writer firmly believes the foregoing principles apply to the vast majority of limestones. However, there are a few limestones in which it is demonstable that the original ooze matrix has recrystallized to coarse spar. These limestones can be identified positively by the criterion that allochems are dispersed widely in a

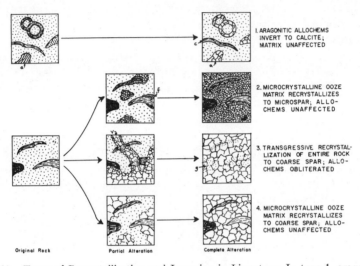

Fig. 39.—Types of Recrystallization and Inversion in Limestone. In type 1, aragonitic allochems (e.g., certain types of shells, and aragonitic algal masses or intraclasts "a") recrystallize or invert to calcite. Consequently their delicate structure is destroyed and replaced by homogeneous mosaic of sparry calcite, in appearance exactly like sparry calcite resulting from cementation. Calcitic fossils "c" and the matrix of rock remain unaffected. In type 2, microcrystalline ooze matrix of rock crystallizes to microspar. During partial stages of alteration, it is common for microspar to begin as radial fringes encrusting allochems "f," and also to occur as irregular patches scattered randomly throughout ooze. Contact between unrecrystallized ooze and microspar is usually gradational. When recrystallization is complete (usual case), entire matrix is converted to microspar but allochems are unaffected (unless they were aragonitic). This is most common type of recrystallization. In type 3, coarse spar develops by recrystallization of both ooze and allochems, and spar transects allochems and matrix indiscriminately. Recrystallization may advance along massive fronts or work out as replacement veinlets "v." When recrystallization is complete, entire rock is converted to coarse, mosaic sparry calcite ("pseudosparite") and allochems may be completely obliterated or visible as faint ghosts "g." This process is apparently identical in mechanism with dolomitization, except that replacing mineral is calcite, not dolomite. This type of recrystallization seems rare. In type 4, original ooze matrix recrystallizes to coarse sparry calcite. This differs from type 3 in that allochems are not affected. In partial stages, irregular patches of spar may occur scattered throughout ooze; these simulate the appearance of "poorly winnowed" limestone. After process is completed, main evidence for recrystallization lies in fact that allochems are too loosely packed (floating) in sparry calcite; rock could not have been deposited that way. This type of recrystallization is felt very important by some workers, but the writer feels that it is of small volumetric importance.

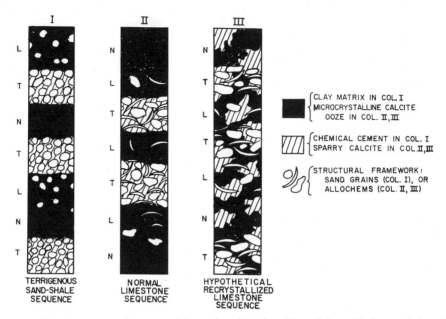

Fig. 40.—Occurrence of sparry calcite as function of packing. Column II shows relations in great majority of limestone sequences: sparry calcite occurs where allochems are relatively tightly packed (strata labeled "T"). No sparry calcite occurs where allochems are loosely packed and "floating" in calcite ooze matrix (strata "L"), and no spar occurs in strata of homogeneous ooze ("N"). This demonstrates pore-filling origin for sparry calcite in most limestones, because relations are exactly those that occur in terrigenous sand-shale sequences (column I). Here also chemical cements (e.g., calcite, quartz) occur only where sand grains are relatively tightly packed, and cement is inconspicuous in clayey strata. Clay forms a rock in its own right, just as does micrite in column II. If spar usually developed by recrystallization of microcrystalline calcite ooze, occurrence of spar should be completely independent of packing as shown in column III. Here the spar would occur as irregular patches even in midst of large areas of ooze. Limestone sections like column III do occur, but in the writer's experience are extremely rare.

matrix of spar. This is simply again a matter of packing, that is, the allochems are widely enough spaced so that they must have been originally deposited with an ooze matrix filling between them, which ooze has now recrystallized to spar. Rarely this process may be seen in mid-stage (Fig. 38). Recrystallization spar tends to form crystals of much more uniform size (like replacement dolomite) in contrast to cementation spar, in which the crystals in the mosaic show wide size variation. Occasionally ordinary pore-filling spar, especially in some trilobite and ostracod-rich sediments, forms fibrous overgrowths on the fossils possessing enough force of crystallization to actually spread fossils apart and expand the sediment volume to double or more its original bulk, but these examples are easily recognizable. Similar displacive growth of calcite cement also occurs in some sandstones. Patches of spar in otherwise homogeneous micrite (i.e., dismicrite) seemingly offer fuel to the recrystallizationist, but careful examination nearly always shows that the sparry patches are demonstrably burrows or desiccation phenomena, as has been discussed previously. The specimens most difficult to interpret are those which have their pore spaces partially filled with spar, and partially with ooze (Fig. 19). These the writer has termed "poorly washed limestones" and regards them as transitional between types I and II because currents were only strong enough or persistent enough to wash away part of the ooze. Many of the remaining clots of ooze have "fuzzy" boundaries because of their incoherence. Others see in these limestones a patchy recrystallization of the ooze to spar. Inspection of these limestones for geopetal accumulations of ooze helps in deciphering their origin. If clay minerals are present in the ooze, then they should also be present in the spar if it has formed by recrystallization of ooze, and clay should be absent if the spar is a pore-filling.

Summary.—Recrystallization is a very important process in some formations and in some localities; but its over-all volumetric importance in limestone is considered minor. Like turbidity currents, recrystallization tends to become a panacea which one falls back on to explain any phenomenon he can not readily understand. In studying samples, one most always keep in mind the possibility of recrystallization, but unless this writer sees definite evidence of recrystallization such as ultra-loosely packed allochems in a sparry cement, or transection of allochems, or gradational recrystallization of matrix, he assumes the simpler view that the spar is primary. Even in the recrystallized rocks one must have started with one of the basic eleven rock types outlined, and recrystallization merely adds a qualification to the system. This is no more serious than recognizing that original limestone fabric may be partly replaced by calcite as well as dolomite or silica.

TERRIGENOUS CARBONATE ROCKS, OR CALCLITHITES

Some mention should be made of those terrigenous rocks that are made up largely of fragments of older limestones eroded from outcrops in a source land. Although these have the mineral composition of limestone (or dolomite), they should really be considered as terrigenous rocks and hence form a rock clan fully equivalent in rank to the orthoquartzite, arkose, or graywacke. Thus the clan triangle composition for sandstones proposed by Folk (1954, 1956) should be expanded to a tetrahedron to include these terrigenous carbonate rocks. These rocks are here termed "Calclithites" alluding to the fact that they are a rock made up of fragments of older calcareous rock. A calclithite, then, is defined as a terrigenous rock whose silt-sand-gravel fraction contains more than 50 per cent carbonate rock fragments. The grain size of these rocks ranges from siltstones (rare) through sandstones (fairly common) to conglomerate (most common). A good example of a calclithite-type sandstone is the Miocene Oakville formation of central Texas, a sandstone made up of 80–90 per cent grains of Cretaceous limestone and fossil fragments. Another example is the Collings Ranch conglomerate (Ham, 1954) a Pennsylvanian conglomerate made up chiefly of fragments of Cambro-Ordovician limestone. The rocks are usually the product of intense orogeny, particularly in its early phases before the sedimentary cover has been stripped off the source area. They require rapid erosion and short transport, otherwise chemical weathering and abrasion would destroy the soft limestone fragments. Consequently, calclithites are typically stream or alluvial fan deposits from rugged terranes.

CONCLUSIONS

Whether the new names and the symbolic shorthand are accepted or not, the writer is convinced that the rock types for which these names and symbols stand are basically valid although the limits may become modified through more research. Using these types, proportionate lithologic diagrams can be prepared for limestone formations, as shown in Figure 41. With time, the environmental significance of each of the eleven rock types will become clearer and they can be used in deciphering the reasons for lateral and vertical changes which have taken place in limestone sections.

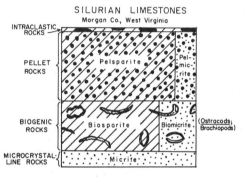

Fig. 41.—Proportionate lithology diagram, Silurian limestones, Morgan County, West Virginia (based on research in progress). This type of diagram is example of graphic lithologic summary, using classification here proposed; summaries like this can be used in comparing formations or tracing lateral and vertical stratigraphic changes in limestone types. Total area of diagram equals 100 per cent; area shown for each rock type is proportional to its volume in section studied. Pelsparite is most common rock type, with Ostracod Biosparite second.

Is this classification too complicated? To the outsider looking in, it undoubtedly appears so—but all classifications look complex to people who have not had to work with the rocks involved. But to the geologist who must struggle with limestones continually, this classification will soon appear too simple. Dozens of different subtypes of biomicrite could be established, dependent upon proportion of fossils to ooze, grain size and amount of breakage of fossils, type of organism, bedding, hardness, color, and impurities. This classification merely established the eleven major limestone types; in any one stratigraphic problem the worker may well be forced to make detailed subdivisions of these major types in order to establish trends or correlations, just as an igneous petrographer mapping a pluton that is entirely granite finds it useful to subdivide his rocks on percentage of quartz, hornblende:biotite ratio, color, or crystal size.

Although limestones are fundamentally simple rocks, the subject of limestone petrography is not something that one can plunge into immediately with "cookbook" in one hand and calculating machine in the other; it still has far more of the characteristics of an art than a science in identity of obscure features such as borderline allochems and recognition of recrystallization. The petrography of fossils (Johnson, 1951) and even of algae alone is a vast subject in itself. Interpretations of some problematical slides can not be proved, but must rest on the observers "feelings" and past experience; different "experts" may come to opposite conclusions. Such controversies are not unknown in geology, however, and are the seasonings that add zest to the subject.

REFERENCES

BUEHLER, EDWARD J., 1948, "The Use of Peels in Carbonate Petrology," *Jour. Sed. Petrology*, Vol. 18, pp. 71–73.

CAYEUX, LUCIEN, 1935, *Les Roches Sedimentaires de France, Roches Carbonatees*. Masson et Cie., Paris.

CUMINGS, E. R., AND SHROCK, R. R., 1928, "Niagaran Coral Reefs of Indiana and Adjacent States and Their Stratigraphic Relations," *Bull. Geol. Soc. America*, Vol. 39, pp. 579–620.

FOLK, ROBERT L., 1952, "Petrography and Petrology of the Lower Ordovician Beekmantown Carbonate Rocks in the Vicinity of State College, Pennsylvania," Ph.D. Dissertation, The Pennsylvania State College.

———, 1954, "The Distinction between Grain Size and Mineral Composition in Sedimentary Rock Nomenclature," *Jour. Geol.* Vol. 62, pp. 344–59.

———, 1956, "The Role of Texture and Composition in Sandstone Classification," *Jour. Sed. Petrology*, Vol. 26, pp. 166–71.

———, AND WEAVER, CHARLES EDWARD, 1952, "A Study of the Texture and Composition of Chert," *Amer. Jour. Sci.*, 250, pp. 498–510.

GRABAU, A. W., 1913, *Principles of Stratigraphy*. Seiler, New York.

HAM, WILLIAM E., 1954, "Algal Origin of the 'Birdseye' Limestone in the McLish Formation," *Proc. Oklahoma Acad. Sci.*, Vol. 33, pp. 200–03.

———, 1954, "Collings Ranch Conglomerate, Late Pennsylvanian, in the Arbuckle Mountains, Oklahoma," *Bull. Amer. Assoc. Petrol. Geol.*, Vol. 38, pp. 2035–45.

HATCH, F. H., RASTALL, R. H., AND BLACK, MAURICE, 1938, *The Petrology of the Sedimentary Rocks*. 383 pp. George Allen and Unwin, London.

ILLING, LESLIE V., 1954, "Bahaman Calcareous Sands," *Bull. Amer. Assoc. Petrol. Geol.*, Vol. 38, pp. 1–95.

JOHNSON, J. HARLAN, 1951, "An Introduction to the Study of Organic Limestones," *Colorado School Mines Quar.*, Vol. 46, No. 2.

KRYNINE, PAUL D., 1948, "The Megascopic Study and Field Classification of Sedimentary Rocks," *Jour. Geol.*, Vol. 56, pp. 130–65.

SANDER, BRUNO K., 1951, "Contributions to the Study of Depositional Fabrics (Rhythmically Deposited Triassic Limestones and Dolomites)," translated by ELEANORA BLISS KNOPF, Amer. Assoc. Petrol. Geol. Originally published in 1936.

UDDEN, JOHANN A., 1914, "Mechanical Composition of Clastic Sediments," *Bull. Geol. Soc. America*, Vol. 25, pp. 655–744.

WENTWORTH, CHESTER K., 1922, "A Scale of Grade and Class Terms for Clastic Sediments," *Jour. Geol.*, Vol. 30, pp. 377–92.

CLASSIFICATION OF CARBONATE ROCKS OF MARINE ORIGIN[1]

N. J. SANDER[2]
New York City, New York

ABSTRACT

Carbonate classifications in current use are to varying degrees subjective and interpretative. Objectivity in classification is highly desirable for an organization working with carbonates in widely separated localities because of the degree of uniformity in nomenclature and judgments concerning some of the properties of reservoir rocks that can be attained thereby. The classification presented here attempts objectivity and succeeds partly by using measured particle size as the basis for subdivision into major categories. Less objectively, it provides a method of distinguishing degrees and kinds of alteration of rock texture by the several processes of diagenesis. A still greater degree of subjectivity is introduced unavoidably in the identification of the grains and crystals comprising the carbonate. Nevertheless, following the procedures prescribed leads even the inexpert to an accurate if not complete name for any carbonate, and any number of observers will arrive at essentially the same designation. Not a part of the classification, but included for completeness, is a discussion of the genetic implications of the texture and composition of carbonate rocks.

INTRODUCTION

During recent years several classifications of carbonate rocks have been published. These classifications set up criteria for erecting and discriminating categories that together comprise all or a large part of the spectrum of the marine carbonates. Ham and Pray (1962) give an excellent discussion and comparison of nine of the classifications. These authors weigh most judiciously the relative merits of descriptive and genetic concepts of categorizing, and show that similarities between the classifications based on either one of these approaches are far more numerous than their differences.

In view of the existence of these papers by distinguished colleagues, a certain temerity is required to publish another discussion of the same subject. However, in the course of preparation of a manual for company geologists, none of the published material was found to be completely satisfactory for world-wide use. The favorable response accorded the following classification led to its preparation in a condensed form for general dissemination.

Those familiar with the literature of carbonate classification will note many similarities between the scheme presented here and that published by Powers (1962). The resemblances are not fortuitous, for both Powers and the writer studied the carbonate reservoirs of Saudi Arabia under the guidance of the late R. A. Bramkamp who first classified Arabian carbonates (Brankamp and Powers, 1958). His work, although subsequently much amended, laid a firm foundation for all later studies of these rocks.

Because petroleum geologists who work as a team must have a mutually comprehensible precise technical vocabulary, the classification proposed here is essentially descriptive. This approach is mandatory, for all concerned must use the same criteria to assign a specimen to its category, and all must be able to infer from the name alone the range of characteristics comprising that category.

Therefore, the name given a rock should be based on easily observable and, insofar as practicable, measurable physical properties. On the other hand, although subjective interpretation concerning the genesis of a carbonate must not be implicated in naming it, carbonates, like all sedimentary rocks, were formed in specific environments, and their physical characteristics in many places can be related to this environment, as suggested on Table III.

CLASSIFICATION BY PARTICLE SIZE

Among the properties of carbonates, grain or crystal size, shape, and spatial arrangement are most important in defining texture. In turn, because texture is related closely to porosity and permeability, use of particle size as a basis of classification tends to group rocks with similar reservoir properties. More significant is the fact that the size of crystals and grains can be measured objectively and more or less accurately. A micrometer in the eyepiece of a low-power (not more than 50×) binocular microscope permits

---

[1] Manuscript received, August 16, 1965; accepted, May 31, 1966. Published by permission of the American International Oil Company.

[2] American International Oil Company, 555 Fifth Avenue.

not only mensuration of the larger of these particles in the two dimensions visible on polished surfaces, thin sections, and peels, but also measurement and appraisal of interrelationships in the three dimensions exposed on the broken surfaces of rocks. Because categorization by particle size is a simple way to achieve relative objectivity in nomenclature, it is used as the basis of the classification proposed here.

The size scale for grains is that proposed by Wentworth (1922); the scale for crystals is similar, but is modified in consideration of the limited range of size attained by crystals in most sedimentary rocks.

Categories of grain size (Tables I, II, III, left-hand column).—Calcilutite and calcisiltite are treated here as separate categories primarily because some well-sorted coarse calcisiltite contains oil in producible quantities, whereas calcilutite or mixtures of clay- and silt-size carbonate particles commonly do not. Grains in the coarse silt range of size are only just resolvable under the binocular microscope at 12× magnification. Coarse particles in the clay-size range can be measured conveniently only at a magnification of 250× or more.

The term *lutaceous calcarenite* designates the very common carbonate rocks made up of both mud- and sand-size grains. The category is necessary not only because of the relative ubiquity of lutaceous calcarenite, but also because the fine-grade constituents strongly influence the reservoir properties of this rock type. Commonly where the larger particles of a lutaceous calcarenite are not mutually supporting, the porosity of a lutaceous calcarenite approaches that which would be measured in a rock made up of the finer fraction alone. Ideally, the limit between lutaceous calcarenite and calcarenite should be selected at the stage where mutual support between the larger grains or crystals starts. This stage corresponds with a fine-fraction component of 25 per cent or less, except where the coarser-grade particles are greatly elongate in one dimension and irregularly packed. Because the degree of mutual support is not easily measurable and consequently is subjective, a maximum of 10 per cent calcilutite or calcisiltite matrix is selected as the limit in differentiating calcarenite from lutaceous calcarenite. This choice is convenient, for porosity and permeability rise sharply where the fine fraction makes up less than 10 per cent of the rock mass. The maximum of 10 per cent coarse fraction that distinguishes calcilutite from lutaceous calcarenite is an arbitrary selection, chosen among a potentially unbroken series of lutite/arenite ratios that ordinarily have no relation to reservoir characteristics.

The coarser grades of clastic carbonates need little discussion. The size limits prescribed for calcirudite include Wentworth's "granule" and "pebble" categories, and those for coarse clastic carbonate his "cobble" and "boulder" classes. If mixtures of these coarser grades are found, compound names may be assigned them, on the basis of the principles used to distinguish lutaceous calcarenite from calcarenite.

The name *statobiolith* (Gr. *statos* = fixed, standing; Gr. *bios* = life; Gr. *lithos* = stone) is coined to accommodate those rocks composed mainly of the remains of sessile reef- —or shoal- —building organisms in their position of growth. The name *biolith* used by Folk (1959) and others in this sense is modified to the extent of adding a prefix, for Grabau (1960:1913) defined biolith in a much broader sense as the name of any rock made up of organic material or material formed by the physiological activities of organisms, plant or animal. Furthermore, the three roots which together make up statobiolith indicate its precise meaning, whereas the two constituting biolith do not.

Although not a category based on particle size, and consequently not amenable to measurement, the statobiolith is important in the study of carbonate reservoirs. It is mentioned here because on Tables I-III it is placed for convenience at the base of the column headed "Grain Size." A statobiolith may in itself be a good reservoir, and, more significantly, its recognition commonly aids in locating the more productive parts of some reservoirs.

Categories of crystal size (Tables I and II, right-hand column).—The range of size of each of the five named categories of crystals was selected to insure reasonable consonance with that of the corresponding grain-size class. However, because crystals in carbonate rocks occur commonly in a relatively modest range of dimensions, the size range of a crystal class does not necessarily coincide with that of what may be considered to be the equivalent grain class. Obviously, the discrepancy is greatest in the coarser grades. The class assignment of crystals with discrete boundaries is determined by measurement along the longest visible dimension.

Classification by Component Structure and Origin

The two main categories of particle found in carbonate rocks obviously are crystals and grains. Mud-size particles can not be so distinguished with the low-power (less than $50\times$ magnification) binocular microscope. Consequently, in the following discussion only particles of coarse silt, sand, and larger grades are considered in detail, because only these sizes are sufficiently coarse to permit easy observation of their structure, and to allow ready deductions concerning their origin.

The nearly total objectivity of a classification based on particle size can not be maintained in a diagnosis based on structure and origin, because ascribing an origin to a given particle is in fact a subjective process, an interpretation of observed characteristics. Some genetic implications are also unavoidable, but must be excluded rigorously from the name given the rock. Some grains are difficult to categorize. These should be ignored if they comprise only a small part of the total volume.

Categories of grains (Tables I, II, III, near top).—Grains making up carbonate rocks come from many sources, some of which can be ascertained readily. The origin or nature of certain other grains may be doubtful or indeterminate. To reduce interpretation to a minimum a grain is assigned one of the following six categories only if its type is indisputable.

(1) *Bioclast.*—This category includes essentially all debris derived from the supporting or protective structures of animals or plants, whether whole or fragmentary. Excluded are the minute crystals secreted by certain algae and alcyonarians as a part of their skeletons, but released on death as individual crystals indistinguishable from physicochemically precipitated crystals of the same composition.

(2) *Pellets.*—This name refers to discrete elongate ellipsoidal or rod-shape grains made up almost exclusively of calcilutite and lacking significant internal structure. Most pellets are of fine-grade sand size. Many probably are the feces of mollusks and worms; others presumably result from algal activity (Wolf, 1965), but no specific origin is implied in the term.

(3) *Oöliths.*—The term is used for ovate or spherical grains less than 2 millimeters in diameter along their major axis, and with a distinctive internal structure made up of calcite arranged in concentric layers, radial prisms, or both. A nucleus may be present. If so, it consists commonly of a pre-existing lithified particle of carbonate, a grain of silica, or of some other solid around which calcite is deposited. If this nucleus makes up more than half of the diameter of the oölith, the qualification "superficial" may be used. Grains more than 2 millimeters in diameter with oölitic structure are called *pisoliths*. A rock made up of oöliths is an oölite; if it consists of pisoliths, it is called a pisolite.

(4) *Terrigenous detritus.*—This category designates all grains and debris derived from pre-existing rocks other than carbonates by any process of weathering or erosion. Clays made up of the argillaceous minerals are the lutaceous grade of terrigenous detritus. Most silt and sand grains are quartz. Clay-, silt-, and sand-grade particles are more common in carbonate rocks than the coarser grades of terrigenous detritus.

(5) *Carbonate detritus.*—All grains and debris in carbonate rocks derived from a pre-existing lithified carbonate are called carbonate detritus. Recognition of carbonate detritus may be difficult, and must be based on distinctive color, texture, or fossil content. If uncertainty exists, the name must not be used.

(6) *Colloclasts* (Gr. *kolla* = glue and *klastos* = broken).—As defined here this term includes two sorts of weakly cemented aggregates of mud or fine silt, one autochthonous, the other para-autochthonous: (a) autochthonous, weakly cemented aggregates commonly attaining silt or sand size are in a few places formed on the sea floor in large numbers. These aggregates have a fairly complex internal structure, but their texture is like that of the stratum in which they occur. Those with a weakly lobate outline were called "grapestone" by Illing (1954); (b) sand- and gravel-size grains formed by the tear-

up, transport, and redeposition of fragments made up of silt-, sand-, or gravel-size carbonates weakly cemented prior to movement. These are para-autochthonous and penecontemporaneous with the indurated rock in which they now occur, for they could not have moved far without having reverted to their original state of many discrete particles.

Differentiation between the two types—both accretionary, but one not transported—is not attempted because distinguishing between them may prove difficult. If evidence for their reworking can be adduced, then colloclast may be replaced by the more specific term *intraclast* in the sense given it by Folk (1962) of "the complete spectrum of reworked contemporaneous carbonate sediment." As defined, colloclast and intraclast include some pellets, here considered apart because of their distinctive shape and lack of internal structure.

Categories of crystals.—Carbonates of marine origin composed predominantly of crystals are common. The criteria given in any textbook of mineralogy should be used to determine their composition. As an aid in identification, the stain tests proposed by Friedman (1959) can be employed with confidence. All crystals, except those obviously of organic derivation such as echinoid spines and plates, should be regarded as physicochemical in origin. However, Lowenstam (1955) and Wolf (1965) present evidence suggesting that the small crystals comprising a large part of some carbonate rocks are organic. In general, to distinguish very small anhedral crystals from grains of similar size is not practicable under the magnifications of less than 50 diameters available in the low-power binocular microscope.

Classification by Type and Degree of Alteration

Because the texture of a rock depends not only on the size, shape, roundness, surface characteristics, sorting, and packing of the grains and crystals of which it is composed, but also on the shape and size of the interstices between these particles, it is desirable to distinguish *texture* from *depositional fabric* in order to discriminate among the effects of certain phenomena causing alteration of texture. For this discussion *depositional fabric is defined as the shape, size, spatial arrangement, and mineralogic composition of the particles comprising a sedimentary rock.* This definition omits deliberately the intergranular and intercrystalline spaces that are an inherent component of texture. Further, it includes intentionally a mineralogical criterion, composition, in a list of requisite qualities that logically should be strictly morphic.

This concept of depositional fabric can be used to distinguish two classes of sedimentary carbonates: (1) a class in which, aside from compaction, the depositional fabric of the rock appears to be unchanged from that existing at the beginning of lithification, insofar as can be determined under the low-power binocular microscope; and (2) a class in which diagenesis or some other process of alteration has visibly changed one or several of the attributes of the depositional fabric.

Such a differentiation is valuable to the petroleum geologist, for it can be used not only as a starting point for the study of diagenetic phenomena, but also permits inferences concerning the porosity and the permeability of the rocks assigned each class. Tables I and II are so constructed that the distinction between the two classes is clearly indicated.

Discriminating Class 1 from Class 2 carbonates presents few problems where grains and crystals are coarser than silt size. However, to differentiate the effects of cementation from those of recrystallization in very fine-grade carbonates is much more difficult, even with recourse to the petrographic microscope. The general appearance of the rock is helpful in making the distinction, although it is not an infallible guide. Unaltered, weakly, or moderately cemented calcilutites are commonly dull, muddy- or chalky-looking, and may include fossil fragments, also unaltered. In thin section under magnifications of 200× or more, the very fine calcite particles (at least partly anhedral crystals, presumably chemically or biochemically precipitated and subsequently deposited as detritus) lack sharp boundaries, although the cement between them may be in clearly defined euhedral form. Megascopically, Class 2 fine-grade carbonates are denser, appear harder and more vitreous, and have a subconchoidal fracture. Under the petrographic microscope a thin section of microcrystalline Class 2 carbonate is usually a mosaic of tightly interlocking euhedral or subhedral crystals. The criteria given by Bathurst (1958) for differentiating the two categories of calcite spar must be used with caution.

Categories of alteration.—By definition Class 1 carbonates are subject only to cementation. Three categories of cementation are shown on Table II. Weakly cemented carbonates are those in which less than 5 per cent of the intergranular or intercrystalline space is filled by cement. In carbonates considered moderately cemented, 5–50 per cent of the pore space is filled with cement, and in carbonates classified as strongly cemented more than 50 per cent of the original void space between particles is occupied by calcite or some other mineral. The mineral comprising the cement should be identified. Most commonly it is calcite, but it may be silica, anhydrite, gypsum, barite, siderite, or iron oxide or sulfide.

Class 2 carbonates are divided into four groups based on the degree of alteration, as indicated on Tables I and II. Criteria for distinguishing between moderately and strongly altered carbonates are difficult to formulate because several processes of diagenesis may have affected the original fabric of the rock either simultaneously or consecutively. In general, if more than three-fourths of the original fabric have been changed or destroyed, the rock is considered to be strongly altered. The category, "relic texture only," is established for those strongly altered carbonates in which traces of the original fabric are still discernible, but only as "ghosts" in an essentially completely recrystallized rock. The "original texture obliterated" category comprises those crystalline carbonates in which no trace of an original clastic fabric remains. These rocks are, however, commonly closely associated with similar carbonates in which indications of Class 1 origin are still detectable.

No provision is made in the classification for "primary" dolomite. Sedimentary rocks apparently made up of carbonate crystals formed *in situ* and with no traces of a detrital or diagenetic origin for any significant component may require the establishment of a third class, "primary crystalline carbonate," for their accommodation. Presumably rocks of this kind are formed only under conditions other than those prevailing in "normal" marine environments.

Discussion of Tables I and II

The discriminatory power of a classification depends on the number of parameters used to define its categories. Table I provides the parameters necessary to classify a sedimentary carbonate according to the size and nature of the particles of which it is composed. It also indicates a way to distinguish the several stages of alteration of Class 2 carbonates. Table I does not include a means of differentiating between the several degrees of cementation of Class 1 carbonates.

Table II, in addition to providing categories for indicating the degree of cementation for Class 1 and moderately altered Class 2 rocks, gives empirically determined ranges of porosity and permeability for each of the grades of these carbonates, and indicates the effects of cementation on these ranges. An attempt is made also to suggest the effects of alteration on porosity and permeability. In particular, emphasis is placed on the increase in porosity commonly associated with dolomitization where rhombs of dolomite make up more than 75 per cent and less than 95 per cent of the rock volume. In making use of this criterion for estimating porosity, the mode of packing and the degree of development of faces on the crystals involved must be considered (Robinson, 1966).

Pore volume tends to decrease systematically as cement fills the intergranular space. If relative uniformity in the degree of cementation can be demonstrated horizontally or vertically, estimates of remaining reservoir porosity may be meaningful, and an evaluation of the effects of cementation on permeability may be practicable.

The effects of the alteration of depositional fabric on porosity and permeability are commonly less predictable than those of cementation. Leaching and solution need not be systematic. The effects of dolomitization and recrystallization on porosity may be more nearly consistent, but they commonly do not affect the fabric of the strata in which they occur precisely to the same degree throughout the reservoir. Consequently, optimum porosity and permeability may be developed in only a small part of the total rock volume. Cementation and recrystallization commonly decrease porosity and permeability (except in the restricted range of dolomitization previously dis-

GRAIN SIZE	SEDIMENTARY FABRIC NOT VISIBLY ALTERED (CLASS 1)						SEDIMENTARY FABRIC ALTERED (CLASS 2)			ORIGINAL TEXTURE OBLITERATED	CRYSTAL SIZE
	BIOCLASTS More than 25%	PELLETS More than 25%	OÖLITHS More than 25%	COLLOCLASTS More than 25%	CARBONATE DETRITUS More than 25%	TERRIGENOUS DETRITUS 10 to 50%	MODERATELY ALTERED Less than 75% xystals	STRONGLY ALTERED More than 75% xystals	RELIC TEXTURE ONLY		
CALCILUTITE (<.004 mm.)	BIOCLASTIC* CALCILUTITE (some chalk)	(Origin of mud-size particles is commonly indeterminate. If they are identifiable, specify, e.g. coccolithic chalk.)				MARL & IMPURE CALCILUTITE	PARTLY RECRYSTALLIZED OR DOLOMITIZED CALCILUTITE	Strongly altered carbonates should be classified according to their crystal size. The name, however, should reflect the original nature of the rock. Modifiers are used to show the type of alteration. A qualifying phrase should be added to indicate the presence of solution phenomena, e.g. STRONGLY DOLOMITIZED MEDIUM CRYSTALLINE OOLITIC CALCARENITE WITH SOME OOLITHS LEACHED AND OTHERS REPLACED BY SPAR CALCITE CEMENT.	MICROCRYSTALLINE LIMESTONE OR DOLOMITE WITH RELIC... TEXTURE	MICROCRYSTALLINE LIMESTONE OR DOLOMITE	MICRO- CRYSTALLINE (<.004 mm.)
CALCISILTITE[a] (.004 to .0625 mm.)	BIOCLASTIC* CALCISILTITE	PELLETIC CALCISILTITE	OÖLITIC CALCISILTITE	COLLOCLASTIC CALCISILTITE	DETRITAL CALCISILTITE	CALCISILTITE	PARTLY RECRYSTALLIZED OR DOLOMITIZED CALCISILTITE		FINELY CRYSTALLINE LIMESTONE OR DOLOMITE WITH RELIC... TEXTURE	FINELY CRYSTALLINE LIMESTONE OR DOLOMITE	FINELY CRYSTALLINE (.004 to .06 mm)
LUTACEOUS CALCARENITE[b]	BIOCLASTIC* LUTACEOUS CALCARENITE	PELLETIC LUTACEOUS CALCARENITE	OÖLITIC LUTACEOUS CALCARENITE	COLLOCLASTIC LUTACEOUS CALCARENITE	DETRITAL LUTACEOUS CALCARENITE	ARGILLACEOUS CALCARENITE	PARTLY RECRYSTALLIZED OR DOLOMITIZED LUTACEOUS CALCARENITE		MEDIUM CRYSTALLINE LIMESTONE OR DOLOMITE WITH RELIC TEXTURE	MEDIUM CRYSTALLINE LIMESTONE OR DOLOMITE	MEDIUM CRYSTALLINE (.06 to .25 mm.)
CALCARENITE[c] (.0625 to 2 mm.)	BIOCLASTIC* CALCARENITE	PELLETIC CALCARENITE	OÖLITIC CALCARENITE OR OÖLITE	COLLOCLASTIC CALCARENITE	DETRITAL CALCARENITE	ARENACEOUS CALCARENITE	PARTLY RECRYSTALLIZED OR DOLOMITIZED CALCARENITE		COARSELY CRYSTALLINE LIMESTONE OR DOLOMITE WITH RELIC TEXTURE	COARSELY CRYSTALLINE LIMESTONE OR DOLOMITE	COARSELY CRYSTALLINE (.25 to 1 mm.)
CALCIRUDITE (2 to 64 mm.)	"The name of the organisms making up the abundant bioclasts should be substituted when identifiable - e.g. "Algal", "Coralline", "Orbitoidal", "Nummulitic".	PELLETIC CALCIRUDITE	PISOLITE	COLLOCLASTIC CALCIRUDITE	DETRITAL CALCIRUDITE	COBBLE CALCIRUDITE	PARTLY RECRYSTALLIZED OR DOLOMITIZED CALCIRUDITE		VERY COARSELY CRYSTALLINE LIMESTONE OR DOLOMITE WITH RELIC... TEXTURE	VERY COARSELY CRYSTALLINE LIMESTONE OR DOLOMITE	VERY COARSELY CRYSTALLINE (>1 mm.)
COARSE CARBONATE (>64 mm.)					LIMESTONE CONGLOMERATE	CONGLOMERATIC LIMESTONE	PARTLY RECRYSTALLIZED OR DOLOMITIZED COARSE CARBONATE		RECRYSTALLIZED COARSE CARBONATE WITH RELIC... TEXTURE		
STATOBIOLITH	ALGAL OR CORAL REEF OR BIOHERM, MOLLUSCAN, CRINOID, OR ALGAL BIOSTROME OR BANK. (Not easily determinable in subsurface) OR ALGAL STATOBIOLITH, ETC.						PARTLY RECRYSTALLIZED STATOBIOLITH		CRYSTALLINE CARBONATE WITH INDICATIONS OF ORIGINAL STATOBIOLITHIC NATURE.		

1. BIOCLASTS — All debris made up of the supporting or protective structures of animals and plants, whole or in fragments. Distinction must be made between bioclasts consisting of small benthonic or planktonic organisms, and those comprised by the entire or fragmentary remains of sessile organisms, for their genetic significance may be very different.

2. PELLETS — Elongated ellipsoidal, or rod-shape grains without internal structure made up almost exclusively of calcilutite. When transported they comprise a portion of Folk's "intraclasts."

3. OÖLITHS — Ovate or spherical grains less than 2 mm. in diameter made up of distinctly laminated radial or concentric layers. Larger grains of the same type are called pisoliths. If a nucleus is present making up half the grain or more of the diameter, the term "superficial oölith" may be used. A rock made up of oöliths is an oölite.

4. COLLOCLASTS — Autochthonous or para-autochthonous penecontemporaneous weakly cemented aggregates of calcilutite or calcisiltite with complex internal structure, but lithologically like the rock in which they are found. If they have demonstrably been transported they comprise a portion of Folk's "intraclasts."

5. CARBONATE DETRITUS — Grains and debris derived from a pre-existing lithified carbonate rock.

6. TERRIGENOUS DETRITUS — Grains and debris derived from pre-existing rock other than carbonates.

a. Not more than 10 per cent coarser detritus in a fine-grained matrix.
b. Calcarenite with more than 10 per cent mud or silt-size matrix.
c. Calcarenite with less than 10 per cent mud or silt-size matrix.
d. Rock composed mainly of attached reef or shoal-building organisms in their position of growth.

TABLE I

CLASSIFICATION OF CARBONATES BASED ON PARTICLE SIZE, ORIGIN, AND DEGREE OF RECRYSTALLIZATION

PARTICLE GRADE	CLASS 1 — ORIGINAL FABRIC NOT VISIBLY ALTERED — CARBONATE ONLY OR UP TO 50% TERRIGENOUS CLASTICS					CLASS 2 — ORIGINAL SEDIMENTARY FABRIC ALTERED — LARGELY CARBONATE (High terrigenous content inhibits alteration)			CRYSTAL GRADE	FRACTURE & FISSURE POROSITY
	COMPACTED ONLY	WEAKLY CEMENTED	MODERATELY CEMENTED	STRONGLY CEMENTED	MODERATELY (Recrystallized or dolomitized)	STRONGLY (Recrystallized or dolomitized)	RELIC TEXTURE ONLY	ORIGINAL TEXTURE OBLITERATED		
CALCILUTITE (<.004mm.)	∅ – M to VH; K – M to VL	∅ – M to VL; K – VL to Nil	∅ – VL to Nil; K – VL to Nil	Porosity and Permeability Very Low to Nil	∅ – L to M; K – L to VL (dolomite 6-78%)	∅ – H (78 – 90% dol.) to VL (>95% dol.); K – H (78 – 90% dol.) to VL (>95% dol.)		In examples where they resemble similar rocks showing relic textures, most dolomite showing no trace of the original rock texture, together with similar limestone made up of sparry calcite formed by grain growth, can more or less confidently be assigned to this class. Other carbonates with no remaining indication of origin and made up of crystals of large size are not assignable genetically, but are put in this class arbitrarily. Porosity and permeability usually are very low, but may exist because of solution or intercrystalline voids. So-called "primary" dolomite is arbitrarily included in this class. This is found commonly with evaporites and may truly represent a distinct category of dolomite, deposited as such.	MICROCRYSTALLINE (<.004mm.)	Fissures formed by solution along bedding planes or fractures are not uncommon in limestone, but these and other karstic phenomena are more numerous in rocks lying well above the water table. Where present in oil wells, fissures can be very efficient collecting locales for oil from finer-grained rocks around the fissure. Many oil fields exist because of fracturing in rocks which at least in part are too fine-grained to yield oil in quantity otherwise. Every effort must be made to determine the orientation and number of fractures, as well as their dimensions, and the number filled with calcite or other mineral.
CALCISILTITE (.004 to .0625mm.)	∅ – M to VH; K – M to VL	∅ – M to VL; K – L to VL	∅ – VL to Nil; K – VL to Nil		∅ – M to L; K – M to VL (dolomite 6-78%)	∅ – H (78 – 90% dol.) to VL (>95% dol.); K – H (78 – 90% dol.) to VL (>95% dol.)	Ghosts of pre-existing textures still present do not affect porosity or permeability. If channeling, vugs, or molds are developed as secondary solution phenomena, porosity and permeability may increase from their commonly very low values in this type of rock, which is usually predominantly either dolomite or sparry calcite developed by recrystallization. Intergranular as well as intercrystalline porosity may exist rarely in dolomite with relic texture as indicated in the column on the left.		FINELY CRYSTALLINE (.004 to .06mm)	
LUTACEOUS CALCARENITE[a]	∅ – M to VH; K – M to VL	∅ – M to VL; K – L to Nil	∅ – VL to Nil; K – VL to Nil		∅ – M to L; K – M to VL				MEDIUM CRYSTALLINE (.06 to .25mm)	
CALCARENITE (.0625 to 2mm)	∅ – H to VH; K – L to VH	∅ – M to H; K – M to H	∅ – M to VL; K – M to VL		∅ – H to VL; K – H to VL	If recrystallization is accompanied by preferential solution so that voids replace oölites or fossils, moderate porosity and permeability may be developed. Strong dolomitization, replacing calcarenite, etc. may yield ∅ and K values as indicated under calcilutite. Strongly recrystallized rocks may contain small or large vugs which increase porosity. These are considered to be solution phenomena.			COARSELY CRYSTALLINE (.25mm)	
CALCIRUDITE (2 to 64mm.)	∅ – M to VH; K – M to VH	∅ – M to H; K – M to H	∅ – M to Nil; K – M to Nil		∅ – M to Nil; K – H to Nil				VERY COARSELY CRYSTALLINE (>1mm)	
COARSE CARBONATE (>64mm.)	∅ and K variable but commonly M to H	∅ and K variable but commonly M to L	∅ and K variable but commonly M to VL		∅ and K variable but commonly M to VL					
STATOBIOLITH[b]	∅ and K variable but commonly M to H	∅ and K variable but commonly M to L	∅ and K variable but commonly M to VL		∅ and K variable but commonly M to VL					

a. Calcarenite with more than 10 per cent mud or silt-size matrix.
b. Rocks composed mainly of attached reef- or shoal-building organisms in their position of growth.

KEY

∅ = Porosity
K = Permeability
Nil = 0
VL = Very Low   ∅ = .1 – 3%   K = .1 – 10 md.
L = Low   ∅ = 3 – 8%   K = 10 – 50 md.
M = Moderate   ∅ = 8 – 16%   K = 50 – 300 md.
H = High   ∅ = 16 – 25%   K = 300 – 1000 md.
VH = Very High   ∅ = >25%   K = >1000 md.

The ranges of porosity and permeability indicated in the several categories of rock are those most likely to be encountered, but may not include the entire range of these properties in the category.

TABLE II

CARBONATE TEXTURES AS RELATED TO POROSITY-PERMEABILITY RANGES

PARTICLE GRADE	SIGNIFICANT PORTION BIOCLASTIC	SIGNIFICANT PORTION CARBONATE DETRITUS	SIGNIFICANT PORTION OÖLITIC	SIGNIFICANT PORTION PELLETIC	SIGNIFICANT PORTION COLLOCLASTIC	SIGNIFICANT PORTION TERRIGENOUS CLASTICS
CALCILUTITE (<.004mm.)	The origin of limestone made up of mud and silt-size particles is in general indeterminable. It may be assumed that most particles of mud size form by inorganic chemical processes which cause precipitation from sea water of minute anhedral or subhedral crystals of calcite. Another source, probably of local importance volumetrically, is the fine fraction of detritus produced by disintegration or abrasion of larger fragments of lithified carbonates of any kind. A third source, possibly in some areas producing a considerable volume of fine detritus, is the organisms which secrete aragonite as discrete crystals. These, released on death and disintegration of the organism, could contribute significantly to the mud-size fraction of sediments, in which lithification would change aragonite to calcite. Some chalk, present over very extensive areas, is made up predominantly of the tests of coccoliths and Foraminifera, the coccoliths identifiable only at high magnification (X600). Other chalk is weakly compacted, essentially uncemented calcilutite of chemical or detrital origin. Where cementation or recrystallization modifies the friable and porous character of a chalk, the term is no longer applicable. Some calcisiltite is made up almost exclusively of pellets formed from calcilutite and shaped by the activity of organisms. Others are made up of grains formed by physico-chemical accretion of mud-size particles and others by concentration of silt-size particles because of winnowing by currents. Commonly, silt and mud-size particles occur in variable proportions in a predominantly fine-grained rock. Some silt includes a sizable proportion of broken foraminiferal tests.					
CALCISILTITE (.004 to 0.0625mm.)						
LUTACEOUS CALCARENITE[a] (.0625 to 2 mm.)	The sessile adult forms of many marine animals and plants can live only in a stable and restricted environment, narrowly definite as to temperature, water depth, salinity, turbidity, oxygen and food supply, strength of currents, and other ecological conditions. Consequently, the presence of significant amounts of unworn remains of one or more of these stenobionomic organisms suggests that conditions favorable to their growth existed at no great distance from the site of deposition of the rock being examined. Other fixed plants and animals are more tolerant of fluctuations in environmental factors, and can live under periodic change in their surroundings. The presence of significant amounts of eurybionomic organisms is less diagnostic than that of stenobionomic species, but is still a good indication of the conditions prevailing. The presence of unattached benthos is even less significant, and plankton in most strata have no genetic import.*	The presence of sand or coarser grade grains of allochthonous lithified carbonate as part of the constituents of a rock reflects an environment in which wave and current energy was strong enough to have transported these grains from a site of erosion to the site of deposition. If the CaCO₃ sand is associated with calcilutite or calcisiltite, the sorting action of waves and currents was not great enough at the moment of deposition or immediately thereafter to winnow out finer materials. If nearly all of the rock is made up of sand-size grains, it may be presumed that winnowing did take place, the amount shown by the degree of sorting.	Oöliths are formed in a turbulent environment by accretion of radial or concentric layers of CaCO₃ either around a minute grain of limestone commonly subsequently recrystallized, or around a quartz grain, shell fragment, or other suitable host, since sea water becomes supersaturated with CaCO₃ as a result of heating and agitation. This heating occurs only in shallow water, commonly when free circulation is somewhat restricted. Presumably oöliths in which the crystals are radially disposed are formed in quieter water than concentrically layered oöliths. Along with oöliths may occur calcite (originally aragonite), coated lumps of accretionary material. Once formed, oöliths can be transported like any other grain of similar size.	Most pellets are presumably formed by organisms and represent fecal deposits of these organisms later lithified. If the genesis of the pellet could be tied to the environment in which the organism lived. In any event, as the pellets are made up of silt- and clay-size particles, the organism lived on or in mud high in calcium carbonate. Once formed and slightly hardened, pellets are like any other detrital grain. They can be transported by water currents. Some pellets may be of physico-chemical origin. Presumably an environment of relatively high energy would be involved in the formation of pellets of this type. Some pellets are intraclasts if Folk's definition is accepted.	Composite grains or "colloclasts" are formed by weak cementation of clay-or-silt size previously discrete particles on the sea floor. Commonly, these aggregates form only where water is saturated with CaCO₃, so that heating or agitation which lower CO₂ content causes precipitation of aragonite as a cement. Other colloclasts are produced by algal cementation or other obscure processes which serve to consolidate fine materials. Rocks in which the lobate outline of many may be distinguished can be assigned a shallow-water origin more or less confidently. On the other hand, cemented aggregates where transported are probably indistinguishable in many cases from other detrital grains, and have, in any event, less primary genetic significance. Colloclasts which can be recognized as having been transported may be called "intraclasts."	The existence of terrigenous clastics in a predominantly carbonate sequence indicates that erosion of lands or land-derived rocks was going on at the same time that local carbonate deposition favored carbonate deposition. The more common terrigenous clastics in carbonate rocks are mud and silt. Thick sections of calcareous mudstone and argillaceous limestone exist throughout the temperate and torrid zones. Reefs and biostromes of most sessile animals cannot survive in turbid water from which rapid deposition of terrigenous clastics is taking place. In some localities calcarenite includes modest amounts of sand grains, which are fairly well sorted. If these grains are predominantly quartz, provenance from a beach sand or a sandstone can be postulated. Slightly arenaceous limestone is commonly relatively free of silt and clay but sessile organisms are not abundant in these rocks.
CALCARENITE[b] (.0625 to 2 mm.)						
CALCIRUDITE (2 to 64 mm.)		Very coarse carbonate gravels, cobbles, or boulders have not been moved very far from their site of erosion, save under exceptional circumstances such as submarine mud flow transport or ice rafting.				
COARSE CARBONATE (>64mm.)	Obviously, strongly weathered and rolled fragments of organisms cannot serve as indicators of the growth environment at the site of their deposition.					
STATOBIOLITH[c]	Strata made up predominantly of organisms still in their position of growth offer unmatched opportunities to determine the genesis of the rock unit and the conditions under which it was formed. A knowledge of the ecology of the organisms preserved is of course necessary in order to make the interpretations required. Most of the statobioliths were formed in shallow water, commonly subject to current and possibly wave action, in localities in which little or no terrigenous clastic was being deposited.					

TABLE III
GENETIC ATTRIBUTION OF MARINE CARBONATES

	BIOCLASTIC	Describes all debris (transported discrete fragments) recognizably of organic origin. In addition to corals, algae, crinoids, bryozoa, stromatoporoids, and attached molluscs with shells, Foraminifera and ostracodes are included, although the genetic significance of the presence of these latter is different from that of organisms sessile in the adult state.
	CARBONATE DETRITUS	Includes all sedimented, lithified, and reworked carbonate particles, regardless of the length of time elapsed between their deposition and later reworking.
	OÖLITIC	Comprises both true oöliths with radial or concentric structure, and superficial or "pseudo"-oöliths in which a smooth coating of accretionary nature covers a previously existing grain of any kind. Pisoliths are placed in this class.
	PELLETIC	Designates discrete elongate ellipsoidal grains made up of calcilutite and lacking significant internal structure. Most pellets are of fine-grade sand size. Many are thought to represent fecal deposits of worms and other organisms, in which case their occurrence in any one rock may fall within narrow size limits.
	COLLOCLASTIC	Aggregates of mud or silt-size grains typically possessing surficial irregularities as defined by Illing (1954), with lobate outline and a texture similar to the stratum in which they occur are called "colloclasts" in this classification. The origin is probably sea-floor weak cementation (e.g. grapestone, bahamite). Where transported from their site of formation, these aggregates, usually not over 2 mm. in diameter, can be called intraclasts. Where doubt exists concerning their transport they are better called colloclasts. In any event, to distinguish autochthonous colloclasts from only slightly abraded intraclasts in older rocks is certainly difficult. Other accretional phenomena include algal colloclasts (algae form irregularly concentric rings in a fine-grained sediment), marine grass lumps, and nodules constructed around a shell fragment. These are commonly transported, and hence classed as intraclasts.
	TERRIGENOUS CLASTIC	All material derived by conventional erosional processes from pre-existing igneous, metamorphic, or sedimentary rocks, excluding carbonates. Most clastics associated with carbonates are mud, silt, or sand size.

a. Calcarenite with more than 10 per cent mud or silt-size matrix.

b. Calcarenite with less than 10 per cent mud or silt-size matrix.

c. Rocks composed mainly of attached reef or shoal-building organisms in their position of growth.

* Some benthonic Foraminifera, for example, are in life restricted to relatively narrow ranges of temperature, salinity, and depth, but at death may be scattered much more widely, so that thanatocoenoses with which paleoecologists are concerned exclusively are commonly less diagnostic of environment than is desirable (Fagerstrom, 1964). Therefore, in environmental studies distinction must be made between the existence of the complete tests of small benthos not sessile in the adult stage, and the entire or fragmentary remains of sessile organisms.

cussed), whereas leaching and solution increase porosity.

In summary, Table II takes into account three types of porosity: (a) *intergranular* and *intercrystalline porosity*, which depends on the size, shape, and spatial arrangement of the particles comprising the rock; (b) *solution porosity*, which commonly is caused by preferential leaching of one or more of the rock components (fossils, oöliths, mineral crystals) leaving molds, vugs, or voids; and (c) *fracture porosity* created by compression or tension which forms linear openings ranging from a fraction of a millimeter to centimeters in width, and from a few centimeters to hundreds of meters in breadth and height.

Fracture porosity may be augmented by solution along the surfaces formed by the breaking up of the rock with the result that fissures and channels are formed. On the other hand, many fractures are sealed by cement rather than enlarged by solution. Stylolites are presumably formed by solution and pressure (Dunnington, 1954), but the concentration of clay or silt along their traces commonly prevents their constituting an important source of porosity in reservoirs.

Table II does not include a category for the porosity formed by the activities of organisms that disrupt or modify the original texture and commonly leave small openings of regular or irregular shape. Neither does it provide for the naming of the discrete types of solution porosity resulting from the solution of algal mats or of shell banks. Furthermore, it does not attempt to distinguish openings in carbonates which could conceivably be produced by slumping or sliding in relatively unconsolidated rocks without the intervention of organisms.

Nevertheless, the careful use of Tables I and II combined with close study of the specimen under the low-power binocular microscope should provide a precise name for almost any marine carbonate seen by the petroleum geologist. The most abundant category of grain should appear first in the name assigned; *e.g.*, "algal, oölitic, moderately cemented, lutaceous calcarenite"; "pelletic, colloclastic, weakly cemented, argillaceous calcisiltite"; and "partly dolomitized, bioclastic, oölitic, strongly calcite-cemented lutaceous calcarenite." The names are long, and may appear awkward compared with those of other classifications for which abbreviations have been coined, but they are precise, and the names given any one specimen by several individuals trained in the use of the classification should be nearly identical. Abbreviations might be agreed on, but standardizing them and learning them could be difficult; *e.g.*, moderately cemented "biopelarenite"; strongly recrystallized "oölutarenite" with voids in place of oöliths; and strongly dolomitized "detsiltite."

GENETIC ATTRIBUTION OF CARBONATE ROCKS

Table III is a summary of the genetic significance of each of the several types of allochthonous and para-autochthonous grains comprising the bulk of many carbonates, but it does not provide for an analysis of the environmental implications of the total population of heterogeneous grains which make up some carbonate rocks. The sedimentary environment under which Class 1 and some Class 2 carbonates were formed can, however, in many cases be determined from their mineralogic, biotic, and textural characteristics. These together are called the rock "facies" which can be defined as the sum of all of the characteristics of a rock which lead to its recognition as having been formed under the more or less clearly delimited conditions which together constitute a sedimentary environment.

The study of facies involves several disciplines, and it is uncommon that all of the parameters comprising sedimentary environment can be identified with more than a modicum of certitude. For example, as pointed out by Folk (1962, p. 76–80), it is indeed difficult to estimate at what depth many carbonates were formed. On the other hand, criteria for determining water turbulence probably are somewhat more precise. Plumley *et al.* (1962), using textural properties and biotic constitution, set up an "Energy Index" by which carbonates are classified according to the energy level of the water in which they were formed.

In commercial work the study of facies is important for several reasons. (1) The existence of intergranular porosity in Class 1 carbonates stems mainly from their texture. The conditions that produce a particular texture can be related to a specific environment and, from this, deductions can be made concerning the vertical and lateral extent of the facies characteristic of the environment, particularly if the configuration of the sea floor controlling its distribution can be delimited. Thus, study of facies is invaluable in planning the location of development wells in a carbonate reservoir, and also in suggesting sites for additional wildcats. (2) Outstanding among the problems that can be solved in part by study of facies and paleogeography are the orientation and delimitation of biostromal and reefoid barriers and fore-barriers, the two commonly making up the reservoir. Each is characterized by a relatively high energy index, distinct from that of the lagoonal or shelf facies usually present shoreward, and from that of the basinal facies on the seaward side of the barrier (Henson, 1950). (3) In attempting to reconstruct the history of any sedimentary basin a study of facies helps to determine the transgressions and regressions of the sea, and to define lócal and regional movements of the sea floor, both of which may influence petroleum accumulation. Investigation of facies is useful not only in establishing the sequence of events in one locality, but also in correlating between localities, because the facies of any lithologic unit formed in the sea is a response to, and is in exact balance with, the epeirogenic and orogenic conditions existing at the moment of its deposition. Any change in this condition, no matter how slight, will be reflected in the facies. Though this is an oversimplification, facies and structure are but the two sides of a single coin.

It is beyond the scope of this discussion to analyze the environmental significance of the several groups of fossils commonly found in carbonate rocks, but it may be assumed that in general the presence of algal fragments in quantity suggests relatively shallow water because these plants grow only in the presence of light. The same may be said for the several types of corals. Many other animals and plants found as fossils were also restricted in the range of environmental conditions under which they could exist, but the fact always must be considered that the fossil assemblage (thanatocoenose) may be much different, because of transport, action of scavengers, and relative frequency of preservation, from that which existed in life at a particular site.

Nevertheless, fossils offer one of the most precise means of determining sedimentary environment if used with discrimination. Nearly all organisms of a sedentary habit are restricted in the range of light, temperature, salinity, oxygenation, turbidity, and energy which they can tolerate, thus offering by their presence as fossils some indication that the environment in which they lived existed at the same time and, commonly, at no great distance from the site of the burial of their remains. If these remains have been transported any considerable distance, they are generally worn and broken, and their environmental significance can be appropriately discounted.

Conclusions

A classification of carbonate rocks of marine origin based on particle size is practicable. Its use is recommended where a comparison of the characteristics of carbonates from many localities is required, for it is sufficiently objective to permit direct collation of the work of many individuals whose knowledge of carbonates may range widely in degree and kind. The means provided in the classification for distinguishing types and amounts of alteration, primitive as they are, are not a part of many other classifications, although the importance of diagenesis to reservoir studies is obvious.

A weak point in the classification is the omission from the name given the carbonate of those grain constituents making up less than 25 per cent by volume of the rock mass. This omission may, in some cases, be deleterious to a study of facies. Under such circumstances, either the rule may be relaxed or a qualifying phrase appended to the designation.

The classification was prepared mainly to standardize within a large and far-flung organization the analysis and description of carbonate rocks. If it serves as a firm foundation for the more advanced and more universally applicable scheme which replaces it, the ends for which it was devised will have been achieved.

Selected References

Bathurst, R. G. C., 1958, Diagenetic fabrics in some British Dinantian limestones: Liverpool and Manchester Geol. Jour., v. 2, p. 11–36.

Beales, F. W., 1958, Ancient sediments of Bahaman type: Am. Assoc. Petroleum Geologists Bull., v. 42, p. 1845–1880.

Bissell, H. J., and G. V. Chilingar, 1958, Notes on diagenetic dolomitization: Jour. Sed. Petrology, v. 28, p. 490–497.

Bramkamp, R. A., and R. W. Powers, 1958, Classification of Arabian carbonate rocks: Geol. Soc. America Bull., v. 69, p. 1305–1318.

Campbell, C. V., 1962, Depositional environment of Phosphoria Formation (Permian) in southeastern Bighorn basin, Wyoming: Am. Assoc. Petroleum Geologists Bull., v. 46, p. 478–503.

Dunham, R. J., 1962, Classification of carbonate rocks according to depositional texture: Am. Assoc. Petroleum Geologists Mem. 1, p. 108–121.

Dunnington, H. V., 1954, Stylolite development postdates rock induration: Jour. Sed. Petrology, v. 24, p. 27–49.

Fagerstrom, J. A., 1964, Fossil communities in paleoecology: their recognition and significance: Geol. Soc. America Bull., v. 75, p. 1197–1216.

Feray, D. E., E. Heuer, and W. G. Hewatt, 1962, Biological, genetic, and utilitarian aspects of limestone classification: Am. Assoc. Petroleum Geologists Mem. 1, p. 20–32.

Folk, R. L., 1959, Practical petrographic classification of limestones: Am. Assoc. Petroleum Geologists Bull., v. 43, p. 1–38.

—— 1962, Spectral subdivisions of limestone types: Am. Assoc. Petroleum Geologists Mem. 1, p. 62–84.

Friedman, G. M., 1959, Identification of carbonate minerals by staining methods: Jour. Sed. Petrology, v. 29, p. 87–97.

—— 1964, Early diagenesis and lithification in carbonate sediments: Jour. Sed. Petrology, v. 34, p. 777–813.

Ginsburg, R. N., 1957, Early diagenesis and lithification of shallow-water carbonate sediments in south Florida: Soc. Econ. Paleontologists and Mineralogists Spec. Pub. 5, p. 80–100.

Grabau, A. W., 1960; 1913, Principles of stratigraphy: v. 1, New York, Dover Pub. Inc., 581 p.

Ham, W. E., and L. C. Pray, 1962, Modern concepts and classifications of carbonate rocks: Am. Assoc. Petroleum Geologists Mem. 1, p. 2–19.

Henson, F. R. S., 1950, Cretaceous and Tertiary reef formations in the Middle East: Am. Assoc. Petroleum Geologists Bull., v. 34, p. 215–238.

Illing, L. V., 1954, Bahaman calcareous sands: Am. Assoc. Petroleum Geologists Bull., v. 38, p. 1–95.

Le Blanc, R. J., and J. G. Breeding, 1957 (eds.), Regional aspects of carbonate deposition: Soc. Econ. Paleontologists and Mineralogists Spec. Pub. 5, 178 p.

Leighton, M. W., and C. Poindexter, 1962, Carbonate rock types: Am. Assoc. Petroleum Geologists Mem. 1, p. 33–61.

Lowenstam, H. A., 1955. Aragonite needles secreted by algae and some sedimentary implications: Jour. Sed. Petrology, v. 25, p. 270–272.

Murray, R. C., 1960, Origin of porosity in carbonate rocks: Jour. Sed. Petrology, v. 30, p. 59–84.

Nelson, H. F., C. W. Brown, and J. H. Brineman, 1962, Skeletal limestone classification: Am. Assoc. Petroleum Geologists Mem. 1, p. 224–252.

Pettijohn, F. J., 1957, Sedimentary rocks: 2d ed., New York, Harper and Bros., 718 p.

Plumley, W. J., G. A. Risley, R. W. Graves, Jr., and M. E. Kaley, 1962, Energy index for limestone interpretation and classification: Am. Assoc. Petroleum Geologists Mem. 1, p. 85–107.

Powers, R. W., 1962, Arabian Upper Jurassic carbonate reservoir rocks: Am. Assoc. Petroleum Geologists Mem. 1, p. 122–192.

Rich, M., 1964, Petrographic classification and method of description of carbonate rocks of the Bird Spring Group in southern Nevada: Jour. Sed. Petrology, v. 34, p. 365–378.

Robinson, R. B., 1966, Classification of reservoir rocks by surface texture: Am. Assoc. Petroleum Geologists Bull., v. 50, p. 547–559.

Sarin, D. D., 1962, Cyclic sedimentation of primary dolomite and limestone: Jour. Sed. Petrology, v. 32, p. 451–471.

Stehli, F. G., and J. Hower, 1961, Mineralogy and early diagnesis of carbonate sediments: Jour. Sed. Petrology, v. 31, p. 358–371.

Wentworth, C. K., 1922, A scale of grade and class terms for clastic sediments: Jour. Geology, v. 30, p. 377–392.

Weyl, P. K., 1960, Porosity through dolomitization: conservation of mass requirements: Jour. Sed. Petrology, v. 30, p. 85–90.

Wolf, K. H., 1965, Petrogenesis and paleoenvironment of Devonian algal limestones of New South Wales: Sedimentology, v. 4, p. 113–178.

DOLOMITE IN PERMIAN LIMESTONES OF WEST TEXAS[1]

WILLIAM A. CUNNINGHAM[2]
San Angelo, Texas

ABSTRACT

This paper presents the results of a series of calcium and magnesium determinations made on well samples from the Permian limestones of West Texas. With the exception of some isolated areas, these limestones have a high magnesium content, and probably should be classified as dolomite limestones. No widespread relationship exists between magnesium content and the porosity of the limestone.

There is wide variation of opinion regarding the relative amounts of limestone and dolomite in the so-called "Big lime" of the Permian basin of West Texas. A few petrographic studies have been made, and some differentiation attempted on the basis of microscopic examination and rate of reaction with 10 per cent hydrochloric acid. Such "determinations" as the latter are, at best, mere approximations. Hence, a series of chemical analyses was started in order to clarify, if possible, some of the confusion and differences now existing.

In making this study on the basis of chemical analysis alone, it is realized that, while the relative percentage of calcium carbonate and magnesium carbonate are easily determined, the physical form in which these constituents exist in the original formation is not determined. True dolomite is a definite mineral composed of equi-molecular quantities of calcium and magnesium carbonate, or a weight ratio of 54.26 per cent $CaCO_3$ and 45.74 per cent $MgCO_3$. It is realized that very commonly impurities decrease these percentages and that in very few rocks is the exact 1:1 molecular ratio encountered in natural deposits.

The term "dolomite" is generally used to designate any limestone which contains appreciable quantities of magnesium carbonate, though such terminology is ambiguous. Concerning the nomenclature of dolomite and related deposits, Twenhofel says:[3]

---

[1] Manuscript received, August 3, 1935.

[2] Chemical engineer, University Lands, 805 San Angelo National Bank Building. Introduced by Hal P. Bybee.

[3] W. H. Twenhofel *et al.*, *Treatise on Sedimentation*, 1st ed. (1926), pp. 251–52. Williams and Wilkins, Baltimore, Maryland.

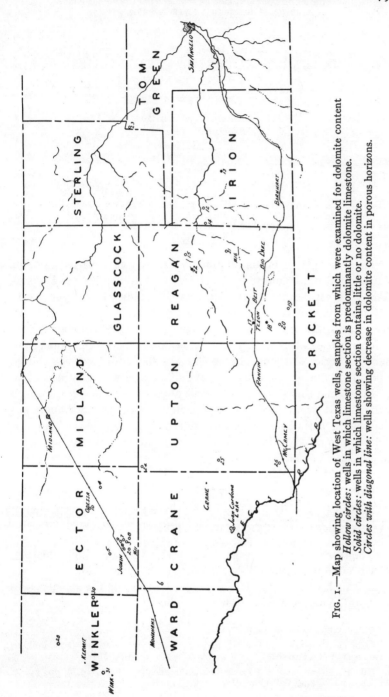

FIG. 1.—Map showing location of West Texas wells, samples from which were examined for dolomite content
Hollow circles: wells in which limestone section is predominantly dolomite limestone.
Solid circles: wells in which limestone section contains little or no dolomite.
Circles with diagonal line: wells showing decrease in dolomite content in porous horizons.

Dolomite limestones are those which are dominantly composed of dolomite. There are probably all gradations between calcite and dolomite limestones, but the occurrences are few in which the calcite and the dolomite occur in approximately equal proportions. Limestones which have less dolomite than justifies their assignment to dolomite limestones may be designated dolomitic limestones. Magnesium carbonate also exists in limestones in other forms than as dolomite, being either in solid solution in calcite or in isomorphous mixture therewith. Calcite is also miscible with dolomite, in each case the mineralogical properties of the mineral with which the mixing takes place not being affected. Limestones in which magnesium carbonate occurs not in the form of dolomite may be termed magnesium limestones.

Calcium and magnesium determinations have been made on lime samples from 35 wells. Approximately 150 analyses have been made, ranging from 1 to 18 samples from each well. All analyses were made by standard analytical methods.[4] In the absence of petrographic evidence to the contrary, and for the purpose of simplification, all magnesium found to be present in the limestones examined has been assumed to be in the form of the carbonate. Likewise, all magnesium carbonate has been assumed to be in the form of dolomite. The latter assumption is undoubtedly open to question, since the two carbonates may conceivably be present as a physical mixture of the two and not as the combined mineral.

The criterion for comparison used in this study is the molecular ratio of calcium carbonate to magnesium carbonate. This is the ratio obtained when the weight percentages of the two compounds are divided by their respective molecular weights. For example, pure dolomite is composed of 54.26 per cent calcium carbonate and 45.74 per cent magnesium carbonate, by weight. Since the molecular weights of these two compounds are, respectively, 100 and 84.3, the molecular ratio is $\frac{54.26}{100} : \frac{45.74}{84.3} :: 1 : 1$.

Actual dolomite percentage is easily calculated by dividing the percentage of magnesium carbonate in the sample by 45.74. However, this datum is not reliable as a basis of comparison because it is affected by the presence of impurities such as sand, pyrite, and shale in the sample. The use of the molecular ratio eliminates any error from this source since only acid-soluble constituents are considered.

In Table I are listed the data from the most important wells examined.

[4] F. P. Treadwell and W. T. Hall, *Qualitative Analysis*, Vol. II, 6th ed. (1924), p. 87.

TABLE I*

Interval in Feet (Depth)	Per Cent $CaCO_3$	Per Cent $MgCO_3$	Molecular Ratio $\frac{CaCO_3}{MgCO_3}$	Per Cent Dolomite	Description and Remarks
CRANE COUNTY					
1. Gulf, Edwards 1-D. Public School Land, Block 18, Section 21, 330 feet from S. and E. lines. Elevation 2,657 feet. Top Blaine limestone, about 2,910 feet					
2,885–90	49.8	31.0	1.35	68	Brown, Whitehorse
3,070–75	52.3	32.4	1.36	71	Light gray, crystalline, some calcite
3,352–62	50.2	36.4	1.16	80	Dark gray, some black shale
3,500–10	55.8	37.0	1.27	81	Gray, granular, crystalline. 1 boiler of water per hour at 3,500–10 feet
3,950–57	55.4	37.8	1.23	83	Soft, white
4,250–52	51.8	41.5	1.05	91	White
4,576–81	51.2	39.5	1.09	86	Dark gray, oil-stained
ECTOR COUNTY					
2. J. S. Cosden, Connell 1-A. Public School Land, Block B-16, Section 13, 1,320 feet from N. and E. lines. Elevation, 2,847 feet. Top Blaine limestone, 3,200 feet					
3,200–02	51.3	38.6	1.12	84	Light gray
3,248–55	51.9	40.0	1.10	88	Light gray
3,290–95	46.3	37.6	1.04	82	Light gray; some gray sand and shale
3,330–38	51.1	40.8	1.05	89	Light gray; considerable pyrite
3,360–69	53.1	42.3	1.06	93	Light gray, hard
3,402–09	49.1	38.6	1.07	84	Cream-colored dolomite, soft
3,443–50	32.9	26.4	1.05	58	Dark gray-brown, sandy, oil-stained; top of pay 3,436 feet
3,495–98	52.3	42.4	1.04	93	Medium gray
3,547–50	56.2	39.2	1.21	86	Mottled gray-brown, some pyrite; oil-stained; some quartz
3,598–3,603	54.1	37.3	1.22	82	Dark gray
3,648–53	47.8	38.5	1.24	84	Medium gray
3,703–08	55.3	38.2	1.22	84	Light gray
3,750–56	55.9	39.7	1.18	87	Very light gray
3,798–3,805	58.1	36.0	1.36	79	Light gray
3,855–68	52.8	38.2	1.16	84	Medium gray
3,892–99	54.0	41.1	1.11	90	Medium gray-brown
3,925–40	56.9	42.6	1.13	93	Light gray-tan
3,975–87	54.5	42.3	1.09	90	Fine light gray
3. Cosden, University 1-B. University Land, Block 35, Section 2, 1,650 feet from N. and 440 feet from E. lines. Elevation, 2,890 feet. Top Blaine limestone, 3,260 feet					
3,120–30	54.9	31.6	1.46	69	Brown; some anhydrite, Whitehorse
3,304–12	52.5	39.0	1.13	85	Very light gray
3,575–80	54.9	43.9	1.05	96	Very light brown; oil-soaked
4. Davis, Hendrick 1. T. & P., T. 2 S., Block 42, Section 46, Center of SW. ¼. Elevation, 2,878 feet. Top Blaine limestone, 4,665 feet					
4,565–83	47.2	26.6	1.50	58	Brown, crystalline; considerable sand
4,700–10	38.1	32.4	0.99	69	Medium gray, sandy, some pyrite
4,825–35	54.7	42.5	1.08	93	Gray-brown
5. Duffey et al., Rutledge 1. Public School Land, Block B-15, Section 7, 990 feet from S. and 330 from E. lines. Elevation, 2,967 feet. Top Blaine limestone, 3,590 feet					
3,630–50	53.3	35.2	1.28	77	Medium gray; crystalline; some calcite
3,740–50	51.8	24.3	1.8	53	Medium gray; much calcite
3,895–3,910	51.9	41.7	1.05	91	Light gray, compact; some pyrite
4,140–60	56.0	38.5	1.23	84	Brown, crystalline ("sugary"), some calcite

* Unless otherwise stated, all samples examined and listed are Blaine in age. Serial numbers at left preceding names of wells correspond with locations in Figure 1.

TABLE I (Continued)

Interval in Feet (Depth)	Per Cent $CaCO_3$	Per cent $MgCO_3$	Molecular Ratio $CaCO_3$ / $MgCO_3$	Per Cent Dolomite	Description and Remarks
6. Gulf, Connell 1-B. Public School Land, Block B-16, Section 1, 440 feet from E. and 2,310 feet from N. lines. Elevation, 2,898 feet. Top Blaine limestone, 3,160 feet					
3,340–50	47.4	32.9	1.22	72	Light gray, sandy
3,480–90	81.6	18.7	3.68	41	Brown; oil-stained (showing oil and gas 3,482 feet)
3,695–3,700	67.2	36.4	1.55	80	Dark gray, oil-soaked
7. Penn et al., Kloh 1. T. & P., T. 3 S., Block 44, Section 7, 330 feet from S. and 2,310 feet from E. lines. Elevation, 2,925 feet. Top Blaine limestone, 3,410 feet					
3,255–75	57.8	20.3	2.40	44	Brown; some anhydrite, considerable SiO_2, Whitehorse
3,425–40	53.2	37.9	1.18	83	Light brown and gray; some SiO_2; showing gas
3,455–70	51.2	40.7	1.06	89	Very light gray; some SiO_2
3,500–15	54.4	37.9	1.20	83	Brown; some SiO_2
3,570–85	49.9	37.7	1.12	83	Mixed light gray and light brown
3,680–90	50.9	38.8	1.10	85	White
3,725–30	56.3	41.2	1.15	90	Brown, very hard; some pyrite and some calcite; top of pay 3,720 feet
8. Penn-Llano et al., Hogen 1. T. & P., T. 3 S., Block 44, Section 18, 330 feet from N. and W. lines. Elevation, 2,914 feet. Top Blaine limestone, 3,485 feet					
3,420–35	52.4	32.2	1.37	70	Brown limestone; little anhydrite; Whitehorse
3,520–30	50.7	36.9	1.16	81	Dark gray
3,625–35	51.7	37.8	1.15	83	Medium gray; some calcite
3,830–40	53.2	41.0	1.10	90	Fine white, crystalline; hole full sulphur water at 3,832–41 feet
9. Simms-Phillips, University A-1. University Land, Block 35, Section 1, 2,970 feet from N. and 440 feet from E. lines. Elevation, 2,915 feet. Top Blaine limestone, 3,290 feet					
3,318–27	54.8	29.9	1.54	65	Medium gray
3,442–50	51.2	38.9	1.11	85	Gray
3,605–11	55.9	43.9	1.07	96	Gray
10. Stanolind, Cowden 1. T. & P., T. 2 S., Block 43, Section 26, 330 feet from S. and 1,320 feet from W. lines. Elevation, 2,942 feet					
3,935–45	53.4	42.5	1.06	93	Core, medium gray
11. Texas, Cosden-Connell 5. Public School Land, Block B-16, Section 24, 330 feet from S. and E. lines. Elevation, 2,843 feet. Top Blaine limestone, 3,382 feet					
3,400–08	52.7	29.8	1.49	65	Mixed light and dark gray; some calcite
3,415–18	50.9	35.2	1.22	72	Brown
3,550–55	49.4	36.4	1.14	80	Light gray; some pyrite
3,654–60	53.4	37.2	1.21	81	Dark gray
3,871–78	53.6	38.3	1.18	84	Dark brown, crystalline; oil-soaked; some calcite
3,993–4,004	69.1	24.4	2.38	53	Light brown, crystalline; some calcite; hole full sulphur water

IRION COUNTY

12. Beasley et al., Sawyer 1. H. & T. C., Block 24, Section 3032. Top Blaine limestone 1,530 (?) feet

1,610–20	60.8	13.5	3.8	30	Light gray; crystalline
1,784–88	48.3	33.7	1.21	74	Mixed brown and gray; oil-stained, soft

13. Benedum and Trees, Sugg 1. H. & T. C., Block 28, Section 3081, 1,560 feet from S. and 925 feet from W. lines. Elevation, 2,376 feet. Top Blaine limestone, 1,165 feet

TABLE I (Continued)

Interval in Feet (Depth)	Per Cent $CaCO_3$	Per Cent $MgCO_3$	Molecular Ratio $\frac{CaCO_3}{MgCO_3}$	Per Cent Dolomite	Description and Remarks
1,630–40	47.9	37.4	1.08	81	Brown, very crystalline; some quartz and some sandy shale
1,725–45	82.9	Trace			Medium and light gray; some calcite and pyrite
1,900–10	95.5	None			Light gray; crystalline
1,990–2,000	89.0	None			Light and dark gray; crystalline, much calcite

14. Kirby et al., Sawyer 1. H. & T. C., Block 24, Section 3047, 150 feet from center of W. line. Elevation, 2,525 feet. Top Blaine limestone, 1,815 feet

Interval in Feet (Depth)	Per Cent $CaCO_3$	Per Cent $MgCO_3$	Molecular Ratio $\frac{CaCO_3}{MgCO_3}$	Per Cent Dolomite	Description and Remarks
1,988–2,000	44.7	33.2	1.13	73	Light gray; some gray shale and pyrite; sandy
2,108–19	54.7	40.0	1.15	88	Hard, white, much calcite; showing oil
2,240–48	54.4	43.4	1.06	95	Medium gray; oölitic; water
2,290–95	55.6	34.1	1.38	75	White, crystalline
2,314–20	56.0	33.4	1.41	73	Dark cream-brown
2,402–07	55.4	38.5	1.21	84	Dirty white; very crystalline
2,500–07	53.5	35.4	1.31	77	Dark gray, mottled crystalline
2,593–2,605	41.5	31.8	1.10	70	Dark gray; crystalline, some sand, shale, and pyrite

REAGAN COUNTY

15. Amerada et al., Sawyer 1. T. & P., Block 1, Section 125, 990 feet from E. and 330 feet from S. lines. Elevation, 2,527 feet. Top of Blaine limestone, 2,350 feet

Interval in Feet (Depth)	Per Cent $CaCO_3$	Per Cent $MgCO_3$	Molecular Ratio $\frac{CaCO_3}{MgCO_3}$	Per Cent Dolomite	Description and Remarks
2,365–70	55.9	31.2	1.51	68	Light gray; oil-soaked; showing gas, 2,380 feet
2,650–60	58.6	31.4	1.57	89	White, crystalline
2,825–40	48.6	36.2	1.13	79	Very light gray; crystalline; hole full salt water, 2,850 feet

16. Beasley, Sawyer 1. Robinson Survey, Section 8, 1,320 feet from N. and 3,960 feet from E. lines. Elevation, 2,586 feet. Top Blaine limestone, 2,240 feet

Interval in Feet (Depth)	Per Cent $CaCO_3$	Per Cent $MgCO_3$	Molecular Ratio $\frac{CaCO_3}{MgCO_3}$	Per Cent Dolomite	Description and Remarks
2,250–60	53.4	31.8	1.41	70	Cream-colored
2,373–85	48.0	33.1	1.22	72	Hard smoky gray, some SiO_2
2,475–83	57.5	31.1	1.56	68	Medium gray, crystalline; some shale; oil, 2,479 feet
2,493–97	49.7	37.8	1.11	83	Soft, dirty white
2,615–20	49.2	39.1	1.06	85	White; some pyrite
2,678–80	47.7	34.8	1.16	76	Cream-colored
2,810–20	42.9	32.3	1.12	71	Medium gray, crystalline; 500 feet sulphur water at 2,815–20 feet

17. Big Lake Oil Company, University 179. University Land, Block 2, Section 10, 660 feet from S. and E. lines. Elevation, 2,690 feet. Top Blaine limestone, 2,980 feet

Interval in Feet (Depth)	Per Cent $CaCO_3$	Per Cent $MgCO_3$	Molecular Ratio $\frac{CaCO_3}{MgCO_3}$	Per Cent Dolomite	Description and Remarks
3,003–07	51.9	28.0	1.56	61	Dark gray and brown, crystalline
3,052–54	87.5	8.3	8.9	19	White, crystalline
3,105–09	89.4	Est. 3–5			Light gray
3,163–68	80.5	Trace			White, crystalline
3,208–13	91.8	Trace			Hard, white, crystalline; porous; showing gas, 3,213 feet
3,238–48	92.5	Trace			Light gray, crystalline; oil-soaked
3,290–3,303	92.4	Trace			Gray, crystalline; porous; some SiO_2; 1,800 feet salt water, 3,305 feet
3,385–93	94.8	Trace			Medium gray; honey-combed; some SiO_2
3,495–3,503	94.7				Medium gray

18. California, University 1. University Land, Block 8, Section 13, 330 feet from N. and 2,310 feet from E. lines. Elevation, 2,861 feet. Top Blaine limestone, 3,215 feet

TABLE I (Continued)

Interval in Feet (Depth)	Per Cent $CaCO_3$	Per Cent $MgCO_3$	Molecular Ratio $CaCO_3 / MgCO_3$	Per Cent Dolomite	Description and Remarks
3,215–20	50.2	33.1	1.28	72	Medium gray
3,415–22	79.4	Trace			Light gray-to-white

19. California, University 3. University Land, Block 7, Section 14, 2,328 feet from N. and 2,310 feet from W. lines. Elevation, 2,821 feet. Top Blaine limestone, 2,980 feet

2,980–90	22.4	14.3	1.33	31	Light gray, granular, very crystalline, sandy; much SiO_2
3,075–82	43.3	33.8	1.08	74	Medium gray; "sugary" crystalline

20. Gulf-State, Campbell 1. University Land, Block 1, Section 1, 330 feet from N. and W. lines. Elevation, 2,911 feet. Top Blaine limestone, 3,165 feet.

3,240–50	34.0	22.9	1.25	50	Medium gray; crystalline, sandy

21. Humble-Dunning, Bar S Ranch 1. T. & P., Block 1, Section 93, 660 feet from S. and E. lines. Elevation, 2,480 feet. Top Blaine limestone, 2,110 feet

2,200–10	50.1	29.8	1.42	65	Brown, some calcite
2,296–2,300	51.6	37.6	1.16	82	Hard, medium gray

22. Simms, Sawyer Cattle Co. 1. C. A. Glass Survey, Block 1, Section 12, 150 feet from S. and 1,250 feet from W. lines. Elevation, 2,570 feet. Top Blaine limestone, 2,465 feet

2,558–70	40.3	19.9	1.71	44	Medium gray; considerable SiO_2
2,670–77	51.3	36.8	1.17	81	Medium gray
2,855–58	51.0	39.4	1.09	86	Hard, medium gray; crystalline
2,975–81	48.6	37.2	1.1	81	Cream-colored; some pyrite
3,056–68	91.2	Trace			Dark gray; crystalline; porous; hole full water at 3,060 feet
3,110–17	61.8	32.2	1.62	71	Dark gray, crystalline
3,193–99	86.2	Est. 3–5			Light gray; semi-porous; hole full salt water at 3,195 feet
3,248–99	85.0	Trace			Light gray
3,366–70	41.6	Trace			Dark gray; some pyrite; much SiO_2

TOM GREEN COUNTY

23. Roxana, Clark 1. G. C. & S. F., Block A, Section 5, 330 feet from N. and 1,320 feet from W. lines. Elevation, 2,224 feet. Top Blaine, 710 (?) feet. Top Clearfork, 1,340 feet

1,345–50	50.0	37.6	1.12	82	Medium gray to cream-colored, sandy, some pyrite; showing oil at 1,355 feet
1,535	50.1	39.1	1.08	86	Medium gray; some quartz
1,650	52.5	40.6	1.09	89	Medium gray; some quartz
1,930	48.4	37.2	1.10	81	Medium gray; some shale
2,095	51.5	40.1	1.08	88	Brown, crystalline
2,250	53.2	38.8	1.16	85	Gray, cream-colored, crystalline, soft
2,450	49.4	36.6	1.14	80	Light gray; some calcite and pyrite; hole full water at 2,470 feet
2,600	50.8	40.4	1.06	88	Gray and brown; some gray shale; hard
2,765	51.2	40.6	1.06	89	Light gray
2,845	55.0	42.6	1.09	93	Very light gray; crystalline; soft, hole full water at 2,835 feet
2,950–55	55.9	41.8	1.13	92	Medium gray; much calcite
3,050–52	54.6	43.6	1.06	95	Very light gray, sandy

UPTON COUNTY

24. Broderick and Calvert, Homer National Bank 1. T. & P., T. 4 S., Block 42, Section 22. Elevation, 2,890 feet. Top Blaine limestone, 4,710 feet

4,720–26	70.9	24.5	2.44	54	White, some calcite
4,819–23	53.1	34.8	1.29	76	Light gray

TABLE I (*Continued*)

Interval in Feet (Depth)	Per Cent CaCO₃	Per Cent MgCO₃	Molecular Ratio CaCO₃/MgCO₃	Per Cent Dolomite	Description and Remarks
25. Humble, Damron 1. G. C. S. D. & R. G. N. G., Block F, Section 192, 330 feet from N. and W. lines. Elevation, 2,756 feet. Top Blaine limestone, 2,870 feet					
2,949–57	48.2	28.7	1.42	63	White, some calcite and pyrite
2,968–77	54.3	31.5	1.45	69	Cream-colored "grainy" calcite and pyrite
3,065–70	54.7	32.4	1.42	71	Dark cream, crystalline; some calcite, soft
3,100–10	55.0	35.2	1.32	77	Brown, oil-stained, soft
3,250–55	55.9	32.0	1.47	70	White, crystalline, soft
3,320–35	52.9	38.2	1.17	84	Mixed white and dark gray, very hard
26. Marland, Burleson 3. T. C. Jones Survey, Section 100. Elevation, 2,595 feet. Top Blaine limestone, 2,030 feet					
2,070–85	33.9	23.7	1.86	52	Gray
2,111–13	53.5	42.2	1.07	92	Gray, porous; good showing oil
2,210–15	56.4	30.7	1.55	67	Gray and brown, porous; oil, 2,235–50 feet
2,260–65	71.4	29.2	2.06	64	Gray and brown, porous
2,300–08	55.7	38.7	1.21	85	Gray and white; some calcite; flowing by heads at 2,298 feet
2,360–70	55.9	39.7	1.19	87	Gray, porous, oil-stained
2,510–20	55.4	38.6	1.21	85	Gray, tight; some calcite
2,705–15	56.9	36.5	1.31	80	Gray, considerable calcite; soft
2,960–70	57.5	40.9	1.18	90	Light gray
3,120–30	54.8	43.9	1.05	96	Brown, porous
3,320–30	54.3	42.3	1.08	93	Gray and brown
3,455–65	54.3	42.1	1.08	92	Brown
3,640–45	50.1	43.5	1.04	95	Medium gray
3,920–25	70.2	18.8	3.14	41	Medium gray; much calcite
4,040–45	84.0	12.7	5.6	28	Brown and gray; some calcite
4,160–70	55.3	41.4	1.13	91	Brown and gray; much calcite
4,490–4,506	52.6	41.4	1.07	91	Dark gray; hole full sulphur water at 4,500 feet

WARD COUNTY

27. Rector, Monroe 1. H. & T. C., Block 33, Section 62, 660 feet from SW. and SE. lines.					
4,654	81.5	Est. 5			Brown Delaware limestone core
4,658	79.5	Est. 5			Black Delaware limestone core

WINKLER COUNTY

28. Cranfill Bros. *et al.*, Bashara 1. Public School Land, Block 77, common corner of sections 10, 11, 22, and 23					
3,345	52.2	38.7	1.14	85	
29. Gibson & Johnson, Leck 1. Public School Land, Block 74, Section 3					
3,070	52.3	43.8	1.01	96	Gray, very porous
30. Tidal, Hill *et al.*, Amburgy 1. P. S. L., Block B-7, Section 24. Elevation, 2,995 feet. Top Blaine limestone, 3,850 feet					
3,951–55	55.4	35.1	1.33	77	Light gray and cream-colored
4,195–99	57.5	28.5	1.7	62	Dark gray; sandy
4,426–30	47.7	35.7	1.13	78	Dark gray; sandy; oil-stained
4,778–84	54.5	38.9	1.18	83	Medium gray, crystalline; 2BWPH at 4,782 feet
31. Westbrook, Hendricks 1. P. S. L., Block B-5, Section 42					
2,881	52.1	39.6	1.11	87	White
3,030–36	52.6	40.7	1.09	89	Light gray

TABLE I (Continued)

Interval in Feet (Depth)	Per Cent CaCO₃	Per Cent MgCO₃	Molecular Ratio CaCO₃/MgCO₃	Per Cent Dolomite	Description and Remarks

MISCELLANEOUS

REAGAN COUNTY

Skelly, Grayson-University 1-D. University Land, Block 8, Section 33, 1,980 feet from N. and 660 feet from W. lines. Elevation, 2,891 feet

3,086–92	59.4	35.4	1.4	77	Core: oölitic; some oil

Outcrop sample. E. part of Survey No. 328, J. M. Archer, Grantee. Fossiliferous limestone from Permian on Terrell County side of Pecos River

	91.4	5.4	14.4	12	

CROCKETT COUNTY

Stanolind, Todd 1. G. C. & S. F., Block UV, Section 67, 1,320 feet from N. and E. lines.

7,334	52.6	40.4	1.1	88	Ellenburger core

McMan and Amerada, Powell 1 (Powell field)

2,495–2,500	50.6	31.6	1.35	69	Gray mottled, crystalline
2,633–40	54.3	40.8	1.12	89	Brown, "sugary" crystalline; top of pay 2,530 feet

PECOS COUNTY

Shell, Smith 13-A — Yates field "pay"

1,712–17	51.5	42.1	1.03	92	Brown, crystalline, porous, oil-soaked

As previously stated, for ease of calculation and for purposes of comparison, all magnesium present in the limestone is assumed to be in the form of dolomite. Any calcium in excess of the amount required for a 1:1 molecular ratio with the magnesium is assumed to be in the form of pure limestone. Thus, if any sample has a $CaCO_3:MgCO_3$ ratio of 1.25:1, it would contain about 88 per cent dolomite and 12 per cent limestone.

The following points of particular interest have been brought out by these analyses.

1. Most of the limestone is highly dolomitic. The samples examined have been so widely scattered and so uniformly high in magnesium content that it appears safe to say that, with the exception of the area described under the following paragraph 3, the most of the Blaine limestone in the Permian basin probably contains at least 75 per cent dolomite and should be classified as dolomite limestone.

2. The "Brown lime" in the lower part of the Whitehorse-Cloudchief as well as the upper 100 feet of the Blaine limestone is, in general, more calcareous than the remainder of the limestone section. There is some anhydrite in the lower Whitehorse, hence the increased

calcium carbonate content might be due to the leaching action of the gypsum solution on the previously formed dolomite.

$CaSO_4$ (in solution) $+ MgCO_3 . CaCO_3 = 2CaCO_3 + MgSO_4$ (in solution)

3. Only four wells—the Big Lake Oil Company's University No. 179, the California Company's University No. 1, and the Simms Oil Company's Sawyer Cattle Company No. 1 in Reagan County, and Benedum and Trees' Sugg No. 1 in Irion County—penetrated limestone sections which contained very little dolomite. In the first two of the Reagan County wells and in the one Irion County well the uppermost limestone section was found to be somewhat dolomitic, but practically no magnesium was found lower than 150 feet in the Blaine limestone section. In the Simms Oil Company's Sawyer Cattle Company No. 1 the top 400 feet of the limestone section is highly dolomitic, but the lower 350–400 feet is practically pure limestone. Other wells, located near these wells, encountered the normal, highly dolomitic limestone section.

4. There appears to be no uniformity in the percentage of dolomite in the limestone at points of porosity, that is, points at which oil, gas, or water were encountered. However, in some wells there was a noticeable decrease in the amount of magnesium in the zones of apparent porosity. That such decreases are not consistent throughout the basin is shown by the following data.

A. Reagan County. The Simms Oil Company's Sawyer Cattle Company No. 1 has almost pure limestone at 3,055 feet and at 3,190 feet; at both points the well encountered a hole full of water. The "lime" above 3,000 feet and 3,115 feet is highly dolomitic; there is little or no magnesium present in the section below 3,200 feet.

B. Reagan County. Beasley's Sawyer No. 1 had a showing of oil at 2,479 feet; at 2,480 feet there is a slight decrease in magnesium content of the limestone.

C. Ector County. The Texas Company's Cosden-Connell No. 5 has a calcium carbonate-magnesium carbonate ratio of 2.38 at 4,000 feet, at which depth a hole full of water was encountered. The ratio above that point averages about 1.2.

D. Ector County. The Gulf Production Company's Connell 1-B has a jump in the ratio to 3.7 at 3,490 feet which is about the middle of the "pay." At 3,350 and 3,700 feet, which are above and below the "pay," the ratios are respectively 1.22 and 1.55.

E. Upton County. Marland's Burleson No. 3 shows a ratio of 1.55 at 2,215 feet and 2.06 at 2,260 feet. Physical appearance of these and intervening samples were the same; oil was encountered at

2,235–2,250 feet. At 2,300 feet the ratio drops to 1.21, though the well was flowing by heads at 2,298 feet. This well was drilled deep into the limestone section below the pay horizon, and between 3,900 and 4,100 feet penetrated a compact limestone in which the ratio increased to a maximum of 5.6 and which contained no oil, gas, or water.

A hole full of sulphur water was encountered at 4,500 feet; the "lime" at this point has a ratio of 1.07—almost pure dolomite. The lower part of the limestone in this well may be Clearfork in age.

These few samples show very well that there is apparently no widespread relationship between the porosity and the dolomite content. However, such a relationship might exist in certain very limited areas, which could be determined only by more detailed study.

Sufficient samples and cores are not available from wells which have penetrated deep into the limestone to furnish much information about the lower part of the section. A number of Ordovician tests have gone completely through the limestone, but most of them have been drilled with rotary equipment, the cuttings from which are of no value for analytical purposes. One cable-tool well, Marland's Burleson No. 3, in Upton County, drilled about 2,575 feet into the limestone; it has been commented on in section 4-C.

5. In one well examined, Roxana's Clark No. 1, in Tom Green County, the limestone was not Blaine, but Clearfork in age. It is particularly interesting to note that throughout the 1,700 feet of section examined the "lime" is more uniformly highly dolomitic than is the Blaine limestone.

6. The color of the so-called "Brown lime" in the Blaine appears to be due primarily to staining with oil or sulphur compounds.

7. Microscopic distinction between pure limestone and highly dolomitic limestone is very uncertain. Though not an absolute criterion, the rate of reaction in cold, 10 per cent hydrochloric acid is much more accurate.

From the standpoint of chemical possibilities there are several ways in which dolomite *might* be formed, though the conditions under which such formation will take place may be unknown or unobtainable. Some of these are here discussed.

1. *Direct precipitation from solution.*—The conditions affecting the solubility of calcium carbonate and magnesium carbonate are very similar. Three factors affect the solubility of both compounds very markedly: the amount of dissolved carbon dioxide, the amount of dissolved salt ($NaCl$), and the temperature. Of these factors the amount of dissolved carbon dioxide is undoubtedly the most important, though it, itself, is directly dependent on the other two

factors. The mode of influence of carbon dioxide is shown graphically in the following equilibrium equations.

$$CO_2 \text{ (gaseous)} \rightleftarrows CO_2 \text{ (dissolved)}$$
$$CO_2 \text{ (dissolved)} + H_2O + XCO_3 \rightleftarrows X(HCO_3)_2 \text{ (dissolved)}$$
$$\text{(solid)}$$

Thus when the partial pressure of the carbon dioxide in the atmosphere is increased, the amount of dissolved carbon dioxide is increased; this in turn increases the amount of calcium or magnesium carbonate dissolved in the form of the respective bicarbonate. An increase in temperature of the solution will cause a decrease in the amount of dissolved carbon dioxide, thus precipitating some of the dissolved calcium and/or magnesium bicarbonate. This is further complicated by the fact that the solubility of calcium and/or magnesium carbonate *increases* as the salt content increases up to 5 per cent to 10 per cent, after which point it *decreases* as the salt content is increased.

Although these various factors tend to change the solubilities of the carbonates in the same direction, magnesium carbonate is, in general, about four to six times as soluble, molecularly, as calcium carbonate under the same conditions.

In view of the foregoing facts it is obvious that a saturated solution from which calcium and magnesium carbonates are being deposited due to a change in temperature, carbon dioxide content, or salt content will precipitate much more magnesium carbonate than calcium carbonate. Also, the complete evaporation of such a solution would tend to precipitate a "limestone" which would be predominantly magnesite. It appears, therefore, that if there is to be a primary deposition of equi-molecular quantities of calcium and magnesium carbonates from a solution saturated with both compounds, both must be added—as by the influx of a "fresh" water carrying both calcium and magnesium carbonate—in just such quantities that equi-molecular amounts must be precipitated in order to bring about equilibrium conditions with respect to the solubilities of the two carbonates. Even though we assume such ideal conditions we are assured only of the fact that the two carbonates are precipitated and have no reason to believe that they are deposited always in the form of dolomite.

2. *Change of precipitated limestone to dolomite.*—Originally deposited calcium carbonate may be converted to dolomite by a partial replacement with magnesium carbonate. Under certain conditions[5]

[5] F. W. Clarke, "Data of Geochemistry," 5th ed., *U.S. Geol. Survey Bull.* 770 (1924), p. 566.

a magnesium sulphate solution will react with calcite to form dolomite. The reaction probably takes place as follows.

$$2CaCO_3 + MgSO_4 \rightarrow CaCO_3 \cdot MgCO_3 + CaSO_4$$

Under different conditions this reaction is reversible,[6] that is, gypsum solutions will react with dolomite to form limestone and magnesium sulphate. This latter reaction may explain the more highly calcareous condition of the "Brown lime" in the lower Whitehorse section and of the upper part of the Blaine limestone.

There is also the possibility that when a solution saturated with magnesium bicarbonate remains in contact with precipitated calcium carbonate sludge for a considerable period of time dolomite may be formed according to the equation

$$CaCO_3 + Mg(HCO_3)_2 \rightarrow CaCO_3 \cdot MgCO_3 + H_2O + CO_2$$

Replacement has also been assumed to take place according to the following equation.

$$2CaCO_3 + MgCO_3 \rightarrow CaCO_3 \cdot MgCO_3 + CaCO_3 \text{ (in solution as bicarbonate)}[7]$$

3. Deposition as a mixture of $Mg(OH)_2$ and $CaCO_3$ (or $Ca(OH)_2$) with subsequent carbonation and conversion to dolomite. Irving[8] states:

If the increase in alkalinity results from increase in free base alone the magnesium is rapidly precipitated as the pH rises above 10. Calcium precipitation likewise occurs, but more slowly. If the precipitation occurs from the addition of carbonates (changing the pH then by electrolysis), magnesium precipitation follows the pH curve, indicating its occurrence as a function of the hydroxyl ion exclusively. Calcium is precipitated chiefly by increase of carbonate ion concentration. The explanation is of course in the fact that $CaCO_3$ is relatively insoluble, $MgCO_3$ relatively soluble. $Ca(OH)_2$ on the contrary is quite soluble as compared with the insoluble $Mg(OH)_2$. There is a possibility of salt waters naturally approaching a pH value of 10, permitting the precipitation of both calcium and magnesium. The conditions would determine the ratio of calcium to magnesium. In waters whose alkalinity is produced by free carbonate, the calcium would far exceed magnesium in the precipitate. Were the alkalinity induced by a process removing carbonates and leaving free base, magnesium precipitation would correspondingly increase.

[6] Ibid.

[7] W. H. Twenhofel, op. cit., p. 348.

[8] L. Irving, "The Precipitation of Calcium and Magnesium from Sea Water," Jour. Marine Biol. Assoc. United Kingdom, Vol. 14 (1926), p. 441; in Twenhofel, Treatise on Sedimentation, p. 338.

On the basis of the foregoing statement it is easy to visualize a group of conditions whereby the formation of dolomite might occur. For example, a large inland sea of high salinity and alkalinity might have fresh water, carrying $Mg(HCO_3)_2$ and $Ca(HCO_3)_2$ in solution emptying into it; the fresh waters would tend to float on top of the water of higher gravity. As the waters mix, the magnesium would normally tend to be precipitated as $Mg(OH)_2$ and the calcium as $CaCO_3$. However, conditions might exist wherein dolomite would be precipitated directly according to the equation

$$Mg(HCO_3)_2 + Ca(OH)_2 \rightarrow CaCO_3 \cdot MgCO_3 + H_2O + CO_2$$

Even though the dolomite is not precipitated directly, later shifts might occur which would bring the fresh water in contact with the previously precipitated magnesium hydroxide—calcium carbonate sludge, when changes might take place in the following manner.

$$Mg(OH)_2 + Ca(HCO_3)_2 = CaCO_3 \cdot MgCO_3 + 2H_2O$$
solid   in solution   solid

$$Mg(OH)_2 + Mg(HCO_3)_2 = 2MgCO_3 + H_2O + CO_2$$
solid   in solution   solid

The subsequent combination of the more or less intimately mixed $CaCO_3$ and $MgCO_3$ to form dolomite can be assumed to have taken place, but the exact procedure of this change we do not know.

It is realized that we have little or no experimental results which will corroborate many of these hypotheses. However, as already stated, they are not given as explanations of how dolomite has been formed, but are merely chemical relationships which *might* take place *if* proper conditions of temperature, pressure, concentration, et cetera, were realized. Neither do the possibilities given here constitute all the conditions under which dolomite could be formed or all hypotheses proposed by other investigators of the problem of dolomite formation.

The highly dolomitic nature of the "limestone" from which much of the West Texas oil is derived may be closely connected with some of the troubles being encountered in acid-treated wells in this area. In the laboratory examination of the samples the following facts of particular interest in this connection have been noted.

1. Weak acid (less than 10 per cent HCl) had little effect on the cuttings without being heated.

2. 6 N acid (about 22 per cent) decomposed most of the cuttings readily.

3. 15 per cent acid reacted with the cuttings readily but required heating to bring about complete disintegration.

4. Oil-soaked and oil-stained cuttings were wetted by the acid with considerable difficulty. Addition of small amounts of nitric acid and heating aided very materially in increasing the rate of dissolution of such cuttings.

5. Many of the samples contained considerable material which was entirely insoluble in acid. Though some of this material was sand, much of it appeared to be a very finely divided, flocculent silt. This is probably the material which tends to choke many of the wells after acid treatment.

DOLOMITIC MOTTLING IN PALLISER (DEVONIAN) LIMESTONE, BANFF AND JASPER NATIONAL PARKS, ALBERTA[1]

F. W. BEALES[2]

Toronto, Ontario, Canada

ABSTRACT

Dolomitic mottling of the Palliser limestone is the result of local replacement of original limestone during diagenesis. Apparently this replacement was controlled by local variations in effective diffusion porosity in the sediments.

INTRODUCTION

The origin of dolomite in almost all its forms remains a problem. A progressive alteration of limestone to dolomite can be demonstrated in some suites of rocks and this has been engaging the writer's attention for some years. The investigation reported here was undertaken in the hope that the dolomitic mottling shown by some limestones (Figs. 1–2) might represent an arrested stage in the process of complete dolomitization. If such were the case, mottled dolomitic limestones could contribute ideas that would assist in the solution of some of the broader problems.

For reasons of availability and convenience the dolomitic mottling of the Devonian Palliser formation of southwestern Alberta was selected for more detailed study. F. G. Fox (1951, p. 838) has already commented upon this mottling as follows.

> Dolomitization apparently occurred prior to lithification of the rocks. A number of thin sections have been studied by de Wit,[4] (4. R. de Wit, personal communication) who believes the dolomite bodies to have been algal remains. Algal colonies, like those of other organisms, may be expected to have considerable original porosity, and such colonies when enclosed and preserved in fine marine muds might readily provide the loci for replacement. The writer is in agreement with de Wit, since his idea is in harmony with the observable facts of the occurrence of the dolomite.

No new general principles have been discovered during the present investigation but something has been learned about the nature of the mottling. As it is not planned to continue this study at present, a brief report is made here.

This discussion on dolomitic mottling in limestones is not merely of local interest since closely similar rocks of differing ages and from other areas have been commented upon by several authors.

Dixon and Vaughan (1911) described the pseudobreccias in the Carboniferous limestones of Gower, South Wales. In these pseudobreccias dolomitization affected the less recrystallized parts of a patchy, recrystallized, fine-grained, foraminiferal limestone. Differential weathering of the surface of the dolomitic and calcareous parts produces a conspicuous mottling. The authors noted that

---

[1] Manuscript received, June 8, 1953.

[2] University of Toronto. The Geological Survey of Canada and the California Standard Company of Canada made possible the field work, and the writer is deeply indebted to the officers of both organizations for numerous favors.

gastropod shells and foraminiferal tests in the unrecrystallized parts were particularly susceptible to dolomitization.

Wallace (1913) and Birse (1928) discussed mottled Ordovician limestones from Manitoba. Wallace attributed the mottling to contemporaneous alteration of original limestone by the agency of sea water and noted that selective alteration of fucoids apparently best explains the "form" of the mottling. Birse pointed out that the mottling is irregular, dendritic, and anastomosing in both dimensions, and generally contains a cavity or a darker core within each patch or ramification. Cavities noted by Birse approximated $\frac{1}{8}-\frac{1}{16}$ inch in diameter and visible lengths on exposed surfaces measured from $\frac{1}{4}$ to 2 inches. Birse considered that Wallace's fucoidal theory did not fit the facts and regarded the mottling as a diagenetic feature, the tubular cavities acting as centers from which dolomitization began.

Van Tuyl (1916) described several types of mottled dolomitic limestones, among which the Tribes Hill (Beekmantown) limestone of New York appears to bear the closest resemblance to the Palliser limestone considered here. He attributed the mottling to replacement of original limestone presumably by the agency of downward diffusing magnesian salts derived from sea water that may at times have been hypersaline by comparison with the recent oceans. He noted that replacement commonly selected fucoids, gastropods, and other organic remains.

DESCRIPTION OF MOTTLING

The Palliser formation (Beach, 1943; de Wit and McLaren, 1950; and Fox, 1951) is Upper Devonian in age and in southwest Alberta consists of a massive cliff-forming unit of dolomitic limestone. The cliffs form majestic features in the Banff and Jasper National Parks where most of the field work for this study was done. The formation is uniform both in thickness (800–1,000 feet), and lithologic character, over wide areas, and the mottling (Figs. 1–2) is present almost throughout.

The mottling is caused by a difference in composition and crystallinity which has been developed irregularly throughout the rock. The character of the mottling suggests diagenetic replacement of part of the limestone by dolomite, involving the development of slightly coarser crystallinity in the altered part, and obliteration of original structures. The dolomitic component of the rock generally appears darker on weathered surfaces (Figs. 1–3). However, it may, less commonly, appear lighter (Fig. 4), according to the texture of the calcareous part and the degree of weathering of the rock.

As the dolomitic and calcareous surfaces normally show marked difference on weathering, the mottling is commonly very conspicuous. On such weathered surfaces the crystalline dolomitic part commonly stands in relief above other parts of the limestone (Figs. 1–3). Even where this relief is not strongly developed the mottling is emphasized by fine channelling that develops on the margin of

Fig. 1.—Irregular dolomitic mottling showing concentration along bedding planes.

Fig. 2.—Same rock as in Figure 1 but bedding plane surface.

Fig. 3.—Example of dolomitic mottling showing "worm burrow" from which mottling appears to have spread into surrounding rock.

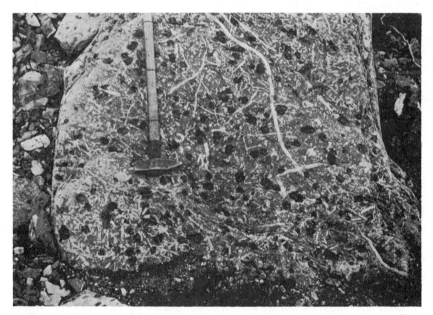

Fig. 4.—Water-worn slab of Palliser limestone showing replaced "worm burrows." Dark spots are rain splashes.

the dolomitic area; presumably this channelling is caused by the solution of the calcite in the heterogeneous carbonate of the margin, resulting in loosening of dolomite crystals. Although the contact between altered and unaltered limestone is normally sharp, microclusters and larger patches of dolomite crystals are disseminated in the calcareous part of the rock.

In the lowest part of the formation the mottling is seamy and irregular following some bedding planes in preference to others and spreading irregularly into the adjacent rock to form an irregular meshwork of partial dolomitization. Higher in the formation the mottling becomes much more pronounced and may in places simulate cuneiform writing, covering and overlapping on the surface of the rock. In parts of the succession dolomitization appears to have selected "worm burrows," and "algal" colonies, resulting in a branching and anastomosing network of replacement in the rock (Fig. 4). This replacement may parallel or cut across the bedding planes. The completeness of the alteration has obliterated the original textures of these structures, making identification doubtful. Straight, or sinuous, unbranching, sub-cylindrical structures are here referred to as "worm" tubes. Branching, sub-cylindrical structures appear to be pseudomorphs after algae in some places. Relatively unaltered algal limestone containing remains of branched and unbranched colonies occurs in the Palliser formation and in the underlying Devonian formations. Certain other fossils, such as gastropods and thin-shelled brachiopods, are susceptible to dolomitization and invariably incorporated in the mottling. Others, such as thick-shelled brachiopods and crinoid ossicles, persist as recrystallized calcite. The limestone filling in shells is commonly altered even where the shell material itself remains as calcite. Where mottling pseudomorphs organic structures, irregular invasion of the surrounding rock is present. Near the top of the formation the mottling is more irregular, consisting of seams and patches following the bedding in part and cutting across primary structures to join other dolomitic material. The uppermost beds of the formation are commonly more calcareous, less massive, and show less pronounced mottling.

The mottling commonly accounts for 10–40 per cent of the rock but some layers show a greater concentration of dolomite than others, resulting in a conspicuous banding. The dolomitic part of the rock may rarely extend to almost 100 per cent but paler areas of slightly coarser crystallinity and small irregular patches of limestone suggest dolomitization by a similar process to that operating in the bulk of the rock.

Petrographically the unaltered limestone appears to be of two main types.

1. Abundant, fine-grained, relatively uniform limestones contain very few macrofossils but much fine fossil débris, ranging from about 100 to 500 microns in size. The fossil débris consists of calcispheres about 100 microns in diameter with walls 3.5 microns thick, also minute entire shells and fine fragments and flakes of other organic material. A few scattered crinoid ossicles of very much larger size, ranging from 1 to 3 mm. seem everywhere to occur. Why such

relatively coarse material should come to rest embedded in limestone of average crystallinity of about 10 microns is not clear. Perhaps the finer branches and stems of crinoids accumulated sufficient gas bubbles on decay to be floated far and wide from their habitat of growth. Rotting of skeletons on the sea floor and within the sediments as they accumulated has without doubt rendered unrecognizable much organic material.

2. The second major group of Palliser limestones differs little in composition but much in texture. These are the lime-sands and lime-grits that consist of well rounded grains of fine-grained limestone embedded in a groundmass of similar texture. Some of these "grains" may be invertebrate fecal pellets, and in such cases the rock would more correctly be classified as a pellet-limestone. However, well formed, ovoid, or sausage-shaped pellets are rare and a penecontemporaneous clastic origin appears more likely for the bulk of the rock. Reworked, and consequently disrupted and distorted, pellets may be an important component among the clastic grains of some of the lime-sands and lime-grits. The average crystallinity of the limestone is of the order of 5 microns. The limestone grains range from 0.2 to 2 mm. according to the coarseness of the specimen. This rock may commonly be partly recrystallized so that only relict grain structure is visible. On the other hand, dolomitic mottling may invade the matrix only, thereby framing the grains and rendering the clastic nature of the limestone much more conspicuous. Scattered, recrystallized macrofossils, probably mainly species of *Cyrtospirifer*, are embedded in the rock but the limestone grains and matrix as a whole show little or no other signs of recognizable organic material.

Thin sections reveal that the mottling of Palliser formation limestones is the surface expression of the replacement of original limestone by dolomite. When the alteration is incomplete, isolated euhedral crystals of dolomite occur, generally ranging in diameter from about 25 to 150 microns, while coarser examples range from 75 to 250 microns (Fig. 5). Where dolomitization, responsible for the mottling, selects definite primary features in the limestone such as bedding laminae, "worm" tubes, or "algae," the edge of the dolomitic part is sharp (Figs. 6 and 7). Where dolomitization has progressed irregularly through the rock a more diffuse margin prevails, showing a gradual decrease in dolomite content away from the main center of alteration (Fig. 8).

Traces of organic structure other than dolomitic pseudomorphs of macrofossils are very rare within the dolomitic patches. Recognizable organic débris is much more common, but not very abundant, in the calcareous parts.

CONDITIONS OF DEPOSITION

The Palliser formation in this area was laid down on the floor of a widespread shelf sea.

1. The deposit is very uniform, fine-grained, and of wide extent.
2. Current-bedding and intraformational conglomerates were noted in the basal beds following the shallow waters of Alexo time but thereafter massive

beds, showing no trace of shallow-water features, predominate, and lime-sands and lime-grits become much less common.

3. Macrofossils are relatively uncommon for a shelf limestone but this apparent absence may be partly due to the rotting of skeletons subsequent to deposition upon the sea-floor. Recognizable shelly and crinoidal material has been observed to grade laterally into fine-grained limestone in which a very faint trace of the original entire outline is still visible.

4. Small, thin-shelled calcispheres and minute bivalves that are commonly the only recognizable organic material in the upper beds are probably of pelagic origin.

5. Algae appear to have been abundant and may account for much of the carbonate precipitation.

6. Insoluble residues are generally small and invariably fine-grained. They consist of a little mud associated with fine silt grade material that is commonly largely authigenic.

7. Steady subsidence over a long period of time with the accumulation of organic calcareous muds is implied.

DIAGENESIS OF PALLISER LIMESTONES

The dolomitic mottling is considered to be a product of diagenesis of the limestone because it follows or truncates primary structures and it replaces or partly replaces fossils. Further, its extreme irregularity and the marked difference in crystallinity from other parts of the rock are not otherwise readily explained.

Dolomitization appears to have taken place early in diagenesis since bedding laminae and other primary structures control the alteration rather than joints and fractures. Mottling is uniform over vast areas and great thicknesses of rock. If, as is likely, sea water was the source of the magnesium it would penetrate most readily at and shortly after deposition of the sediments. Small, infilled cracks and fractures are seams of recrystallized calcite cutting sharply across limestone and dolomite alike; stylolites likewise cut the rock indiscriminately, suggesting that alteration took place shortly after deposition.

The crystallinity of the matrix in some specimens of lime-sand and lime-grit (Fig. 6) is slightly coarser than that of the grains. The matrix might therefore be expected to be slightly more resistant to dolomitization. However, in the mottling on such beds dolomitization invariably selects the matrix in preference to the grains which may be completely unaltered. This suggests that dolomitization took place at a time when the grains were still embedded in relatively porous mud.

The conditions and mechanism of replacement remain unsolved problems. In consequence it is still necessary to consider a variety of possible processes that may have resulted in the formation of dolomite even though some of them are not favored. For example, precipitation of $MgCO_3$ from sea water followed by recrystallization with $CaCO_3$ to form dolomite has commonly been thought of as a possible process. However, it is not at all clear why the presence of precipitated $MgCO_3$ should be more favorable than magnesium ions in promoting recrystallization, particularly since increased temperature and pressure appear to be unnecessary.

Thus the magnesium, which is considered to have been derived from sea water, may have been retained within the sediments by either (1) precipitation of $MgCO_3$ contemporaneously with the accumulation of the lime-muds and followed by recrystallization, or (2) penetration of magnesian solutions with associated replacement of some of the calcite.

The cessation of the dolomitization short of complete alteration of the limestone could be due to recrystallization having used up all the $MgCO_3$ in the first case or, in the second, by restriction of permeability, for example, by compaction and recrystallization in the steadily accumulating sediments, thereby prohibiting diffusion of further magnesium ions.

Another suggestion really combines both the foregoing possibilities: it is that $MgCO_3$ or a related compound was precipitated in the surface layers of the newly formed lime-muds perhaps by the agency of organic bases. A very close control would be necessary to favor precipitation of $MgCO_3$ but not $CaCO_3$. This control, also possibly derived from organic decay products, may have been on the pH of the sediments by maintaining an alkalinity low enough to keep $CaCO_3$ in solution but nevertheless permitting precipitation of dolomite or of $MgCO_3$ and recrystallization of dolomite.

Alternatively, a colloidal control may have operated. Goetz (1921) pointed out that colloidal $CaCO_3$ and $MgCO_3$ are stabilized by certain organic colloids and under suitable conditions may then be co-precipitated and recrystallized to form dolomite. Conditions in the lime-muds may have favored precipitation of colloidal $MgCO_3$.

The skeletons of certain groups of organisms contain a strongly magnesian calcite (Clarke and Wheeler, 1917; Johnson, 1951) and the possibility thus arises that the mottling may derive much of its magnesium from this source. Certain algal skeletons are particularly magnesian and the algal contribution to the Palliser limestones may well be considerable. However, original textures and structures have been so largely altered that it is impossible to evaluate this point. The algal component may equally have been largely of externally precipitated non-magnesian calcite or aragonite. It is not probable that the recrystallization of $MgCO_3$ from organic magnesian calcite or from syngenetic precipitation would provide an adequate explanation of the mottling for the following reasons.

1. Mottling is pronounced in some parts of the section but not in others equally dolomitic. This suggests that there is not much movement and segregation of dolomite subsequent to its formation, and favors an origin by diagenetic alteration for the dolomite. The mottling is not an original sedimentary feature and if

segregation is not marked it is unlikely that the magnesium was derived from a primary precipitate of $MgCO_3$ settling with the other sediments on the sea floor except where dolomite is disseminated in the sediments. Even under such conditions it could be due to solutions penetrating and precipitation taking place evenly in the rock, mottling developing where certain channels provided easier routes.

2. More heavily altered parts of the limestone may approach 100 per cent dolomite but some of these dolomitic beds show very faint traces of mottling. This indicates two stages of dolomitization though the fact that such beds are overlain and underlain by much less altered limestone suggests the two phases were not greatly separated in time. However, the mottling so closely resembles the other parts of the rock that no difference can be detected apart from the very slight difference in weathering. If the mottling represents a segregation of $MgCO_3$ originally present, a process that came to a stop because of the limited supply of $MgCO_3$, how was the rest of the rock converted to dolomite? The uniformity of the rock is against two totally different mechanisms producing the dolomite. A more likely explanation is that the diffusion gradient of magnesium ions was interrupted for a time when partial dolomitization had taken place and then reestablished, permitting completion of the process.

3. The matrix and grains of the mottled lime-sand and lime-grits of the Palliser formation commonly appear very similar in composition and if dolomitization were due to recrystallization it would be expected to affect both alike. Very commonly, however, the matrix only is altered and the grains remain unchanged; this is highly suggestive of dolomitization by diffusion of magnesian salts subsequent to deposition but prior to consolidation of the matrix.

It appears that dolomitization selected the course of least resistance over a considerable period of geological time. If certain bedding laminae offered the easiest route for the penetration of dolomitizing solutions these were altered, together with cross-cutting penetration perhaps along lines of compactional shears or stretching. Where organic matter or the activity of burrowing organisms rendered parts of the rock more susceptible, these were the parts replaced. Alteration took place only where solutions could penetrate, precipitation of dolomite taking place most readily where effective diffusion porosity permitted a magnesium ion gradient to be set up. This gradient would depend on the ability of ions to enter the sediments by diffusion from the overlying sea water and on their removal from solution with the ultimate formation of dolomite.

Presumably the formation of dolomite would be more rapid where magnesian calcite was being replaced rather than pure calcite. This, and the known preferred alteration of aragonite relative to calcite, may explain the common replacement of organic remains.

If the process of dolomitization and associated recrystallization increases the effective diffusion porosity, channels of less restricted ionic movement, hence more rapid diffusion, might develop along organic and other susceptible trends in the rock.[3] This could cause more complete alteration, and perhaps explains why the contact between dolomitic mottling and residual calcareous rock is commonly so sharp.

Appreciable quantities of magnesium could penetrate several feet into the sediments in a matter of tens of years (Garrels, Dreyer, and Howland, 1949, p. 1827). The transfer of magnesium ions would stop when dolomitization was complete or when compaction, or recrystallization, or the thickness of sediments deposited above, interfered with the diffusion gradient.

CONCLUSIONS

The Palliser formation, laid down as limestone that was possibly magnesian, was subsequently altered to dolomitic limestone at an early stage in diagenesis. Secondary alteration and recrystallization produced the dolomitic mottling now so conspicuous in the rock. This dolomitization began from more susceptible centers possibly resulting in a trigger effect in that the effective diffusion porosity of these areas was increased, thus facilitating the diffusion of the magnesian solutions which first affected the centers and then spread into the surrounding rock. The variability in selection at different horizons suggests that effective porosity was the controlling factor rather than original composition. Thus in the lower beds alteration is localized largely along certain bedding laminae and spreads irregularly from these; higher in the succession "worm burrows" and "algae" are most affected. This localization of alteration, coupled with the fact that disseminated dolomite does not seem to have segregated strongly, favors magnesian solutions reacting with calcium carbonate being responsible for the formation of dolomite rather than recrystallization of a primary magnesian precipitate. In short, dolomitization follows the easiest route and this route varies in different parts of the succession, thus explaining the variation in detail of the mottling.

From this and other studies the writer is strongly impressed by the relationship that commonly appears to exist between dolomitization and organic remains in the limestone. A positive link between the two could operate in a variety of ways.

1. By control of the pH of the sediments due to products of organic decay. This pH value would vary as the cycle of decay progressed and might therefore start the process and bring it to an abrupt close.

2. Organic catalysts may assist the process. Goetz (1921) has shown how organic colloids could assist in bringing about co-precipitation of $CaCO_3$ and $MgCO_3$. Although the writer does not favor this process for the open sea, it may be important in the controlled environment in sediments on the sea floor.

3. Heeger's test for dolomite depends on the presence of ferrous ion which suggests that its formation took place in a reducing environment. Further, the

---

[3] For a more detailed discussion of the relationship between ionic diffusion and the permeability and porosity of rocks see Garrels, Dreyer, and Howland (1949).

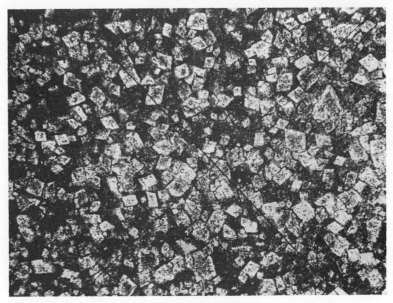

Fig. 5.—Isolated euhedral crystals of dolomite largely replacing fine-grained limestone. Mag. ×68.

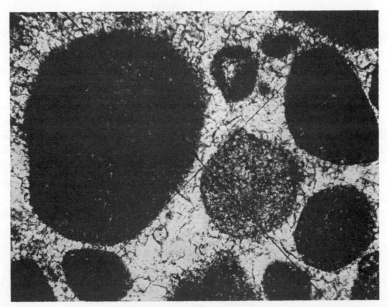

Fig. 6.—Mosaic of interlocking anhedral dolomite crystals replacing groundmass of lime-grit. Mag. ×68.

Fig. 7.—Mosaic of interlocking subhedral dolomite crystals replacing fine-grained limestone. Shows sharp contact between dolomitic and calcareous parts. Mag. ×68.

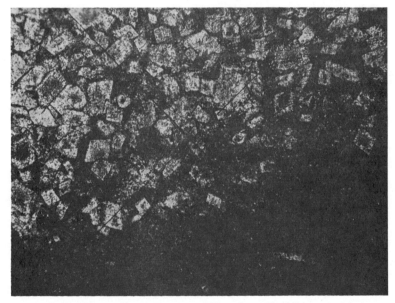

Fig. 8.—Dolomitic mottling grading laterally into fine-grained limestone. Mag. ×68.

presence of the ferrous ion itself may have a pseudocatalytic effect. It is possible that ferrous ions first react with calcium carbonate to displace calcium and form an ankeritic compound which in turn is converted to dolomite by the replacement of the ferrous ion by a magnesium ion. Ferrous ions thus liberated might react again with residual calcium carbonate. Some such indirect process of dolomitization is suggested by the lack of success that has thus far attended experimental attempts at the synthesis of dolomite.

4. Aragonite and perhaps magnesian calcite, both common constituents of invertebrate skeletons, are more susceptible to replacement than pure calcite.

REFERENCES

BEACH, H. H., 1943, "Moose Mountain and Morley Map-Areas, Alberta," *Geol. Survey Canada Mem. 236*, pp. 1–74.

BIRSE, D. J., 1928, "Dolomitization Processes in the Palaeozoic Horizons of Manitoba," *Trans. Roy. Soc. Canada*, Sec. IV, Vol. XXII, pp. 215–21.

CLARKE, F. W., AND WHEELER, W. C., 1917, "The Inorganic Constituents of Marine Invertebrates," *U.S. Geol. Survey Prof. Paper 102*, pp. 1–56.

DEWIT, R., AND MCLAREN, D. J., 1950, "Devonian Sections in the Rocky Mountains between Crowsnest Pass and Jasper, Alberta," *Geol. Survey Canada Paper 50-23*.

DIXON, E. E. L., AND VAUGHAN, A., 1911, "The Carboniferous Succession in Gower," *Quar. Jour. Geol. Soc. London*, Vol. LXVII, pp. 447–571.

FOX, F. G., 1951, "Devonian Stratigraphy of Rocky Mountains and Foothills between Crowsnest Pass and Athabaska River, Alberta, Canada," *Bull. Amer. Assoc. Petrol. Geol.*, Vol. 35, No. 4, pp. 822–43.

GARRELS, R. M., DREYER, R. M., AND HOWLAND, A. L., 1949, "Diffusion of Ions through Intergranular Spaces in Water-Saturated Rocks," *Bull. Geol. Soc. America*, Vol. 60, pp. 1809–28.

GOETZ, VAN C., 1921, "Über die Entstehung des Dolomits der Muschelkalkschichten nördlich des Lothringer Hauptsattels und über den Einfluss von kolloiden Phasen auf die Bildung von Dolomit überhaupt," *Geol. Rundsch.*, Bd. 12, pp. 138–50.

JOHNSON, J. H., 1951, "An Introduction to the Study of Organic Limestones," revised ed., *Quar. Colorado Sch. Mines*, Vol. 46, No. 2, pp. 1–185.

VAN TUYL, F. M., 1916, "The Origin of Dolomite," *Iowa Geol. Survey*, Vol. XXV, pp. 251–427.

WALLACE, R. C., 1913, "Pseudobrecciation in Ordovician Limestones in Manitoba," *Jour. Geol.*, Vol. XXI, pp. 402–21.

DOLOMITES IN SILURIAN AND DEVONIAN OF EAST-CENTRAL NEVADA[1]

JOHN C. OSMOND[2]
Salt Lake City, Utah

ABSTRACT

The Sevy dolomite (500–1,600 feet thick) unconformably overlies Laketown dolomite (Middle Silurian) and is tentatively correlated with the Lone Mountain formation of central Nevada. It is aphanic and homogeneous and represents very slow deposition near sea-level. *Halysites* is found near the base. In a belt marginal to the miogeosyncline during Sevy time, Silurian and Ordovician dolomites and quartzites were eroded. The widespread sandy member at the top of the Sevy is derived from erosion of the Eureka (Middle Ordovician) quartzite.

The Simonson dolomite (500–1,600 feet thick) is characterized by heterogeneous types of dolomite strata, pronounced lamination, conspicuous mottling, and erosional disconformities. It contains stromatoporoid biostromes and bioherms, tubular corals and bryozoans, and Middle Devonian brachiopods including *Stringocephalus*. Periodically the sea floor was exposed and subjected to erosion and reworking of the sediments. Toward the west limestone becomes prominent, and the Simonson is correlated with the Nevada formation. The upper contact of the Simonson is a narrow zone of interbedded dolomite and limestone which passes upward into a thick limestone section.

The evidence of increasing depth of water from the Sevy to the limestone above the Simonson is believed to represent eastward transgression of the Devonian sea.

In eastern White Pine County the miogeosyncline subsided less than in the Confusion basin at the south and southeast or the Roberts Mountains basin at the west. This is known as the North Snake Range "positive area." Simonson sediments are proved to converge over this area which persisted during Sevy and Simonson time.

INTRODUCTION

GENERAL STATEMENT

In the Great Basin the middle Paleozoic deposits form a thick sequence of carbonate rocks. This section has been skillfully subdivided by Merriam (1940, pp. 10–16) at Lone Mountain in central Nevada and by Nolan (1935, pp. 17–21) at Gold Hill in west-central Utah.

The area between these established local standard sections is discussed in this paper. The units studied are the Sevy dolomite and Simonson dolomite defined as Devonian by Nolan.

Stratigraphic and petrographic studies form the basis for interpretations of the depositional environments of these units and their paleogeographic relationships. Correlation of these units with the section at Lone Mountain indicates thickening and facies changes toward the west.

---

[1] Submitted in partial fulfillment of the requirements for the degree of Doctor of Philosophy in the Faculty of Pure Science, Columbia University, New York. Manuscript received, February 15, 1954.

[2] Gulf Oil Corporation, Salt Lake City, Utah.

The helpful guidance of Professor Marshall Kay is sincerely appreciated. Field expenses and transportation were provided very generously by grants from the Kemp Memorial Fund and the Lamont Geological Observatory, Columbia University. John Harms, Rush Valentine, and George Thomas assisted greatly in the field work. P. F. Kerr, P. K. Hamilton, and A. N. Strahler made helpful suggestions on the analytical procedures. C. W. Merriam, John Imbrie, and R. L. Batten kindly identified parts of the fauna. The aid extended by many associates at Columbia and geologists in the area is also much appreciated. Nancy Osmond graciously encouraged the writer and typed and edited the text.

LOCATION

The work was done in the mountains of the Basin-and-Range Province in east-central Nevada and west-central Utah (Fig. 1). Ely, Nevada, is the center of the area which includes White Pine County and adjacent parts of Nye and Lincoln counties, Nevada, and Millard, Juab, and Tooele counties, Utah.

The area was in the Millard belt of the Cordilleran geosyncline during lower and middle Paleozoic time and was inundated during most of this time (Kay, 1951). The carbonate and orthoquartzite shelf deposits indicate an absence of

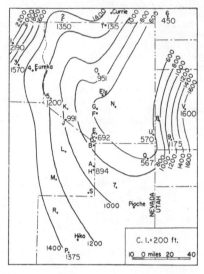

Fig. 1.—Isopachous map of Sevy dolomite.

adjoining tectonic source lands. These conditions did not continue into Carboniferous time.

DESCRIPTION OF AREA

In this part of the Basin-and-Range Province (Nolan, 1943), the fault-block mountains trend north-south and are 5–10 miles across. They are separated by alluvium-filled valleys 15–20 miles wide. The mountains are primarily tilted Paleozoic sediments, but the structures within the ranges are complicated by folding and thrusting. Tertiary sediments and volcanic rocks lie on the older strata and are variously tilted.

The distribution of mountains and valleys provides much better north-south continuity of exposures than in the east and west directions. The filled valleys create gaps between outcrops. Unfortunately, the stratigraphic continuity inherited from the trends of the Millard belt also runs north-south; hence, the east-west gaps conceal important facies changes transverse to the geosynclinal trend.

LOWER PALEOZOIC SECTION

The Cambrian sequence in east-central Nevada approaches a thickness of 7,000 feet (Wheeler and Lemmon, 1939, pp. 13 and 31). A very thick lower quartzite is overlain by shale and argillaceous limestones. The Lower Ordovician is represented by about 3,000 feet of Pogonip limestone (Hintze, 1951, p. 38), overlain by 400 feet of Middle Ordovician Eureka quartzite (Webb, 1953, Pls. 2 and 3). Upper Ordovician and Silurian dolomites attain a thickness of more than 1,000 feet in most of the area. Middle and Upper Devonian carbonate rocks aggregate 4,000 feet.

LOCATIONS OF MEASURED SECTIONS

The summer of 1951 was devoted to measuring eight sections along 53 miles of the Egan Range south of Ely. This provided a familiarity with the units so excellently exposed there and served as a basis for correlation over a larger area during the summers of 1952 and 1953. Efforts to correlate with one of the established sections were most successful with the one at Gold Hill, Utah. The sections measured are listed with corresponding letters on the index map (Fig. 1).

- A. Trough Spring Canyon, South Egan Range
- B. Adams Ranch, South Egan Range
- C. Six-Mile Spring, South Egan Range
- D. Nine-Mile Canyon, South Egan Range
- E. Lund, South Egan Range
- F. South Ward Peak, South Egan Range
- G. Rowe Creek Canyon, South Egan Range
- H. South End of Egan Range
- J. Blackrock Canyon, White Pine Range
- K. Cherry Spring, White Pine Range
- L. Blue Eagle Mountain, Grant Range
- M. Bordoli Ranch, Quinn Canyon Range
- N. Cooper Wash, Shell Creek Range
- O. Steptoe, North Egan Range
- P. Pahranagat Pass, Northern Pahranagat Range
- Q. Lime Mountains, South of Snake Range
- R. Southern Worthington Range
- S. Timber Mountain, Seaman Range
- T. Indian Springs, Northern Cherry Creek Range
- U. Big Springs Ranch, Southern Snake Range
- V. Kings Canyon, Confusion Range

In addition to the measured sections the following numbered locations are significant.

1. Roberts Creek Mountain, Roberts Mountains (Merriam, 1940, p. 31)
2. Pearl Peak, Southern Ruby Mountains (Sharp, 1942, p. 660)
3. Lone Mountain (Merriam, 1940, p. 18)
4. Brush Peak, Fish Creek Range (Merriam, 1940, p. 18)
5. Standard Oil of California and Continental's Meridian 1
6. Sevy (Dewey) Canyon, Deep Creek Mountains (Nolan, 1935, p. 18)
7. Silverhorn, Fairview Range (Westgate and Knopf, 1932, p. 16)
8. Standard Oil of California and Continental's Burbank 1
9. Burbank Hills (Rush, 1951, p. 45)

The exposures at Lone Mountain and Sevy Canyon (Gold Hill district, Utah) were examined with the purpose of establishing correlations of these important sections with strata of the intervening territory. Pearl Peak, Brush Peak, and Silverhorn also were examined.

SEVY DOLOMITE

GENERAL DESCRIPTION

The Sevy dolomite was named by Nolan (1935, p. 18) from exposures in Sevy Canyon (loc. 6, Fig. 1) in the Deep Creek Mountains south of Gold Hill, Utah.

The Sevy dolomite is remarkably homogeneous throughout the area of outcrop. The typical rock is a well-bedded mouse-gray dolomite in layers 6 to 12 inches thick and weathers to a very light gray. It is of exceedingly dense texture and has a conchoidal fracture. In most of the beds a faint lamination parallel to the bedding is visible, in part at least, because of slight differences in color in adjoining laminae. A few beds of darker dolomite occur near the top of the formation, and locally there are present beds containing tiny nodules of light-colored chert.

The Sevy dolomite is present in east-central Nevada where the brown to dark gray aphanic dolomite weathers to the characteristic whitish gray. As the rock is composed of crystals too small to be seen by the unaided eye it is an aphanic dolomite (DeFord, 1946, p. 1923). The exposures are steep slopes with steps and risers developed by strata 2–3 feet thick. These beds break into large splinter-shaped fragments which have a distinctive clinking sound underfoot. The thickness of the Sevy ranges from 500 to 1,600 feet. Fossils are very rare.

SUBDIVISIONS

The Sevy dolomite was recognized at every location examined. Two subdivisions were made on the basis of lithic change (Fig. 2). Quartz sand grains are scattered through several hundred feet of the Sevy, but they are most concentrated near the top, forming a sandy member which is a consistent datum in east-central Nevada. Beneath this member at the two southernmost locations there is a cherty argillaceous dolomite member (Fig. 24).

CORRELATION WITH LONE MOUNTAIN FORMATION

The Lone Mountain formation was redefined by Merriam (1940, pp. 13–14) from the exposures on Lone Mountain, 17 miles west of Eureka, Nevada. The unit is massive, blocky dolomite which weathers light mouse-gray. "A characteristic feature of its upper portion is an alternation of heavy beds of light mouse-gray and darker smoke-gray members." No diagnostic fossils were reported but the relation to the overlying Lower Devonian Nevada limestone is transitional.

The Sevy is tentatively correlated with the Lone Mountain formation on the basis of striking lithologic similarity and comparable position in the stratigraphic sequence.

The correlation with the Lone Mountain formation can be extended into southeastern California. In the northern Panamint Range, Death Valley, California, the Hidden Valley dolomite is subdivisible into three units (McAllister, 1952, pp. 15–17). The lower unit, number 1, contains Silurian fossils and seems comparable with the Roberts Mountains formation which underlies the Lone Mountain formation. Unit 2 is devoid of diagnostic fossils, is massive, very light gray, and is correlated with the Lone Mountain formation. The unit is coarse-crystalline, but this difference is attributed to recrystallization. The upper unit, number 3, contains a Lower Devonian fauna and has brownish weathering, silty and sandy beds at the base. This sand may be related to the sandy member in the top of the Sevy, and unit 2 is probably related to the Sevy.

LOWER CONTACT

At Gold Hill the Sevy is underlain by fossiliferous Laketown dolomite (Silurian, "Niagaran"). Similarly at Lone Mountain the Lone Mountain formation is underlain by Roberts Mountains formation (Silurian, "Niagaran").

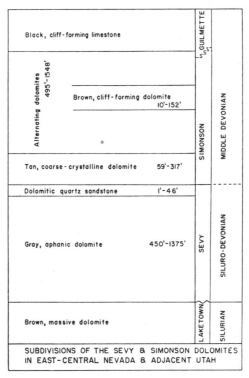

FIG. 2.—Subdivisions of Sevy and Simonson dolomites. Thickness in feet.

At the locations in east-central Nevada the Sevy is underlain by dark, massive, cherty, fossiliferous dolomites comparable with these Silurian formations.

The contact between the Laketown and the Sevy is unconformable at Gold Hill (Nolan, 1935, p. 18). At Kings Canyon, Utah, the top of the Laketown exhibits a few feet of relief with beds having been removed by pre-Sevy erosion. In the Ruby Range (Sharp, 1942, p. 660) Lone Mountain unconformably overlies Lower Ordovician Pogonip. However, Sharp mentions that toward the south the Lone Mountain becomes coarse-crystalline, brown-weathering dolomite with black chert. This description seems to fit the Roberts Mountains formation and may alter the significance of the unconformity at this place.

At all locations examined the base of the Sevy was represented by a sharp contact or marked transition with the underlying dark dolomites. There is no evidence of interbedding of the distinctly different lithologic types.

The contact represents a surface of erosion with varying relief. In some places sharp relief, carbonate-fragment conglomerates, and missing units provide clear evidence; but at other places, bleaching of the underlying units and incorporation of their fine detritus in the basal part of the Sevy create an apparent transition. The magnitude of this unconformity can be determined only by subdivision and regional study of the Silurian strata. This has not been the object of the present

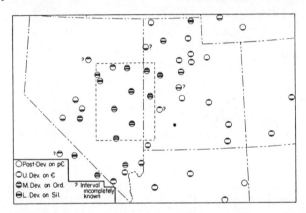

FIG. 3.—Map illustrating magnitude of regional unconformity which underlies Sevy dolomite in east-central Nevada.

study, but some consideration has been based on the following observations and perusal of Figure 3.

At Steptoe, Blackrock Canyon, Lund, Trough Spring Canyon, and Kings Canyon the top of the massive dark dolomite contains abundant dark chert and over-silicified fossils. Similar units are reported near the top of the Roberts Mountains formation in the type section (Merriam, 1940, p. 12) and the Laketown dolomite at Gold Hill (Nolan, 1935, p. 17). At Pahranagat Pass and in the Lime Mountains the dolomite contained abundant chert, but no fossils were found. Beneath the Sevy at Indian Springs, Blue Eagle Mountain, and Big Springs Ranch is tan-gray, medium- to coarse-crystalline, massive dolomite quite different from the cherty member which underlies this unit at the other places. Nolan mentions the presence of the tan-gray medium-crystalline unit at the top of the Laketown at Gold Hill and its absence locally as the result of pre-Sevy erosion. These relationships are evidence of a regional unconformity beneath the Sevy dolomite.

UPPER CONTACT OF SEVY DOLOMITE

Above the Sevy is the tan, massive, coarse-crystalline basal member of the Simonson dolomite. The contact between the whitish aphanic dolomite steps of the Sevy and the cliff of the basal Simonson is clear and sharp. One to 3-foot interbeds of aphanic dolomite, like the main body of the Sevy, occur as high as 60 feet up in the coarse member of the Simonson and account for about 10% of the thickness of this interval. Because of this interbedding there does not seem to have been a major break in deposition at this position in the section. At Lone Mountain, the Lone Mountain formation is transitional into the overlying Nevada formation (Merriam, 1940, pp. 22–23).

PETROGRAPHY

Color.—The color of the rocks was determined by use of the Rock Color Chart (Goddard *et al.*, 1951). The value (lightness) of the color was found to be much more useful in comparing the dolomites than the chroma (saturation) or hue. The value on the weathered surfaces ranged from 6 to 9 (1 is black; 9 is white). On fresh surfaces the value varied from 3 to 6. The average value differential is 3, and the weathered surfaces are considerably lighter. The olive-gray fresh surfaces are the result of internal absorption of light caused by the exceedingly small, transparent grains.

Texture.—The Sevy dolomite is a homogeneous aphanic rock. The crystal size is typically less than .01 mm. (silt size and smaller in the Wentworth scale). This is in agreement with W. E. Ham, who experimentally found the .03-mm. crystal size to be the aphanic-phaneric boundary (in DeFord, 1946, p. 1923). Exceptions to this very small size of the textural components are the clastic quartz and dolomite sand grains. Recrystallization adjacent to fault zones also creates larger crystals. Toward the west and southwest this recrystallization may be on a regional scale.

The interlocking mosaic of dolomite anhedra is made of irregularly shaped, closely packed, clear crystals. The sizes range from less than .002 mm. to .009 mm. with most of the particles about .005 mm. in the largest dimension. The contacts between the crystals appear especially dark in thin sections and probably represent concentrations of the impurities.

Scattered subhedra in the large sizes represent recrystallization of the smaller anhedra. This process has developed individual subhedra disseminated through the rock as well as in local zones along fractures.

The anhedra do not exhibit original clastic forms but appear to have fitted together by adjustive processes of solution and precipitation. Originally they could have been spheroidal, rhomboidal, or any other shape but probably were no larger than now. In this well sorted sediment, dolomite sand grains are associated with quartz sand grains; and it seems reasonable to assume that dolomite of silt size and smaller was deposited with the insolubles of comparable sizes.

Insoluble residues.—About 2% of the total weight of the rock is insoluble residue. The percentage is higher in the sandy and argillaceous members. The residue is mostly clay-size particles with some silt-size quartz. There is a notable absence of bitumen. This agrees with the paucity of fossils and indicates a lack of organic material which is instrumental in the dolomitization of many calcareous sediments (Linck, 1937, p. 281).

Williams (1948, p. 1157) noted that there are negligible insoluble residues in the Laketown dolomite in the Logan Quadrangle, northeastern Utah. Sediment derived from the erosion of this unit, which is 95% dolomite, would likewise be a dolomite low in insolubles.

A differential thermal analysis was made of the insoluble residue. The residue contained 10% angular silt-size quartz which was separated by elutriation from the material to be analysed. The D.T.A. curve showed no indication of either crystalline quartz or clay minerals. The insoluble residue is assumed to consist of a small amount of authigenic silica and rock flour. The last category might include feldspars, micas, and chert. This would agree with a carbonate-rock source supplying its insoluble residues.

Primary structures.—Although the over-all appearance of the Sevy reveals only bedding, on closer scrutiny faint lamination may be found. The lamination is developed by the presence of very thin laminae (1 mm.) of silt and very fine-grained clastic quartz and dolomite. These laminae are separated by much greater thicknesses of homogeneous aphanic dolomite. The laminae show evidence of current control in small-scale scouring and cross-lamination (Fig. 4).

In addition to the even laminae there are roiled (contorted) laminae and plastimorphic fragments that evince the plastic nature of the original mud. Plastimorphic fragments are torn from a plastic parent sediment such as the Sevy muds and are neither rounded by wear nor angular by breakage (Fig. 5). They are irregular in outline, ordinarily small; but, if large (a few mm.), they are flat. They are best preserved and probably recognized only where entrapped in another plastic or highly viscous sediment of contrasting color.

CHERTY ARGILLACEOUS MEMBER

At the two southernmost locations (Worthington Range and Pahranagat Pass) the part of the Sevy underlying the sandy member is different from all other sections. At these places there is an argillaceous dolomite phase with chert nodules. This member is 101 feet thick in the Worthington Range, where the chert is red and brown, and 202 feet thick at Pahranagat Pass, where the chert is black (Fig. 24).

The argillaceous dolomite is aphanic, olive-drab on the fresh surface, and weathers to yellow-gray. It forms rounded blocks in contrast to the splinters and angular rhombs more characteristic of the Sevy.

The chert nodules are irregularly shaped, individual or coalescing, and 1–3 inches high on the average with maximum height of 6 inches. All the chert has yellow-brown to reddish tan rinds $\frac{1}{5}$–$\frac{1}{2}$ inch thick. The silica for the chert was probably derived from the argillaceous matrix.

SANDY MEMBER OF SEVY DOLOMITE

Definition.—In the upper few tens of feet of the Sevy dolomite occur laminae

Fig. 4.—Exceptionally well developed lamination from Sevy dolomite. Note silt and very fine-grain quartz and dolomite in lower laminae. Cross-lamination in upper left and lower middle of field. Granule of dolomite in lower right has caused depression of underlying dolomite. (×1.5)

Fig. 5.—Etched surface of sandy member of Sevy dolomite showing well rounded quartz sand grains (black), zones of slightly different color and crystal-size dolomite, and plastimorphic fragments of dolomite (some contain quartz sand grains). Roiled sedimentary breccia indicating plastic state of deposit when deformed. (×1.5)

and strata containing more than 25% quartz sand. Common occurrence of concentrations of this magnitude differentiates the sandy member from identical dolomite below, which contains negligible sand.

Description.—The sandy zones typically weather brown in contrast to their whitish gray associated beds. A few of the dolomite strata in this member are dark gray. The sand is concentrated along laminae and stylolites. There are also strata in which the sand grains "float" in the dolomite matrix.

The sand is typically fine-grained with coarse-grained and clay-size extreme fractions. The grains are rounded to well rounded and frosted but with only a moderate degree of sphericity (.7–.9), according to the standards cited by Petti-

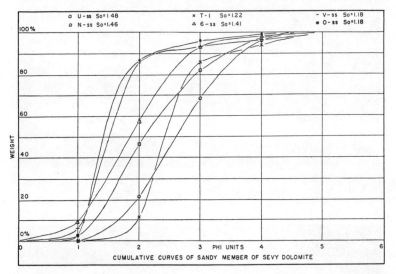

Fig. 6.—Cumulative curves of sandy member of Sevy dolomite, north and east of Egan Range.

john (1949, pp. 50–52). The grains are not secondarily enlarged, but in some places part of the unit is quartzitic with clear siliceous cement. Thin sections of the sandy zone reveal well rounded dolomite sand grains in the same size range as the quartz grains (Fig. 10).

Spatial relationships.—Quartz sand grains are present as less than 1% of the rock in thin zones through several hundred feet of the Sevy and also in the Simonson above. The sandy member is represented by dolomitic quartz sandstones at the locations in Figure 9 where the thicknesses are indicated in parentheses. The area of prominence is in the southern part of east-central Nevada. Elsewhere, the member is represented by a zone of thin sandy laminae and strata with scattered sand grains. At Gold Hill this horizon was placed in the basal Simonson by Nolan (1935, p. 19), but the matrix of this member so resembles the Sevy that it is placed as the upper member of the formation in this report. The thinness of this widespread member indicates that the sediments were subjected to extensive reworking and not merely dumped into the area and rapidly buried.

Quartz sand was not found in this position at Lone Mountain or at Brush Peak. It is not reported by Merriam in the Roberts Mountains. Sand is in a corresponding stratigraphic position in the Sulphur Spring Mountains, northwest of Eureka, and in the base of unit 3 of the Hidden Valley dolomite in the Panamint Range near Death Valley (McAllister, 1952, pp. 15–16).

The sandy member is believed to represent a continuous and correlatable zone through east-central Nevada and not to be confused with other sand zones which

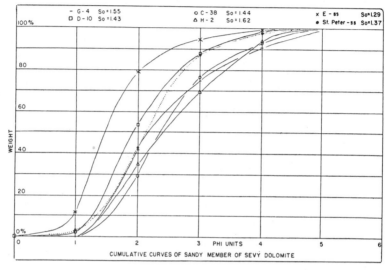

FIG. 7.—Cumulative curves of sandy member of Sevy dolomite, in Egan Range.

occur in the middle Paleozoic carbonates of the Cordilleran miogeosyncline (Merriam, 1951, p. 1508).

Size analyses.—Specimens of dolomitic quartz sandstone were disaggregated with hydrochloric acid; and the residues were washed, dried, and sieved. Figures 6, 7, and 8 and Table I show the results of these analyses. It was found that concentrations of 30% or more quartz sand created a rock which was readily identified as dolomitic quartz sandstone in the field. The maximum concentration determined was 80%.

The cumulative curves of specimens from locations north and east of the Egan Range are shown in Figure 6; from the Egan Range south of Ely, in Figure 7; and from west and south of the Egan Range, in Figure 8. Their most striking feature is their similarity. This uniformity is all the more remarkable since the samples were taken from various positions within the member. The second significant relationship is the high angle of the slope of the curves, indicating a well sorted sediment.

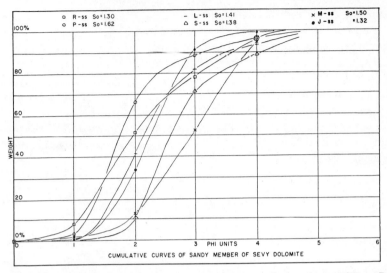

Fig. 8.—Cumulative curves of sandy member of Sevy dolomite, south and west of Egan Range.

Table I shows a comparison of some of the statistics obtained from the size analyses. The maximum-size grade of the sand grains is evenly distributed geographically between coarse and medium. All values relate to the Wentworth scale. The minimum size is clay size; however, the material smaller than $\frac{1}{16}$ mm. was not subdivided into silt and the smaller clay sizes. These particles constitute less than 5% of the insoluble residue in most samples.

The modal-size grade is very fine sand for one sample, fine sand for 9, and

TABLE I. SIZE ANALYSES OF SANDY MEMBER OF SEVY DOLOMITE

	Location	Max.-Size Grade	Modal-Size Grade	Median-Size (mm.)	Sorting (Payne, 1942)	So	QD_ϕ	Sorting Index (Dapples, 1953)
North of Egan Range (north to south)	6 T O	Coarse Medium Coarse	Fine Fine Fine	.275 .185 .375	Fair Fair Good	1.41 1.22 1.18	.500 .276 .233	.052 .028 .024
East of Egan Range (east to west)	V U N	Coarse Coarse Medium	Fine Fine Fine	.350 .235 .163	Fair Fair Fair	1.18 1.48 1.46	.233 .577 .551	.026 .058 .055
Egan Range (southward from Ely)	G E C D H	Medium Coarse Medium Coarse Medium	Medium Medium Fine Medium Medium	.210 .355 .195 .260 .190	Fair Fair Fair Fair Fair	1.55 1.29 1.44 1.43 1.62	.630 .377 .525 .511 .699	.064 .037 .055 .048 .069
St. Peter sandstone (average)		Coarse	Fine	.230	Fair	1.37	.450	.045
West of Egan Range (north to south)	J L M	Coarse Medium Coarse	Fine Medium Very fine	.210 .225 .130	Good Fair Poor	1.32 1.41 1.50	.399 .500 .590	.040 .051 .056
South of Egan Range (north to south)	S R P	Medium Coarse Coarse	Fine Fine Fine	.165 .290 .255	Fair Fair Fair	1.38 1.30 1.62	.468 .403 .699	.053 .042 .071

medium sand for 5. Four of the latter samples came from the Egan Range. The median size varies from .130 to .355 mm. with an average value of .239 mm. The geographic distribution is uniform.

The coefficient of sorting (So) as described by Trask (1932, pp. 71–72) varies between 1.18 and 1.62 and averages 1.40. According to Trask's classification these values indicate a well sorted sandstone. Hough (1940, p. 26) and Stetson (Stetson and Upson, 1937, p. 57) agree that most near-shore marine sediments of sand grade have sorting coefficients between 1 and 2 with an average of 1.45. This does not imply a near-shore depositional environment for the sandy member but provides some basis for evaluating the coefficients of sorting.

Sorting, according to the standard used by Payne (1942, p. 1707), is only fair for 15 of the 18 samples. This method of determining sorting seems too rigorous for what appears to be a well sorted sand in the Sevy and also for the generally recognized well sorted St. Peter sandstone of the Mid-Continent Ordovician.

Recently Dapples and others (1953) have discussed the petrographic and lithologic attributes of sandstones. These writers attempt to relate the physical properties of sandstones to a genetic classification.

The statistic of sorting index introduced in this article is easily attained and may provide a basis for comparison of the degree of sorting of various sands. This value is the ratio of the number of phi (Wentworth) classes in the modal two-thirds of the sample to the square of the number of phi classes in the entire size range. Sorting indices calculated for the quartz sand residues of the sandy member range from .03 to .07 and place this unit in the quartzose category of their classification. The reason for the divergence between the index and the coefficient is the consideration of the extremities of the curve which are included in the index calculations.

The quartile deviation in phi units (Krumbein, 1936, p. 104), calculated from phi numbers ($-\log_2$ diam. in mm.), is useful in comparing the spread of the cumulative curves. This quartile deviation indicates the number of Wentworth size grades in half of the distribution from the first to the third quartile (25%–75% of the cumulated weight).

The quartile deviation (phi) varies from .233 to .699, with an average of .481. This means that for the average sample less than one Wentworth size grade spans the distribution of the central 50% of the curve. This indicates a well sorted sediment when considered in combination with the shortness of the coarser end of the curve. Authigenesis may be involved in the length of the smaller-size segment of the curve.

Figure 9 shows the quartile deviations and lines delineating areas of equal values. The samples with the greatest dispersion occur southeast of those with less dispersion. The maximum, modal, and mean sizes do not correspond with this pattern but vary randomly throughout the area. The sorting of the sediments is better outward from the southeast which probably means that the sands came from this direction: those travelling the farthest became better sorted.

Comparison with St. Peter sandstone.—The cumulative curve for the St. Peter sandstone is given in Figure 7 and corresponding data in Table I. This information is from the excellent study by Thiel (1935, p. 584).

The median diameter varies from .114 to .350 mm. with an average of .222 mm. The So varies from 1.15 to 1.93 with an average of 1.39, and the quartile deviation was calculated by the present writer to vary from .20 to .96, with an average value of .49. There is a marked similarity between these figures and those for the sandy member of the Sevy. This condition does not necessarily reflect similar origins, but it indicates that the sand in the Sevy is comparable with a well known example of a well sorted sandstone.

Heavy minerals.—Heavy minerals were extracted from some of the samples,

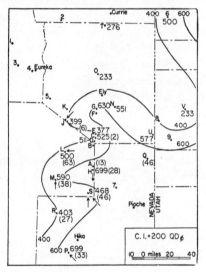

Fig. 9.—Map showing lines of equal sorting of sandy member and current directions (arrows) as indicated by cross-lamination. Thicknesses of dolomitic quartzites given in parentheses.

but the quantities were exceedingly small. The most common non-opaque minerals are zircon (some blue), tourmaline, and garnet. Minor constituents are rutile, hornblende, diopside, brookite, and tremolite. The opaque fraction, hematite, magnetite, leucoxene, and allophane, formed nearly 50% of the total.

The sampling technique was not aimed at a heavy-mineral analysis, and the suite of minerals found seemed of little value for correlation. However, this evidence seems to indicate a second- or later-cycle sandstone.

Cross-lamination.—Planar cross-lamination (McKee and Weir, 1953, p. 387), with a low angle and up to 2 feet long, was developed at six of the locations (Fig. 9). The outstanding development was at the south end of the Egan Range (Fig. 11), where the current seems to have moved southward. At the other locations cross-lamination was rare and poorly developed. At Blackrock Canyon (J) the

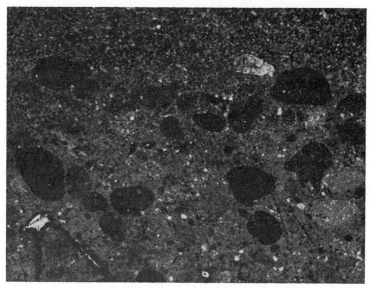

Fig. 10.—Rounded dolomite sand grains associated with quartz sand grains (white) in sandy member of Sevy dolomite. (Photomicrograph; ×30)

current seemingly came from the northeast; at Blue Eagle Mountain (L), from the east; at Bordoli Ranch (M), from the south; at Timber Mountain (S), from the south and southeast; and at Pahranagat Pass (P), from the southeast.

The three southern readings agree in indicating currents from the south; the

Fig. 11.— Planar cross-lamination of dolomitic quartz sandstone in sandy member of Sevy dolomite. View eastward at south end of Egan Range shows exceptionally well developed structure.

other three readings, in the central part of the area, show currents from the north and east. These estimates are based on field observations and not on statistical analyses.

Conclusions.—Two factors have combined to bring about the deposition of such a uniform, widespread sand in the Sevy dolomite. These are sorting during transportation and a well sorted source containing grains of a limited size.

A source area probably existed at the south as indicated in a general way by the increasing proportion of quartz sand in the dolomite section, cross-lamination, presence of the underlying cherty argillaceous member, and the distribution of the quartile deviations (and So). This area might be in southern Nevada or adjacent Arizona. At Goodsprings, Nevada, Middle Devonian rests on Middle Cambrian (Hazzard and Mason, 1953, p. 649), and in the Grand Canyon region Upper Devonian is on Upper Cambrian (McKee, 1945, Pl. I). Sand also was probably derived from a belt trending from the southwest corner of Utah to the southeast corner of Idaho.

These sources were gentle upwarps along the eastern hinge of the Millard belt. Erosion on this type of area would remove the Silurian and Upper Ordovician dolomite blanket and progressively expose Eureka and Swan Peak quartzites. The sands would be carried to the river mouths and dropped into carbonate muds. They would then be subject to currents parallel with the shore and to landward winds. Finer débris, if present, might stay in suspension and be carried farther from shore.

Webb (1953, pp. 93–104) has studied the Eureka quartzite and the grain-size distributions within it. The cementation of the rock necessitated analysis from thin-section examinations. He obtained numerical counts of the grains in the various size grades. These data were converted to weight distribution to be compared with sieving analyses (Greenman, 1951). The mean size at Sunnyside, 3 miles north of the Trough Springs location, and in the Confusion Range is 0.88 mm. (very fine sand). From Webb's binocular estimates the modal-size grade, which varies from fine to coarse, appears to be predominantly medium sand. These values are larger than the modal and maximum sizes of the sands in the Sevy. Thus the sandy member of the Sevy dolomite could have been derived from erosion of the Middle Ordovician quartzites peripheral to the Millard belt.

The sandy member of the Sevy does not possess regressive characteristics similar to those of the Eureka. Instead, the distribution of sorting values and thicknesses indicates slow southeastward and eastward transgression of the sand-depositing environment.

SEVY ISOPACHOUS MAP

Figure 1 is an isopachous map of the Sevy dolomite; thicknesses at locations 1 and 3 are Merriam's measurements of the Lone Mountain formation; the thickness at location 2 is from Sharp (1942, p. 661). The thickness of the Lone Mountain formation in the Meridian well (loc. 5) is from Donovan (1951, p. 48). The

Lone Mountain thicknesses are in agreement with the isopachous pattern of the Sevy and substantiate the correlation of the two names.

The dominant feature of the map is the area of thinning along the Utah-Nevada boundary in White Pine County, shown also in the isopachous maps of the members of the Simonson dolomite. It has been referred to informally as the north Snake Range "high."

The Sevy is thin in the Egan Range between Ely and Lund and in the Snake Range. The base is not exposed in the Shell Creek Range at location N. In the Gold Hill district (Nolan, 1935, p. 19) the Sevy is 450 feet thick in Sevy Canyon and may be 600 feet thick less than 3 miles south. If this is primary thickening and not the result of measurements across faults, the area of thinner sections is smaller than indicated on the map, and another such area would be north of location 6.

This area of thin sections is the effect of differential deposition (convergence) and not of subsequent erosion. It is therefore a relatively more positive area in a region of subsidence. The interbedded upper contact and the presence at all locations of the upper sandy member prohibit the possibility of erosion. The area subsided less than its surroundings. Such conditions were present in the Ordovician where the Eureka quartzite was either eroded away or never deposited in the northern part of the Snake Range (Christiansen, 1951b, p. 75).

The maximum thickness measured of the Sevy is 1,600 feet in Kings Canyon in the Confusion Range of western Utah. West of Eureka, Nevada, the Lone Mountain formation thickens to 2,190 feet in the Roberts Creek Mountains (Merriam, 1940, p. 32). The thinnest measured section of the Sevy is 567 feet in the Lime Mountains, an east-west-trending range south of the Snake Range.

Outward from the thinner area, the Sevy thickens at the rate of 40 feet per mile into the Confusion basin on the east, as shown in the cross section (Fig. 25). The Sevy becomes thicker toward the south, west, and northwest about 10 feet per mile. In the northwestern part of the map the Sevy thickens into the Roberts Mountains basin.

REGIONAL EXTENT

Equivalents of the Sevy dolomite may be present in northwestern Utah in the Desert Range and Lakeside Mountains. Deposits of this age are absent in the Raft River Mountains in the corner of the state (Stokes, 1952, p. 1300). Sloss (1954, p. 365) mentions the possibility of extending the Sevy into the Lemhi Range of south-central Idaho, 230 miles north of the Lakeside Mountains. The Sevy may also be represented as far east as the Wasatch Mountains of Utah.

Farther south, in the Canyon Range, 4,750 feet of undifferentiated limestones and dolomites overlie the Cambrian Ophir formation (Christiansen, 1951b, p. 9). This section contains Eureka quartzite (Webb, 1953, p. 80) and may include strata as high as the Laketown. Christiansen's description of the youngest rocks as "light gray, dense dolomites" might indicate the presence of Sevy 72 miles east of Kings Canyon.

The Sevy is missing in southern Nevada in the Muddy Mountains (Longwell, 1952), at Frenchman Mountain (McNair, 1952, p. 46), and at Goodsprings.

At these locations, peripheral to the miogeosyncline, post-Sevy strata lie on the Lower Paleozoic. Farther east and southeast the magnitude of this unconformity increases (Fig. 3).

The Sevy is represented by 1,500 feet of dolomite in central Nevada and by 700 feet of dolomite in the northern Panamint Range near Death Valley, California.

In the Tonopah Quadrangle of southwestern Nevada, Silurian and Devonian rocks have been eroded from the Manhattan geanticline (Nolan, 1943, pp. 172–173; Ferguson and Muller, 1949, pp. 4 and 51). A remnant of Devonian limestone in the San Antonio Mountains in this quadrangle lies with slight angular discordance on Ordovician cherty argillaceous sediments and may indicate a pre-Devonian uplift in the area.

The Sevy may pass into argillaceous and silty rocks in the western part of Nevada. In the Toquima Range several hundred feet of Silurian calcareous shales with graptolites directly overlie the Ordovician (personal communication, Marshall Kay, 1953). The western limit of the miogeosynclinal carbonate sediments corresponds with a major zone of overthrusts (Roberts, 1949, p. 1917; Kay, 1952, p. 1269).

The Sevy is less extensive than the underlying "Niagaran" formations and also less extensive than the Devonian strata above. The extent definitely seems related to the miogeosyncline and to have been limited by its eastern hinge, the Wasatch line. No facies identified as near-shore were found. These may have been eroded away, if they were ever present. It seems possible that the near-shore muds were similar to those deposited many miles offshore. Following this line of reasoning, absence east of the hinge line is the result of non-deposition.

FAUNA

Fossils are extremely rare in the Sevy; and, where present, they are poorly preserved. At Nine-Mile Canyon in the Egan Range some silicified crinoid stems, *Syringopora* sp. cf. *S. perelegans* Billings, and *Halysites* sp. cf. *H. catenularia* Linné were found in the lower part of the unit. At Big Spring Ranch a phaceloid disphyllid? was 50 feet below the top of the Sevy.

Merriam reported only a poorly preserved *Syringopora*? near the base of the type Lone Mountain formation; McAllister found only favositids near the base of unit 2 of the Hidden Valley dolomite in the Panamint Range, California; and Nolan mentioned "small crinoid stems at a few horizons and several poorly preserved gastropods near the base" of the type Sevy.

The preservation of these fossils could be attributed to their sturdy structure having withstood dolomitization as suggested by McAllister (1952, p. 15). The *Syringopora*, however, has very fine connecting tubules preserved. Also, the very fine laminations in the rock have not been destroyed by recrystallization. It seems likely that very few fossils were ever present in these deposits. The surface of Sevy deposition probably was very inhospitable and supported little life. Corals with phaceloid coralla seem indigenous, but the crinoid columnals might have been introduced from outside the area.

Fossils were found in the upper 10 feet of the Laketown at Kings Canyon, Steptoe, and Blackrock Canyon. *Halysites* and favositids are conspicuous at each location. The fauna from Kings Canyon also includes stropheodontids, *Murchisonia* and a low-spire gastropod (*Eotomaria*?), zaphrentids, streptelasmids, and *Synaptophyllum*. At Blackrock, heliolitids, zaphrentids, streptelasmids, and fragments of *Syringopora*? occur in the fauna. These forms indicate a Silurian age for the dolomite beneath the Sevy at these places. Further study of these strata is being made by R. W. Rush.

AGE OF SEVY DOLOMITE

Nolan (1935, p. 18) classified the Sevy as Middle Devonian in age because (1) it truncates the upper member of the Laketown dolomite at Gold Hill, (2) there are 30 feet of dolomite sand and conglomerate at the base of the Sevy, and (3) the Sevy grades upward into the Simonson which contains Middle Devonian fossils.

The Lone Mountain formation at Lone Mountain, Nevada, 130 miles west-southwest of Gold Hill, was classified as Silurian by Merriam (1940, p. 13). He based his opinion "on the position of the formation between the fossil-bearing Roberts Mountains formation and the Nevada formation."

Merriam (personal communication, 1952) reports Oriskany Lower Devonian fauna at several places in central Nevada and southeastern California. In the Ubehebe district, 211 miles south-southwest of Lone Mountain, this fauna is definitely in the upper part of the dolomite sequence which corresponds with the Lone Mountain formation. This fauna was not found at any of the sections examined for this report.

The Eastern Nevada Geological Association Stratigraphic Committee (1953, p. 146) correlated the lower part of the type Sevy as Silurian and the upper part of the Lone Mountain formation as Devonian.

The lithologic change at the base of the Sevy is in most places sharp and unconformable on Silurian rocks. At the top the Sevy is interbedded with the lower part of the Simonson dolomite which contains Middle Devonian fossils near its top. The occurrence of *Halysites* suggests that at least the lower part of the Sevy was deposited during Silurian time.

Petroleum geologists have reported an Upper Silurian fauna from several localities in the Great Basin. These occurrences are probably in strata which underlie the Sevy.

In light of these conditions a late Silurian age seems probable for the lower part of the Sevy; the upper part may be early Devonian. If a transgressive nature is accepted for the Sevy, it is older in central Nevada than at the east.

ORIGIN OF SEVY DOLOMITE

The unconformity at the base of the Sevy characterizes it as a transgressive deposit. This unconformity and the Sevy span the time from Middle Silurian to Lower or Middle Devonian.

The provenance of the Sevy sediments was an extensive area of low relief underlain by carbonate rocks of Silurian age and older. These were exposed along a wide belt of the eastern shelf of the Cordilleran geosyncline. There were local areas where these sediments were uplifted to a greater extent and consequently were eroded to older and older horizons.

The Silurian a nd Upper Ordovician deposits are almost entirely dolomite overlying Middle Ordovician sandstone. The absence of clay minerals in the Sevy agrees with the absence of shale in this sequence.

Erosion of this sequence on the positive areas peripheral to the shelf provided abundant dolomite sediment and solute. The surface waters of this time probably contained greater concentrations of magnesium than at any other time in the history of western United States. In time, erosion reached the sandstone and provided the quartz sand of the upper member of the Sevy. The Sevy sequence is the inverse of the sequence in the source area.

The Sevy sediments were originally carbonate muds. Evidence of this is shown in the plastimorphic deformation and in the smallness of the particles. The surface of deposition was very near sea-level. It may have been a tidal mud flat or a shallow shelf or bay. In this environment the waters from the source would become further concentrated by evaporation in the partly restricted environments. Aragonite and/or calcite were precipitated and probably were completely dolomitized during the long time before lithification; hence, there are no remnants of an original calcareous mud. Cullis (1904) has given valuable evidence of the stages aragonite-calcite-dolomite, and Skeats (1918, p. 199) has explained the shallow-water contemporaneous replacement of calcite by dolomite. His theory assumes the presence of CO_2 in saturated water and the critical pressure (1 to 4 atm.) at shallow depth where $CaCO_3$ and $MgCO_3$ are equally soluble.

The dearth of organic material would be expected in an environment where deposition was slow and the muds were extensively reworked. The compaction of these muds might have been similar to that of clay deposits (Terzaghi, 1940, p. 89). Hedberg's work (1936) on the compaction of clays and shales discusses the stages through which the deposit passes: dewatering stage, mechanical rearrangement, mechanical deformation, and recrystallization. However, the shape of carbonate particles and the chemistry of the deposits do not permit direct comparison with the compaction of clay minerals.

PETROLEUM CONSIDERATIONS

The Sevy does not possess the qualities of a source bed of petroleum. Its texture is unfavorable for a reservoir rock except where porosity is developed by fracturing. In this case its brittleness would be a vital contributing and sustaining factor. Such fracture porosity developed in the base of the Sevy might provide a reservoir for petroleum derived from the underlying petroliferous dolomites.

The Sevy might act as an impervious barrier to accumulations related to the unconformity at its base. The sandy member contains some bitumen and, being well sorted, might provide a reservoir. However, diagenetic dolomitic cementation has destroyed original porosity at the surface.

SIMONSON DOLOMITE

GENERAL DESCRIPTION

The striking difference between the Sevy dolomite and the overlying Simonson dolomite is the darker color and heterogeneous nature of the latter. The Si-

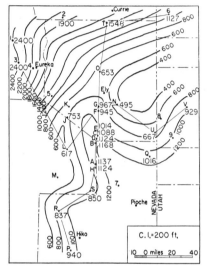

FIG. 12.—Isopachous map of Simonson dolomite showing traverses of cross sections (Figs. 24 and 25).

monson exposures form concave slopes and are less conspicuous than the thick white convex band below or the precipitous cliffs of black limestone above.

The Simonson is readily subdivisible into four members (Fig. 2). The lowest member is a massive cliff-forming unit of tan coarse-crystalline dolomite. Above this is a distinctive sequence of alternating light gray and dark brown dolomite beds. Lower and upper alternating members are separated by a massive, brown, cliff-forming dolomite member. Nolan (1935, p. 19), when first describing the Simonson, wrote "The most striking feature is the very general presence of a fine lamination."

Figure 12 shows the isopachous pattern of the Simonson with values of the Nevada formation used at the three locations in the northwest corner of the map. The structural elements are virtually the same as during Sevy time, except for a

COARSE MEMBER

The oldest member of the Simonson dolomite is a light tan, coarse, crystalline, cliff-forming dolomite. This unit is obvious at all the locations except those near Eureka where it may be represented in the top of the type Lone Mountain formation.

Recrystallization which developed the characteristic texture destroyed most of the primary structures. The present rock is thick-bedded with coarse, crude laminae. At some locations there are a few 3 inch zones of undulating laminae resembling recrystallized stromatoporoid colonies.

The lower part of the member is interbedded with light gray-weathering aphanic dolomites like those of the Sevy. This type of interbedding is also found in the upper part where it grades upward into the alternating member. Strata of tan, coarse-crystalline dolomite do not occur above the lowest brown, fine-crystalline, well laminated bed of the lower alternating member.

Color, texture, insoluble residues, and thickness variations of the coarse member are discussed under those headings for the lower alternating member.

LOWER ALTERNATING MEMBER

The name of this unit is derived from the alternation of brown and gray dolomite beds. The latter, which are whitish gray aphanic dolomite, aggregate much less thickness than the brown beds. There are many types of beds which weather brown. These include finely laminated rocks, accumulations of fossil and sedimentary fragments, intraformational breccia, massive homogeneous strata, and conspicuously mottled beds (Osmond, 1953, p. 1460).

The contrasting colors of these beds create an outstanding striped appearance of the outcrops (Fig. 13). The basic types are referred to as gray beds and brown beds. Table II presents the contrasting characteristics of these beds.

TABLE II. COMPARISON OF BROWN AND GRAY BEDS

Brown Beds	*Gray Beds*
1. Fine to medium crystals	Aphanic
2. Bituminous residues	Silt- and clay-size residues
3. Fossil-bearing	No fossils
4. Rapid lithification	Prolonged plastic stage
5. Erosional upper surface	Non-erosional upper surface
6. Pronounced lamination	Poorly developed, faint, very fine laminae
7. Lateral gradation into limestones	No lateral gradations
8. Sequences of brown beds average 12 feet thick	Beds average 3.4 feet thick
9. Sequences thicker in thicker sections of member	Thickness of beds not related to thickness of member
10. Locally persistent	Locally thin to zero thickness and represented beyond by erosional disconformity between brown beds
11. Several types: laminated, mottled, homogeneous, organic, and sedimentary breccia	One type

Color.—The brown strata possess yellow-red hues which range on the fresh surfaces from lightness values of 2 to 5 and on the weathered surfaces from 3 to 6.

A significant difference in the color of these typically brown strata is found at Gold Hill, Indian Springs, Pearl Peak, and below the *Stringocephalus* zone at Blue Eagle Mountain. At these locations the hues are neutral (shades of gray); other aspects of the color remain the same. This contrast from dark gray strata to the brown is thought to be caused by a difference in the state of the bituminous pigment. Light gray beds are interbedded with both types.

FIG. 13.—Exposure of lower alternating member of Simonson dolomite east of Cooper Wash (N), Shell Creek Range.

The laminated strata are brown at Blackrock Canyon (loc. J). With this exception, the distribution of the dark gray occurrences is west and north of the area. This relationship may shed some light upon the presence of a few hundred feet of very dark gray dolomite above the Lone Mountain formation at Brush Peak (loc. 4), 2 miles southwest of Eureka. The section contains a few thin light gray aphanic beds and is overlain by a 15-foot sandstone in the base of a thick black Nevada (?) limestone sequence.

A thorough study of the coloration of these units might lead to an interpretation of differences in the sedimentary environment or differences in subsequent history that would affect the petroleum possibilities of these strata.

The light gray beds have lightness values which range from 2 to 6 on the fresh surfaces and from 6 to 8 on the weathered surfaces. The average differential is the average of the differences between the values on the fresh and weathered surfaces of each specimen. In the gray beds the average differential is 2.4 and is brought about by the very small grain size; the lesser differential (1.1) for the brown beds is more the effect of the bituminous pigment and to a less degree the crystal size.

The coarse member has lightness values ranging from 3 to 7 on both fresh and weathered surfaces and an average differential of 0.7. This small difference is the result of recrystallization. Some bitumen was trapped in the crystals and is not easily removed by surface leaching. Also, the unit is very porous; and the effect of weathering probably extends to a considerable depth into the outcrops.

FIG. 14.—Typical lamination of brown beds in Simonson dolomite. (Centimeter scale in ½ mm.)

Texture.—The texture of the gray beds is similar to that of the Sevy. The phaneric crystal size of the brown beds is in sharp contrast to the aphanic size of the particles in the gray beds. The crystals of the brown beds range from .05 to .2 mm., which approximates the very fine and fine sand sizes of the Wentworth scale. The texture ranges from uniformly crystalline, interlocking mosaics of dolomite anhedra to dolomite euhedra scattered through a finer crystalline, more calcareous matrix.

Primary structures such as fragments, fossils, and laminae commonly show little or no effect on the dolomitized texture; such structures are present in these rocks as pigmented ghosts. In less thoroughly dolomitized strata the structures are accentuated by a difference in crystal size between the original limestone and the large crystals developed during partial dolomitization.

The coarse member of the Simonson is characterized by crystal sizes ranging from .5 to 1.0 mm. Intercrystalline porosity is well developed in this member, and some enlargement of these voids was observed.

Insoluble residues.—Insoluble residues from the different lithic types in the Simonson were studied, and this information is presented in Figure 15. One of the outstanding features is the presence of bitumen as the major constituent of the residues from the brown beds. This bitumen characteristically shares importance with clastic and authigenic material.

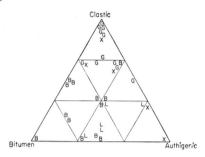

FIG. 15.—Triangular diagram of relative abundance of three types of insoluble residues. B=brown beds, G=gray beds, L=limestone, X=coarse crystalline.

Silt- and clay-size particles constitute the clastic fraction of the residues. Quartz sand grains are very rare in the Simonson; however, they were seen at all locations scattered through one or two of the alternating beds and in the lower part of the coarse member.

The authigenic material consists of euhedral quartz as individual, doubly terminated crystals and aggregates of these. Silicified fossil fragments, chertified veinlets, and dolocastic chert also are included in the authigenic residue.

Residues of the gray beds contain abundant clastic material and very little else. The coarse member yielded residues containing clastic and authigenic material but negligible bitumen. This scarcity of bitumen may be the result of flushing which accompanied recrystallization.

Limestones from the upper alternating member contain residues predominantly of bitumen and authigenic material. The similarity between the residues of the brown beds and those of the limestones seems to substantiate other evidence of their relationship.

The high proportion of clastic residue from the gray beds would be expected if the bitumen were removed during the reworking of the sediment and the clastics from the source were concentrated. Authigenic quartz in the brown beds probably was not formed when the tops of these beds were subjected to erosion, since these crystals were not found in any of the gray-bed residues.

The lithic varieties differ in the percentage of insoluble material as well as in the proportion of the types of residues. The amount of residue in the gray beds ranges from .6% to 11.7% with an average of 3.4% for 12 specimens. This average is distinctly higher than the 1.7% obtained from 15 samples of the brown beds which ranged from .8% to 4.9% insoluble. Such a difference in percentage of insolubles might be explained in the following manner.

A uniform supply of clastic material entered the area of deposition. There it was added to carbonate sediment which was deposited as a time-rhythmic component whose rate of deposition was controlled by fluctuations of the depth of water. As a result, the clastic material appears as a space-rhythmic component vertically through the member (Sander, 1951, p. 132).

The lesser percentage of clastic insoluble in the brown beds also might be the result of bypassing during deposition (Eaton, 1936) or periodic influx of insolubles. Chenoweth (1952, p. 539) also noted the percentage of calcarenite to

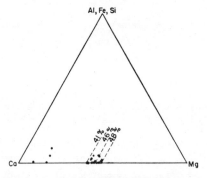

Fig. 16.—Triangular graph showing chemical analyses of 21 samples as determined by spectrographic analysis.

vary inversely with that of shale plus calcilutite in the Trenton of western New York with the latter components increasing toward the source. From a consideration of the insolubles alone, it is difficult to choose between the three possible causes of the higher percentage of insolubles in the gray beds. To agree with other features, as discussed under the origin of the Simonson, the hypothesis of time-rhythmic carbonate deposition is most plausible.

Insoluble residues of 5 samples of the coarse member range from .4% to 1.1% with a very low average of .6%. Residues from 4 specimens of the limestones range from 1.0% to 4.1% and average 2.7%. This small sampling of the limestone falls within the range of the analyses of the brown-bed residues.

Chemical and mineralogical analyses.—Spectrographic analyses were made of 21 samples; and the relative amounts of calcium, magnesium, silica, aluminum, and iron were determined. The procedure followed that described by Nachtrieb (1950, p. 324), and the results are shown graphically in Figure 16.

Seventeen of the specimens were chosen in the field as various types of dolomite; all of these contain between the maximum (48.4%) and the minimum

(40.6%) amounts of magnesium possible in crystals of the mineral dolomite. These deviations from the theoretical dolomite composition (45.65% magnesium) are the result of substitution of calcium for magnesium and magnesium for calcium in the crystal lattice (Palache et al., 1951, p. 211).

Four of the specimens were chosen in the field as limestone, and their analyses verified this. These contained between 9% and 17% magnesium, indicating 20–35% dolomite crystals in the rocks.

None of the 21 specimens contained more than 9% of the combination silica,

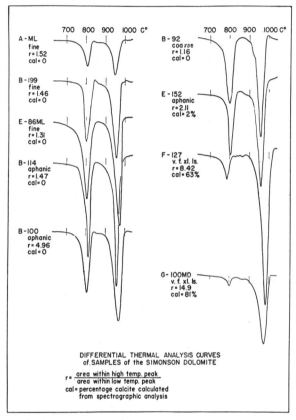

FIG. 17.—Differential thermal analysis curves of samples of Simonson dolomite.

aluminum, and iron; however, the four samples of the gray beds averaged 3.5% of these elements in contrast to the 1.3% average of the other 17 samples. The main constituent of this group in the gray beds is silica with some aluminum and negligible iron. This agrees with the interpretation of the insoluble residues.

Differential thermal analyses were made of 9 of the samples which had been analysed by spectrograph. The D.T.A. curves are shown in Figure 17. The areas within the endothermic peaks were measured with a planimeter. According to Berg (1945, p. 650), the ratio of the area within the high-temperature peak to the area within the low-temperature peak is a quantitative estimate of the amounts of calcite and dolomite in the sample.

Several other factors contribute to this ratio. The presence of soluble salts, the degree of grinding of the sample, and the original crystal size influence the shapes and sizes of the curves (Graf, 1952, pp. 17-19). Also, the extent of substitution in the lattice is so great in some of the samples that the rock calculated from the spectrographic analyses is deficient in calcium.

The data obtained do indicate a general relationship between the ratio of the areas and the ratio of calcite to dolomite. A curve of this relationship could not be plotted from the data because of the uneven distribution of the points.

An area ratio of 14.9 was obtained for sample G-100 MD with 81% calcite; 8.4 for 63% calcite in F-127; 2.1 for 2% calcite in E-152; and 1.5 –1.2 for samples calculated as all dolomite. A single exception to this trend was sample B-100, gray aphanic dolomite, which had an area ratio of 5.0, and, calculated to be all dolomite; this is in contrast to the ratio of 1.47 for B-114 which is brown aphanic dolomite.

Two other methods of mineralogic analysis are those of staining and etching surfaces of carbonate rocks. These methods have an advantage over the D.T.A. method because they show the spatial relationships of calcite and dolomite as well as the percentages.

Etched surfaces (Lamar, 1950) were found to be highly satisfactory. The difference in solubility of calcite and dolomite crystals causes the dolomite and insoluble material to stand in relief above the calcite after the specimen is immersed briefly in dilute acid. The surfaces can then be examined with a binocular microscope, if necessary. This method has the advantages of simplicity of procedure combined with detailed discrimination.

Staining (Rodgers, 1940) is the most satisfactory way to differentiate calcite and dolomite in the thin sections. Most of the specimens which contained both minerals were fine-crystalline, and their relationships in hand-specimen were most satisfactorily studied by etched surfaces.

Lamination.—At first sight the laminae appear to be related to coloration; but on closer examination this is found to be complicated by variations in mineralogic composition and crystal size. Most of the dolomitic rocks have a uniform texture irrespective of the structures. To know the original relationships, it is necessary to study specimens not so completely dolomitized.

The brown beds grade laterally into more calcareous strata at several locations. The lamination of this original calcareous sediment is not accentuated by changes in material but is shown by interlaminar surfaces which represent brief interruptions in deposition. The laminae are identical and show no variation from bottom to top. Some laminated strata grade vertically and probably horizontally into homogeneous rock but ordinarily there is an intervening disconformity.

The porosity along these intercrystalline lamination planes, which are commonly microstylolitic, allowed dolomitization to go upward into the laminae. This process progressed through the laminae. The first dolomitized material became fine-crystalline gray or tan dolomite. Apparently the bitumen was partially removed in the process or was concentrated in the upper part of the laminae. The bitumen was trapped, or the dolomitization could not consummate the replacement of the whole laminae, with the result that the lighter-colored and finer-crystalline dolomite grades upward into darker calcareous dolomite with recognizably larger crystal size (Fig. 14).

All stages from limestone laminae to dolomite laminae are present. The dolomite laminae range from those entirely of the lighter phase to others completely dark. These variations are related to the rate of deposition and burial. This would control the time available for dolomitization and cause variations in the chemistry of the fluids within the laminae. Laminae range up to 5 mm. with an estimated average of 2.5 mm. (.1 inch). Where the lower alternating member is 400 feet thick, there are approximately 48,000 laminae. The uniform periodicity of the depositional changes exhibited in the lamination strongly suggests seasonal control. The 48,000 laminae indicate that the lower alternating member was deposited in 48,000 years plus the indeterminate, probably much longer, time represented by the gray beds and the disconformities between the brown beds.

Sequence of brown beds.—There are two fundamental types of brown beds; these are the laminated and unlaminated (homogeneous) strata. Mottling can form a secondary pattern in either type, and fossil and sedimentary fragments are present in some of both types of strata.

Of 162 gray beds at various locations, 123 of them are overlain by laminated brown strata. In the other 39 instances unlaminated strata were in this position. No definite sequence was recognized above this, but it was rather a random superposition of various types of brown beds with gradational and disconformable contacts.

Bioherms.—In the lower alternating member at Six-Mile Spring, south of Lund, two bioherms were found. A cross section through one of these is given as Figure 18, and Figure 20 is a photograph of the upper surface and its relationships.

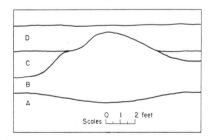

Fig. 18.—Cross section of bioherm in lower alternating member of Simonson dolomite.

The stratum is normally one foot thick with small brachiopod shells (.2 inch, 5 mm.) convex upward along the lamination planes. Within 6 feet laterally this unit expands to more than 4 feet thick and likewise becomes thinner farther along the strike. The plan of these bodies is circular. In the biohermal development there is no recognizable lamination, only a jumbled mass of stromatoporoid heads, some of which are broken.

As shown in Figure 18, unit C overlies the bedded part of unit B but only onlaps the flanks of the bioherm. Unit D is thinner over the bioherm, and subsequent beds show no influence. Adjacent to the bioherm, unit C contains fragments derived from the bioherm; these are much less apparent in unit D.

The top of unit A beneath the bioherm has been depressed, probably by the weight of the bioherm which grew above the general elevation of the surrounding sediments and thus applied greater downward pressure.

It is probable that many such structures are present in the alternating members; however, broader and flatter structures would be more difficult to detect. None would be very thick, as no thick biostromal units were observed other than the brown cliff member.

Deformation within brown beds.—The brown beds are well laminated and exhibit both penecontemporaneous folding and fragmentation (paradiagenetic deformation and inhomogeneity breccia of Sander, 1951, pp. 2 and 26). The folding ranges from gentle arching to overturned folds typically involving a few inches of laminated rock.

Fragmentation of the laminae and the resulting "sedimentary breccia" are common in the brown beds. The fragments are angular and tabular, consisting of part of one laminae, and were found up to 4 inches long.

There may be several causes for the deformation. The instability of underlying muds might cause the hardened laminae to bend and break. Also, channels cut in the top of the brown beds might undercut their flanks causing the slumping or breaking loose of fragments of laminae. Some zones of slippage along laminar planes contain breccia. This slippage is small-scale and of local extent within a bed. Such deformation reflects shifting of the sediments down local slopes.

It is possible that exceptionally strong sporadic currents tore loose fragments of the upper laminae and transported them some distance. Evidence of this is seen in isolated flat fragments up to 4 inches long and $\frac{1}{2}$ inch thick enclosed in otherwise evenly laminated sediments.

Deformation within gray beds.—The deformation within the gray beds is more intensive, more extensive, and of a different nature from that of the brown beds. The contortion within the gray beds is referred to as roiling to connote intense disturbance of a plastic sediment. Roiling takes the form of small overturned complex folds, pulled-apart laminae, and plastimorphic fragments. The very thin darker laminae which outline these structures are commonly faint.

The roiling of these units is probably the result of slumping as the slope of the sediment surface became oversteepened or as currents abated in channels cut in the previously stabilized muds. No regional slope could be detected which might be involved; so it is assumed that the slumping was associated with small local variations in the surface of the sediments.

The oversteepening might also be related to the shrinking of the sediments when subaerially exposed. Another possible cause of slumping would be the removal of aqueous support from a slope which has an angle too high for the repose of the same material in air (Kiersch, 1950, p. 942).

The lower surfaces of the gray beds are the unevenly eroded tops of the brown beds. The upper surfaces of the gray beds are no smoother than the basal surfaces, but they are not the result of erosion. The undulating upper surfaces reflect the plastic nature of these muds during the deposition of superjacent strata. Small irregularities were accentuated by slightly differential loading (Kindle, 1917, pp. 327–32). Thus, the higher places rose still higher during the deposition of the brown laminae above (Fig. 19).

FIG. 19.—Uneven upper surface of gray bed in Simonson dolomite. Relations to lamination in overlying brown bed indicate ridges of gray sediment moved upward during deposition of brown sediments.

Significant physical differences between the gray sediments and the brown sediments are evidenced in their deformation. The gray sediments show every evidence of having been a plastic mud even after their burial by a few feet of the brown deposits.

The brown sediments exhibit various degrees of folding but preserve their integrity until broken. These units hardened before the overlying beds were deposited. The two sediments correspond with the hard limestone (Hartkalke) and soft limestone (Weichkalke) in Pia's classification (1933, pp. 9–10).

Slump megabreccia.—A sedimentary breccia zone 100–200 feet thick occurs between the coarse member and the lower alternating member in Kings Canyon (Fig. 25). The zone persists in this interval for at least 2 or 3 miles.

The matrix is light gray aphanic dolomite similar to the gray beds. This material engulfs randomly oriented angular blocks of brown laminated dolomite up to 4 feet across. In the matrix also are blocks and large plastimorphic fragments of the gray beds. This breccia zone cuts into the top of the coarse member, and this may account for the local thinness of the unit (Fig. 21).

FIG. 20.—Upper surface of bioherm in lower alternating member of Simonson dolomite. Strata thin over bioherm. Note hammer in lower right of field.

The size, shape, orientation, lithic types, and regional setting seem best explained as the results of slumping. The earliest alternating beds may have been deposited on a steep slope dipping into the Confusion basin. The slope became oversteepened, and a large volume of alternating beds slid into the basin. The brown beds were already hardened, but the gray ones were still plastic. The resulting breccia mass spread over part of the basin floor; the slope became stabil-

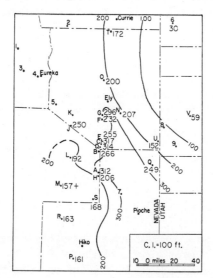

FIG. 21.—Isopachous map of tan coarse-crystalline member of Simonson dolomite.

ized and did not slump again. The cause of the slump might have been a too energetic renewal of subsidence in the basin.

The relief of the sediment surface at the time of the slump exceeded 200 feet. The maximum relief and the angle of the slope would require a regional study to learn the original thickness of the slumped material and the distance of the movement.

Evidence of convergence.—The sections were measured in detail with a 6-foot steel tape where the exposures justified such treatment. From a study of the data the manner of thickening of the lower alternating member was determined. The evidence is presented in Table III.

The table and its interpretation are based on an analysis of the distribution of the gray beds through the lower alternating member. The unit is subdivisible into gray beds and intervals of brown beds between them. The gray beds represent times of erosion and reworking of the brown-bed sediments. The brown beds are the result of normal carbonate deposition. A cyclic repetition of normal deposition and erosion was caused by fluctuations in the depth of water throughout east-central Nevada.

The second line of the table gives the number of beds in the member. This is a rather subjective category since there are innumerable variations in the brown beds; however, consistent criteria seem to have been applied as this number is in general accord with the other figures. The fourth line is also subjective to the extent of involving data from the second line.

TABLE III. DATA ON MANNER OF THICKENING OF LOWER ALTERNATING MEMBER

Locations	P	H	A	B	D	C	E	J	G	N	O	T
Thickness of member (in feet)	249	301	302	331	388	352	435	314	359	128	264* (179)	244
Number of beds	67	114	80	94	139	104	119	105	109	48	52	104
Number of gray beds	16	23	17	22	25	17	22	15	24	10	8	17
$\frac{\text{No. beds}}{\text{No. gray beds}}$ (ratio)	4.2	5.0	4.7	4.3	5.6	6.1	5.4	7.0	4.5	4.8	6.5	6.0
Aver. thickness of gray beds	4.7	2.4	4.8	5.0	2.5	3.1	3.6	4.0	3.6	3.0	2.7	1.4
Range of thickness of gray beds	0.4–17.4	0.1–10.0	1.6–10.8	0.7–12.2	0.3–6.8	0.6–10.7	0.4–8.9	0.3–9.5	0.3–12.8	1.3–3.8	1.0–6.5	0.3–4.0
Aver. thickness of intervals	9.8	10.7	11.4	10.9	12.6	12.1	15.6	12.3	11.9	7.2	14.8	14.0
Range of thickness of intervals	1.2–50.8	0.7–28.0	1.9–35.9	1.8–26.9	2.1–41.9	1.4–35.1	1.8–51.0	1.5–43.8	2.0–32.3	1.7–15.0	3.8–38.9	1.0–35.8
Percentage of thickness of member at N	195	235	236	259	303	275	340	246	281	100	206	191
Percentage of thickness of aver. interval at N	136	149	158	152	175	168	217	172	166	100	206	194

* Only the lower 179 feet of the member were sufficiently well exposed.

The sections are shown in their south-north order with J moved into line from the west and N from the east. The section at N is the thinnest, being on the flank of the north Snake Range positive area; while section E is the thickest, being in the Lund arm of the Confusion basin.

A comparison of these two sections, N and E, illustrates the manner of thickening southward from Ely. The thickness of the member at Lund (E) is 340% of the thickness in the Shell Creek Mountains (N). The average thickness of the intervals between the gray beds at E is 217% of the average thickness of the intervals at N. Thus, the increase in thickness must be explained as the result of thickening of these intervals and of the increase in the number of intervals.

Applying this reasoning in reverse, the thinness of the section on the positive area was caused by less deposition and more extensive erosion. The deposits of some cycles must have been completely removed by subsequent erosion during the accumulation of the member or were never deposited.

The data from the sections between Pahranagat Pass and the Shell Creek Mountains were analysed statistically. The degree of covariance of the thickness of the member and the thickness of the average interval was calculated as the correlation coefficient according to the procedure outlined by Goulden (1939, pp. 65–77).

Values of the correlation coefficient can range from zero for no correlation to plus or minus one for perfect correlation. The value obtained was .9463. This high value was tested for significance by use of the t test, which indicated that less than one time in a thousand would covariance be as great as or greater than that of the present data if the data had been taken at random from thicknesses which were not covariables. This substantiates the observation that the thicknesses of the intervals and the member are related.

The two sections northward from Ely (O and T) indicate that thickening of the intervals alone accounts for the entire thickening of the member in that direction. This would imply greater subsidence during each cycle than at N. The average thickness of the gray beds decreases in this direction. This could be the result of shorter periods of erosion.

BROWN CLIFF MEMBER

The brown cliff member is a dolomite biostrome as much as 152 feet thick without gray aphanic interbeds. The unit forms high cliffs in the Egan Range south of Ely, at Timber Mountain, in the Worthington Range, and on the east in the Lime Mountains and at Big Springs.

The area of greater thickness overlies an area of thickening in the lower alternating member (Fig. 22). Outward from this area the brown cliff member is much thinner. It does not thicken over the Confusion basin in western Utah.

The main contributors to the biostromal nature are stromatoporoids, tubular corals, and bryozoans. Additional remains are those of 2-inch brachiopods and small gastropods which are most common in the upper part.

The unit is massive-bedded with faint mottling. As is common with the brown-weathering dolomites, the rock emits a fetid odor when broken. Small, dark chert nodules were found only at Six-Mile Spring and south end of the Egan Range. Even at these places chert is rare.

The very light tan dolomitized stromatoporoids occur as heads 3–8 inches in diameter. They expanded as they grew upward at various angles. Some zones a few feet thick consist entirely of these heads closely packed and with the spaces between filled by brown dolomite. In other zones they are more widely spaced and associated with tubular corals and bryozoans. These zones attain biohermal character within the member.

UPPER ALTERNATING MEMBER

The upper alternating member is the uppermost part of the Simonson dolomite. It lies between the brown cliff member and the black limestone cliffs above. The upper member resembles the lower alternating member but differs from it in that it is thicker-bedded and the lamination is much less prominent. These ef-

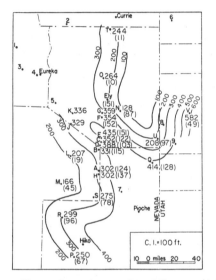

Fig. 22.—Isopachous map of lower alternating member of Simonson dolomite with thicknesses of brown cliff-forming member in parentheses.

fects seem caused by more rapid and more uniform deposition, probably indicative of sedimentation in consistently deeper water.

In the basal few feet of this member there is commonly a 3–5-foot bed with abundant stromatoporoids similar to those in the brown cliff member. Tubulars are important contributors to the sediments of the upper member, and toward the top small identifiable brachiopods occur in the intercalated limestones. Stromatoporoids resembling dark gray or brown golf balls occur near the top of the member at Pahranagat Pass and in the Lime Mountains. They are loosely packed in strata with 65% carbonate matrix. In one stratum the stromatoporoids are brown dolomite in a dark gray limestone, but less than 30 feet away in the same stratum they are dark gray limestone (calcite) in a tan dolomite matrix.

An interesting porosity is conspicuously developed in this member in the southern two sections in the Egan Range. Medium crystalline beds containing tubular fossils are extremely porous with a connected meshwork of tubes as a result of the fossils having been dissolved.

The possibility that the upper alternating member is a facies of the brown cliff member, such as fore-ree for back-reef, has been considered. Such relationship between "reef" and overlying strata has been well illustrated in the Permian of West Texas (Newell *et al.*, 1953, Fig. 53), but the evidence in the case does not prove this relationship with the brown cliff member.

LIMESTONE IN SIMONSON DOLOMITE

Black, laminated aphanic limestone strata occur in the lower alternating member at South Ward Peak, Steptoe, Blue Eagle Mountain, and Indian Springs. At these four locations the limestone beds are a minor, almost unnoticeable, constituent of the top of the member and grade laterally into brown dolomite beds. This gradation is remarkably abrupt. Individual laminae change from dolomite to limestone in $\frac{1}{10}$ inch, and beds 1–3 feet thick change in 3–30 feet through a zone of interlaminated limestone and dolomite.

Of the sections measured, only the one at Blackrock Canyon is northwest of the four mentioned. At Blackrock Canyon the top of the member passed into limestone without the brown cliff dolomite member being present. These relationships indicate an increase in the amount of limestone in the Simonson toward the northwest, and the limestone appears lower in the section in this direction. These lines of evidence agree with the greater amount of limestone in the Middle Devonian at Lone Mountain and with the premise that dolomite sediments are deposited closer to shore than limestone. Limestone occurs interbedded in the top of the upper alternating member at all of the locations.

Upper contact.—At the type locality, near Gold Hill, the top of the Simonson was placed at the base of a persistent dolomite conglomerate bed 100 feet below a 4-foot zone of *Stringocephalus* (Nolan, 1935, p. 19). The overlying Guilmette formation is predominantly dark dolomite with thick limestones and several lenticular sandstones. The dolomite units are mostly unlaminated, thicker than those of the Simonson; and there are zones of abundant tubular corals in the upper part.

The upper contact at the type section is not a striking one and must be related to either the *Stringocephalus* zone or the appearance of limestone if it is to be recognized in east-central Nevada. *Stringocephalus* ranges through a few hundred feet of strata at several locations and is not recognized at all the others; hence there are objections to its use in marking the top of the Simonson.

In east-central Nevada and adjacent Utah the most easily recognized contact is where the dolomite section changes upward into one predominantly of limestone. The Simonson formation, to be most applicable in the field and thus fulfill the requirements of formational rank, should have its upper contact at this lithologic change. The interfacies nature of this contact must be kept in mind.

The limestone section forms prominent cliffs which are the most characteristic topographic expression of the Devonian in the region. These massive units aggregate 1,000–2,000 feet at most of the locations and are representatives of the Guilmette and Devils Gate formations, of Gold Hill and Lone Mountain, respectively.

ISOPACHOUS MAPS

Figure 12, an isopachous map of the Simonson dolomite, shows a thin area in the vicinity of the northern part of the Snake Range along the Utah line. This positive area persisted from Sevy time and is similar to positive areas in the miogeosyncline in northwestern Utah (Stokes, 1952, p. 1300) and south-central Idaho (Sloss, 1954, p. 368). Adjacent to the north Snake Range positive area the Simonson thickens into the Confusion basin. An arm of this basin seems to have extended westward into northern Lincoln County, Nevada. This extension of the basin is referred to as the Lund arm, after the town of Lund, Nevada, which is astride this thicker zone. West of the arm, in the eastern part of Nye County, there is some indication of another zone of thinner Simonson.

The Simonson thickens northward through White Pine County. Nolan gave a thickness of 963 feet for the Simonson at Gold Hill; however, a thickness of 1,127 feet is used on the map to correspond with the interval of dolomite as measured at the other sections. This value is accordant with the trends of thicknesses based on the Nevada sections.

The three thicknesses in the northwestern part of the map are for the Nevada formation and are taken from the literature. At Pearl Peak in the southern Ruby Mountains (location 2 on the map), Sharp (1942, pp. 662–63) describes 1,900 feet of dolomitic section with *Stringocephalus* 660 feet below the top. He notes that "*Stringocephalus* is distributed through several hundred feet" of the Nevada. The fossil also ranges through a few hundred feet at Gold Hill.

The Nevada formation at Lone Mountain (location 3) contains abundant dolomite except in the lower part, and the *Stringocephalus* zone is a few feet thick at the top (Merriam, 1940, pp. 22–25).

These values fit the pattern which would be expected, the Simonson thickening northwestward into the Roberts Mountain basin.

The isopachous map of the lowest member of the Simonson, the coarse-crystalline member (Fig. 21), reveals a general uniformity of thickness except over the northwest-trending arm of the Confusion basin. The thickness at location V, in the basin, has probably been reduced by truncation by submarine slumping.

Figure 22 shows the thickness of the lower alternating member. This unit thickens in the area of the Lund arm of the Confusion basin and thins over the northern Snake Range positive area. In addition, the map shows the thicknesses of the overlying member, the brown cliff member, which also is thicker in the vicinity of the Lund arm. The latter is more than 100 feet thick at all locations where the lower alternating member is more than 300 feet thick, and it is less than 100 feet thick where the underlying member is less than 300 feet thick. The section at Kings Canyon is the sole exception to this relationship.

The isopachous maps of the Simonson and its members conform in showing thinner sections in eastern White Pine County and south of Blue Eagle Mountain (location L) in northeastern Nye County. The Blackrock Canyon section, between these areas, is in a saddle formed by thickening toward the northwest into central Nevada and southeast into the Lund arm.

The isopachous map of the upper alternating member (Fig. 23) is more subjective than the other maps because of the nature of the top of the unit. It dis-

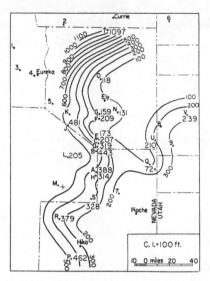

Fig. 23.—Isopachous map of upper alternating member of Simonson dolomite.

plays the same distribution of thicker and thinner areas as the other maps. Most notable is the drastic increase in the rate of westward and northwestward thickening, which is the result of greater and more rapid subsidence toward the west if the depth of deposition is assumed to be the same. The great transgression represented by the dolomite-limestone contact seems a result of rapid subsidence of the whole miogeosyncline. Thus, this contact may be only slightly younger in western Utah than in central Nevada.

These isopachous maps present a picture of the miogeosynclinal floor as being uneven and having local basins and platforms. Actually these are areas which subsided to greater or less amounts than their adjacent parts. The positive areas and basins in the miogeosyncline are gentle features about 40–50 miles across and involving about 1,000 feet of differential vertical movement in east-central Nevada during Sevy and Simonson time.

REGIONAL RELATIONSHIPS

The Simonson dolomite represents the shoreward facies of part of the Nevada limestone of central Nevada. The dolomite facies is present in the Nevada forma-

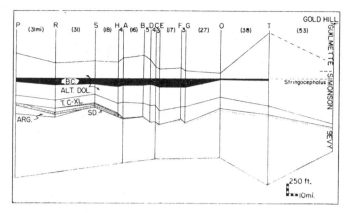

Fig. 24.—South-north cross section of Sevy and Simonson dolomites.

tion at Lone Mountain, beginning about 800 feet above the base and comprising about 30 % of the section up to the *Stringocephalus* zone. In east-central Nevada and at Gold Hill this part of the section is entirely dolomite. This agrees with the idea of off-shore limestone and nearer-shore dolomite facies.

West and northwest from central Nevada this interval passes into a predominantly clastic section. This information is from Gilluly as quoted by Berdan (1953, p. 1394) and Ferguson (1952, p. 73).

Toward the east the Middle Devonian disappears in central Utah by convergence and non-deposition. Upper Devonian seas extended much farther eastward and are represented by dolomites in Wyoming, Colorado, and Arizona and by limestones in western Utah and Nevada.

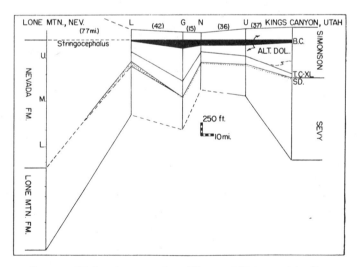

Fig. 25.—West-east cross section of Sevy and Simonson dolomites.

FAUNA OF SIMONSON

Fossils are moderately abundant in the Simonson but without exception are poorly preserved. They have been dolomitized or imperfectly silicified with the exception of those in the limestones at the top of the formation.

The most abundant forms are tubular and represent three distinct types discerned in the field by virtue of their diameters. *Cladopora* is the largest and occurs as irregular anastomosing tubes 5 mm. (.2 inch) in diameter. It is extended parallel with the stratification of the enclosing sediments. This coral is widespread throughout the Devonian of western United States.

The next smaller tubulars are similar to those referred to as *Thamnopora* by Canadian geologists working on the Devonian of western Canada (deWit and McLaren, 1950, p. 10). These bryozoans are straight or gently curved, unbranched rods 2–3 mm. (.1 inch) in diameter. This form is found scattered through zones containing stromatoporoids and, also, as the sole constituent of beds up to 1 foot thick which have been referred to as "spaghetti beds" by workers in the area. Newell *et al.* (1953, p. 148) noted *Cladopora* and tubular bryozoans to be mutually exclusive in the Rader member of the Delaware Mountain group in the Permian of West Texas. This relationship is also essentially true in the Simonson and suggests slight variations in environment.

The smallest tubular form, of unascertained affinities, is less than 1.5 mm. (.05 inch) in diameter and forms laminae of its intertangled meshwork. The last two forms were effective as blankets over the sediment surface and probably acted to stabilize it. The smallest form, being the most fragile and abundant, probably supplied considerable fine débris to the sediments. Stromatoporoids occur as heads in the brown cliff member and as small spheres locally in the upper member. Small cylindrical septate solitary corals and small indeterminate brachiopods also occur in the brown cliff member.

C. W. Merriam kindly identified the following forms from small collections made from the upper limestones of the Simonson. At Steptoe, *Productella*, *Leiorhynchus* sp., *Emanuella*?, and *Styliolina* indicate a Devils Gate or possibly upper Nevada correlation. *Atrypa* cf. *missouriensis* and *Atrypa* sp. from the Lime Mountains could be upper Nevada.

The brachiopod, *Stringocephalus*, was found in Blue Eagle Mountain associated with *Thamnopora* in a zone less than 20 feet thick. This zone occurs in the alternating sequence at a position comparable with the brown cliff member. As stated in the discussion of thicknesses, this member would be expected to be this thin at this location.

The occurrence of *Stringocephalus* approximates the top of the Middle Devonian throughout the Millard belt. At some locations, such as Pearl Peak and Gold Hill, this brachiopod ranges through a few hundred feet of section; as there is an absence of other diagnostic fossils for a thousand feet below and various hundreds of feet above, it is useful and necessary for time correlation.

AGE OF SIMONSON DOLOMITE

The Simonson dolomite contains Middle Devonian fossils including *Stringocephalus* which was found at Blue Eagle Mountain in a thin zone correlated with the brown cliff member. This fossil occurs in the Simonson at Gold Hill and also in the overlying Guilmette formation. *Hexagonaria?* was found in the brown cliff member at Trough Spring Canyon in the Southern Egan Range and is considered to be a Middle Devonian fossil.

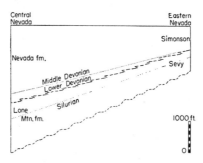

FIG. 26.—Diagrammatic cross section showing proposed correlation of formations and position of Lower Devonian strata.

Below the Simonson, Silurian age is indicated by *Halysites* found in the lower part of the Sevy and in the top of the Laketown, where the form is abundant. The Simonson thus contains Middle Devonian fossils, and the next older are Silurian. Lower Devonian time must be represented by sediments or by non-deposition or erosion in the interval between these zones.

The most obvious explanation would be a widespread unconformity to account for the thinner section and the absence of the Lower Devonian fauna found in central Nevada. No such break was recognized. The Sevy is interbedded with the coarse member of the Simonson which, in turn, is interbedded with the alternating member and so on. No indication of a major break in deposition from Silurian to Middle Devonian time is recognized in the sections.

A possible explanation of the relationships would be that the facies transgress time planes, becoming younger toward the east. In conjunction with this would be differential subsidence as shown by the comparison of thicknesses of the section below *Stringocephalus* at Blue Eagle Mountain (location L) and Lone Mountain (Fig. 25). According to this line of reasoning, the Sevy, which underlies Lower Devonian at Lone Mountain, would be of Lower Devonian age in its upper part at Blue Eagle Mountain and farther east. The Simonson in east-central Nevada would be equivalent to part of the Nevada formation at Lone Mountain and would be Middle Devonian with some Lower Devonian sediments possibly in its base (Fig. 26).

The alternative to this interpretation would be to place the Sevy entirely in the Silurian and for the Middle Devonian to overlap transgressively the Lower

Devonian from west to east. This would necessitate a break between the Sevy and the Simonson or between the lower members of the Simonson; the field evidence does not substantiate this.

ORIGIN OF SIMONSON DOLOMITE

The coarse-crystalline dolomite member at the base of the Simonson is a recrystallized dolomite. This is evidenced by the tessellated (mosaic) texture which has virtually destroyed the primary structures. This member represents a transition between the slow deposition of the Sevy mud flats and the more rapid deposition in deeper water during Simonson time. The coarse crystallinity is probably related to the history of the fluids during a time of change in rates of subsidence and deposition. The base of the member coincides with the top of the sand concentration which may mean a change in geographic relationships at this time.

During deposition of the alternating members the sea was extensive and shallow over a floor which probably sloped slightly toward the west. This surface was not perfectly smooth but was composed of gentle depressions and intervening rounded ridges a few thousand feet across. Sediments were deposited in the depressions and eroded from the higher parts of the sea floor. Stromatoporoids and tubulars probably found the higher areas more to their liking since the water was better circulated. However, their skeletons were comminuted and removed to the depressions where this material was added to the indigenous organic material, sedimentary fragments, and carbonates which were precipitated in the sediments and in the water.

Some of the deposits within the depressions were laminated because of a periodicity of the depositing media and deposition below the profile of equilibrium. Periodic control may have been exercised by sea plants or plankton; by tides, seasons, or other cosmic phenomena; or by atmospheric disturbances such as storms or seiches (Deicha, 1942; Sverdrup, 1946, pp. 538–542 and 965). The beds formed by these deposits were very thin lenses laid disconformably on the underlying strata, especially toward the margins of the depressions. These sites of sedimentation migrated on the sea floor, and the brown sequences are composed of several types of brown dolomite as thin lenses overlapping each other. Some of these pass into limestones where dolomitization was not complete.

Ripple marks, cross-lamination, and oölites are usually considered indicative of shallow marine deposition; however, these structures are conspicuously absent in the Simonson. These sedimentary effects of moving water might be missing for some of the following reasons.

Water of high turbidity (muddy) might have passed so gradually downward into a laminated deposit that there was no distinct water-sediment interface on which these currents could work. The zone between 90% fluid and 70% fluid might be several inches thick and provide a protective cover for the deposits beneath. Another factor is the presence of a considerable volume of sea plants or plankton that would increase the resistance of the water to winds. The configuration of the sea floor of this extensive shallow body of water would dampen the effects of most waves.

The general lack of circulation of the waters in these environments also accounts for the abundance of bitumen in the rocks and for their dolomitization (Linck, 1937, p. 282). The water within the lower parts of the depressions probably became highly concentrated in carbonate salts and organic matter. These factors contributed to the precipitation of calcite both in the water and in the sediment and to its replacement by dolomite.

Periodically, large areas of the sea floor were very close to sea-level, if not exposed. During these times the "just born" brown deposits were subjected to erosion, fragmentation, and reworking. The bituminous content was removed and since the environment was so inhospitable, none was added. The reworked material was distributed unevenly around the surface as thin lenses. In the intervening areas the brown beds were eroded; then the reworked débris was moved, and another part of the top of the brown beds was exposed to erosion. Channels and small-scale cross-lamination were developed by this redeposition. On this mud-flat environment the gray deposits were not allowed to exceed about 20 feet in thickness before they were dispersed. Sporadically the sea-level changed, and these deposits slumped or were torn into fragments of mud (plastimorphic fragments) during the redeposition.

Abruptly the environment reverted to marine, and brown-bed deposition returned. At this time the gray muds were spread over greater areas, but they did not cover the entire sea floor by the time the first brown beds were deposited. Thus, in some sections these intervals are represented by disconformities between brown beds. The cohesiveness of the gray muds was such that they were not mixed with the first brown beds but presented an uneven, unstable floor for these sediments.

The repeated presence of the gray beds in the Simonson is evidence of a major rhythm, and the lamination represents a minor rhythm. These structures are the results of two types of variations in the depositional environment superimposed on an over-all condition. During Simonson time the net effect of crustal movements was increased burial of the top of the Sevy dolomite. Progress toward this net effect was periodically interrupted by decreases or cessations of the rate of burial caused by eustatic changes of sea-level. The gray-bed rhythm is the result of these variations. The minor rhythm, evidenced by the lamination, resulted from local variations in the rate of burial or type of sedimentation during brown-bed deposition.

PETROLEUM CONSIDERATIONS OF SIMONSON

The Simonson and Sevy should be recognizable in drill cuttings from holes in the area. The change from limestone to predominantly dolomite and the occurrence of brown fine-crystalline and light gray aphanic dolomite would characterize the Simonson. The cuttings would develop these colors after immersion in dilute acid, washing, and drying. The coarse-crystalline member at the base of the Simonson, the sandy member at the top of the Sevy, and the non-bituminous homogeneous, aphanic Sevy should also be recognizable and distinguishable from the darker dolomites below.

The petroleum possibilities of the Simonson are better than those of the Sevy. The brown beds of the Simonson contain 3–5% (by volume) bituminous residues indicative of indigenous petroleum. Intercrystalline porosity is well developed in the coarse member; exceptional porosity is also caused by the dissolving of tubular fossils in some beds of the upper member. In addition, porosity might be related to dissolving of the calcitic meshwork of the rock and to fracturing.

The biohermal nature of the brown cliff member and some of the brown beds provides evidence of reef-type traps for accumulations of petroleum. The heterogeneity of the Simonson deposits and the interstratal disconformities are favorable for accumulations. Normal faults are the most evident structures in the Basin-and-Range Province and are potential traps which are worthy of investigation. A few anticlinal structures have been tested and found dry; however, more knowledge of the structural history and pattern of the region will provide a better understanding of the possibilities of all these structures.

CONCLUSIONS

The Sevy and Simonson are miogeosynclinal deposits of dolomite and orthoquartzite in east-central Nevada. Toward the east they thin in a short distance; toward the west they are represented in thick argillaceous and silty limestone deposits. Facies contacts parallel the margins of the miogeosyncline.

The miogeosynclinal surface is composed of shallow basins and gentle positive areas which subsided at different rates. These structural elements persisted through Sevy and Simonson time, the strata converging over the positive areas and diverging into the basins.

Pre-Sevy carbonates and orthoquartzites were uplifted and eroded on the east, providing material for the geosynclinal deposits.

The erosion of the top of the Laketown (Niagaran) was succeeded by transgressive conditions. The sea in eastern Nevada became deeper; and the shore moved eastward across Nevada and Utah during Sevy, Simonson, and subsequent Devonian time.

In the past, thick "monotonous" sequences of carbonate rocks have been lumped together; however, when examined in detail, they reveal characteristic textures, structures, and stratification. When these features are applied to regional studies, paleogeographic relationships can be interpreted and a great variety of sedimentary environments recognized.

BIBLIOGRAPHY

BERDAN, J. M., 1953, "Devonian Ostracode Fauna From Nevada," *Bull. Geol. Soc. America.*, Vol. 64, p. 1394.
BERG, L. G., 1945, "On Area Measurements in Thermograms for Quantitative Estimates and the

Determination of Heats of Reaction," *Compte Rendus (Doklady) de l'Academie des Sciences de l'U.R.S.S.*, Vol. 49, pp. 648-51.
CHENOWETH, P. A., 1952, "Statistical Methods Applied to Trentonian Stratigraphy in New York," *Bull. Geol. Soc. America*, Vol. 63, pp. 521-60.
CHRISTIANSEN, F. W., 1951a, "A Summary of the Structure and Stratigraphy of the Canyon Range," *Guidebook to the Geology of Utah*, No. 6, Intermtn. Assoc. Petroleum Geologists, pp. 5-18.
———, 1951b, "A Summary of the Structural History of the Great Basin Province in Utah," *ibid.*, 68-80.
CULLIS, C. G., 1904, "The Mineralogical Changes Observed in the Cores of the Funafuti Borings," in *The Atoll of Funafuti*, Royal Soc. London, Sec. XIV, pp. 392-420.
DAPPLES, E. C., ET AL., 1953, "Petrographic and Lithologic Attributes of Sandstones," *Jour. Geol.*, Vol. 61, pp. 291-317.
DEFORD, R. K., 1946, "Grain Size in Carbonate Rocks," *Bull. Amer. Assoc. Petrol. Geol.*, Vol. 30, pp. 1921-28.
DEICHA, GEORGES, 1942, "Stratification, Stratification fine et Microstratification," *Comp. Rend. Soc.. Geol. France*, Fasc. 7, pp. 60-61.
DEWIT, R., AND MCLAREN, D. J., 1950, "Devonian Sections in the Rocky Mountains Between Crowsnest Pass and Jasper, Alberta," *Geol. Survey Canada Paper 50-23*.
DONOVAN, J. T., 1951, "Devonian Rocks of the Confusion Basin and Vicinity," *Guidebook to the Geology of Utah*, No. 6, Intermtn. Assoc. Petroleum Geologists, pp. 47-51.
EASTERN NEVADA GEOL. ASSOC. STRAT. COMM., 1953, "Revision of Stratigraphic Units in Great Basin," *Bull. Amer. Assoc. Petrol. Geol.*, Vol. 37, pp. 143-51.
EATON, J. E., 1936, "By-Passing and Discontinuous Deposition of Sedimentary Materials," *ibid.*, Vol. 13, pp. 713-61.
FERGUSON, H. G., 1952, "Paleozoic of Western Nevada," *Jour. Washington Acad. Sci.*, Vol. 42, pp. 72-75.
———, AND MULLER, S. W., 1949, "Structural Geology of the Hawthorne and Tonopah Quadrangles, Nevada," *U. S. Geol. Survey Prof. Paper 216*.
GODDARD, E. N., ET AL., 1951, *Rock Color Chart*, National Research Council, distrib. by Geol. Soc. America.
GOULDEN, C. H., 1939, *Methods of Statistical Analysis*. John Wiley and Sons, Inc., New York.
GRAF, D. L., 1952, "Preliminary Report on the Variations in Differential Thermal Curves in Low-Iron Dolomite," *Amer. Mineralogist*, Vol. 37, pp. 1-27.
GREENMAN, NORMAN, 1951, "The Mechanical Analysis of Sediments from Thin Section Data," *Jour. Geol.*, Vol. 59, pp. 447-62.
HAZZARD, J. C., AND MASON, J. F., 1953, "The Goodsprings Dolomite at Goodsprings, Nevada," *Amer. Jour. Sci.*, Vol. 251, pp. 643-55.
HEDBERG, H. D., 1936, "Gravitational Compaction of Clays and Shales," *ibid.*, Vol. 31, pp. 241-81.
HINTZE, L. F., 1951, "Lower Ordovician Detailed Stratigraphic Sections for Western Utah," *Utah Geol. and Mineralog. Survey Bull. 39*.
HOUGH, J. L., 1940, "Sediments of Buzzards Bay, Massachusetts," *Jour. Sed. Petrology*, Vol. 10, pp. 19-32.
KAY, MARSHALL, 1951, "North American Geosynclines," *Geol. Soc. America Mem. 48*.
———, 1952, "Late Paleozoic Orogeny in Central Nevada," *Bull. Geol. Soc. America*, Vol. 63, pp. 1269-70.
KIERSCH, G. A., 1950, "Small-Scale Structures and Other Features of Navajo Sandstone, Northern Part of San Rafael Swell, Utah," *Bull. Amer. Assoc. Petrol. Geol.*, Vol. 34, pp. 923-42.
KINDLE, E. M., 1917, "Deformation of Unconsolidated Beds in Nova Scotia and Southern Ontario," *Bull. Geol. Soc. America*, Vol. 28, pp. 323-34.
KRUMBEIN, W. C., 1936, "The Use of Quartile Measures in Describing and Comparing Sediments," *Amer. Jour. Sci.*, Vol. 32, pp. 98-111.
LAMAR, J. E., 1950, "Acid Etching in the Study of Limestones and Dolomites," *Illinois State Geol. Survey Cir. 156*.
LINCK, G., 1937, "Bildung des Dolomits und Dolomitisierung," *Chemie der Erde*, Bd. 11, pp. 278-86.
LONGWELL, C. R., 1952, "Basin and Range Geology West of the St. George Basin, Utah," *Guidebook to the Geology of Utah*, No. 7, Intermtn. Assoc. Petroleum Geologists, pp. 27-42.
MCALLISTER, J. F., 1952, "Rocks and Structures of the Quartz Spring Area, Northern Panamint Range, California," *California Div. Mines Spec. Rept. 25*.
MCKEE, E. D., 1945, "Cambrian History of the Grand Canyon Region," *Carnegie Inst. Washington Pub. 563*.
———, AND WEIR, G. W., 1953, "Terminology for Stratification and Cross-Stratification in Sedimentary Rocks," *Bull. Geol. Soc. America*, Vol. 64, pp. 381-90.
MCNAIR, A. H., 1952, "Summary of the Pre-Coconino Stratigraphy of Southwestern Utah, Northwestern Arizona, and Southeastern Nevada," *Guidebook to the Geology of Utah*, No. 7, Intermtn. Assoc. Petroleum Geologists, pp. 45-51.

MERRIAM, C. W., 1940, "Devonian Stratigraphy and Paleontology of the Roberts Mountains Region Nevada," *Geol. Soc. America Spec. Paper 25*.
———, 1951, "Silurian Quartzites of the Inyo Mountains, California," *Bull. Geol. Soc. America*, Vol. 62, p. 1508.
NACHTRIEB, N. H., 1950, *Principles and Practice of Spectrochemical Analysis*. McGraw-Hill Book Company, Inc., New York.
NEWELL, N. D., ET AL., 1953, *The Permian Reef Complex of the Guadalupe Mountains Region, Texas and New Mexico*. W. H. Freeman and Company, San Francisco.
NOLAN, T. B., 1935, "The Gold Hill Mining District, Utah," *U. S. Geol. Survey Prof. Paper 177*.
———, 1943, "The Basin and Range Province in Utah, Nevada, and California," *ibid.*, *Prof. Paper 197-D*.
OSMOND, J. C., 1953, "Mottled Carbonate Rocks in the Middle Devonian of Eastern Nevada," *Bull. Geol. Soc. America*, Vol. 64, p. 1460.
PALACHE, C., ET AL., 1951, *The System of Mineralogy*, Vol. II. John Wiley and Sons, New York.
PAYNE, T. G., 1942, "Stratigraphic Analysis and Environmental Reconstruction," *Bull. Amer. Assoc. Petrol. Geol.*, Vol. 26, pp. 1697–1770.
PETTIJOHN, F. J., 1949, *Sedimentary Rocks*. Harper and Brothers, New York.
PIA, JULIUS, 1933, "Die Rezenten Kalksteine," *Miner. und Petrog. Mitt.*, N.F. Ergänzungsband, Leipzig (cited in Sander, 1951, p. 6).
ROBERTS, R. J., 1949, "Structure and Stratigraphy of Antler Peak Quadrangle, North-Central Nevada," *Bull. Geol. Soc. America*, Vol. 60, p. 1917.
RODGERS, JOHN, 1940, "Distinction Between Calcite and Dolomite on Polished Surfaces," *Amer. Jour. Sci.*, Vol. 238, pp. 789–90.
RUSH, R. W., 1951, "Stratigraphy of the Burbank Hills, Western Millard County, Utah," *Utah Geol. and Mineralog. Survey Bull. 38*.
SANDER, BRUNO, 1951, *Contributions to the Study of Depositional Fabrics* (translated from the German by E. B. Knopf). Amer. Assoc. Petrol. Geol.
SHARP, R. P., 1942, "Stratigraphy and Structure of the Southern Ruby Mountains, Nevada," *Bull. Geol. Soc. America*, Vol. 53, pp. 647–90.
SKEATS, E. W., 1918, "The Formation of Dolomite and Its Bearing on the Coral Reef Problem," *Amer. Jour. Sci.*, Ser. 4, Vol. 45, pp. 185–200.
SLOSS, L. L., 1954, "Lemhi Arch, a Mid-Paleozoic Positive Element in South-Central Idaho," *Bull. Geol. Soc. America*, Vol. 65, pp. 365–68.
STETSON H. C., AND UPSON, J. E., 1937, "Bottom Deposits of the Ross Sea," *Jour. Sed. Petrology*, Vol. 7, p. 57.
STOKES, W. L., 1952, "Paleozoic Positive Area in Northwestern Utah," *ibid.*, Vol. 63, p. 1300.
SVERDRUP, H. U., ET AL., 1946, *The Oceans*. Prentice-Hall, New York.
TERZAGHI, R. D., 1940, "Compaction of Lime Muds as a Cause of Secondary Structure," *Jour. Sed. Petrology*, Vol. 10, pp. 78–90.
THIEL, G. A., 1935, "Sedimentary and Petrographic Analysis of the St. Peter Sandstone," *Bull. Geol. Soc. America*, Vol. 46, pp. 559–614.
TRASK, P. D., 1932, *Origin and Environment of Source Sediments of Petroleum*, Gulf Publishing Company, Houston, Texas, pp. 71–72.
WEBB, G. W., 1953, "Middle Ordovician Stratigraphy in Eastern Nevada and Western Utah," unpublished Ph.D. thesis, Columbia Univ.
WESTGATE, L. G., AND KNOPF, ADOLPH, 1932, "Geology and Ore Deposits of the Pioche District, Nevada," *U. S. Geol. Survey Prof. Paper 171*.
WHEELER, H. E., AND LEMMON, D. M., 1939, "Cambrian Formations of the Eureka and Pioche Districts, Nevada," *Univ. Nevada Geol. and Mining Ser. Bull. 31*.
WILLIAMS, J. STEWART, 1948, "Geology of the Paleozoic Rocks, Logan Quadrangle, Utah," *Bull. Geol. Soc. America*, Vol. 59, pp. 1121–64.

DOLOMITIZATION BY SEEPAGE REFLUXION[1]

JOHN EMERY ADAMS[2] AND MARY LOUISE RHODES[2]
Midland and Amarillo, Texas

ABSTRACT

Bedded dolomites in the Permian Basin were formed by the alteration of metastable limestones by hypersaline brines refluxing from evaporite lagoons. The hot, heavy, highly alkaline, carbon dioxide-free, magnesium-supercharged brines displaced connate waters to provide both a chemically favorable environment for magnesium-calcium exchange and a vehicle for removing displaced calcium. Fossil lagoonal brines and fillings of halite and anhydrite in the dolomite pores offer proof of the brine invasion. Sedimentary dolomites in other areas are commonly associated with evaporites, and for these dolomites a similar origin is postulated.

INTRODUCTION

The stratigraphic column from the mid-Precambrian to the Recent contains many bedded dolomites. In spite of their abundance, stratigraphic span, and geographic range, the origin of these dolomites is a controversial subject. Recent regional mapping of depositional and diagenetic relationships supplies the data for interpreting the origin of some major Permian dolomites. The explanation here advanced may aid in determining the origin of other deposits of sedimentary dolomites.

The abundance and widespread distribution of major sedimentary dolomites indicate that these rocks are products of common, relatively simple natural processes. The environments in which the now dolomitized rocks were deposited, as recorded in the dolomites themselves and in associated sediments, support this conclusion. The environmental range represented is largely identical with that responsible for normal marine limestones, but in addition, includes a restricted lagoon phase. Major dolomite deposits of both normal marine and restricted environments are limited to, or closely associated with, shallow epicontinental shelves or oceanic banks. Thus, all adequately explored major sequences of sedimentary dolomite are known to include or interfinger with calcitic limestones, characterized by abundant shallow-water fossils. Surprisingly high percentages of bedded dolomites are associated with evaporites. Close associations of evaporites, dolomites, and marine limestones are common enough to suggest that restricted lagoon-open ocean couples are essential for the formation of extensive dolomites. This idea is not new, but a workable, theoretical basis for applying it has been lacking. Indeed, Fairbridge's (1957) excellent review and appraisal of the dolomite problem ranks evaporitic control with the least plausible of many discredited dolomitization hypotheses.

Geologists generally agree that most bedded dolomites were produced by post-depositional alteration of limestones. Disagreements arise over the time of alteration and the nature of the physiogeological and physiochemical processes involved. The main problems are: (1) supplying enough chemically active magnesium to complete the dolomitizing process; (2) transporting this magnesium into and distributing it through the host limestones; (3) providing a chemically favorable environment in which approximately half the calcium can be displaced by magnesium; (4) completing the replacement while active introduction of magnesium is in progress; and (5) removing the displaced calcium. A discussion of the chemical processes involved in displacing the calcium by magnesium is beyond the scope of this report. The other problems are essentially geological and yield to geological analyses.

In the more extensive dolomite formations, hundreds or even thousands of cubic miles of bedded limestone were dolomitized. Processes capable of dolomitizing these large masses are more critical than those effective on bodies of limited size. The method of dolomitizing an extensive limestone body after its consolidation and calcitization is difficult to explain. Theories of secondary replacement requiring diffusion of magnesium through solid rock sound plausible for small metasomatic deposits, but not for whole diagenetic environments.

---

[1] Manuscript received, May 19, 1960.
Published by permission of the Standard Oil Company of Texas.

[2] Geologists, Standard Oil Company of Texas.
The manuscript was read and corrections made by W. T. Holzer.

Comparing depositional environments characterized by dolomites with those dominated by non-dolomitic limestones provides an interesting approach. The most obvious difference is the presence or absence of associated lagoonal evaporites. It is readily apparent, where complete depositional sequences are preserved, that limestones closely associated with contemporaneous evaporite lagoons seldom escaped extensive dolomitization. Limestones lacking evaporitic associations at the time of deposition normally remain undolomitized even after hundreds of millions of years. The chemical, geometric, and hydrodynamic reasons for this relationship become apparent on analyzing the two enviromnents.

Modern Limestone Shelves

The intensively studied modern Florida shelf-Bahama Bank area typifies the environments where undolomitized limestones form. These shallow limestone shelves covering 100,000 square miles perch on steep-walled carbonate pedestals dating back to the mid-Mesozoic. Water depths on the limestone banks range to perhaps 1,000 feet, with 95 per cent of the bank floor above the 25-foot level. Discontinuous island chains and marginal reefs line the windward shelf edges. Tides, currents, and storms push great quantities of sea water onto and across the shelves. Waves and currents keep the water on the shelf mixed and almost unstratified and at low tide the excess escapes over the margins of the shallow bank. The shelf limestones are largely of organic origin (Epstein and Lowenstam, 1957). Living plants and animals carpet much of the shelf; other areas are covered by submerged limestone dunes and mud flats (Illing, 1954). Bottom sediments are relatively permeable, and tides flow through as well as over the floor.

The loosely cemented Florida-Bahama limestones possibly resulted from rapid upgrowth of banks and shelf surfaces, or from solution of exposed areas during the Pleistocene lowering of sea-level (Newell, 1955). Limestones in South Florida are only slightly less permeable than a sponge. Some of these porous and permeable limestones have been in contact with sea water for thousands of years. They show no evidences of dolomitization, although composed of metastable aragonite and organically deposited calcite. The only magnesium present is preserved as scattered molecules of $MgCO_3$ in the crystal lattice of the organic calcite. Apparently concentrations of magnesium in modern sea water are too low for direct precipitation of dolomite or active replacement of calcium. Restricted lagoons are lacking and losses due to evaporation are quickly replaced by sea-water inflow and by rainfall. The Recent Florida-Bahama limestones have never been saturated with hypersaline brines. Wells penetrating the pre-Pleistocene bank sections, however, encounter bedded evaporites and dolomites alternating with calcitic limestones. The current cycle of free circulation may eventually be restricted, and evaporites with dolomites may again be deposited.

Ancient Limestone Shelves
Open Marine Environment

Depositional environments on wide-spreading ancient limestone shelves were somewhat more diversiform than those characterizing the Florida-Bahama banks. Many early limestone shelves were hundreds of miles wide and their floors were relatively flat (Fig. 1). Seaward slopes on these shelves commonly averaged less than an inch per mile. Permian deposits in the southwestern United States classically illustrate the wide sedimentary sequential range possible on shallow shelves. Exploration for Permian and older oil fields furnishes opportunities and incentives for studying environmental and facies variations in these Permian rocks on a three-dimensional front.

The early Permian seaway resembled many other limestone shelf-seas recorded in the stratigraphic column. It occupied a partly landlocked embayment extending into the continental interior more than 1,000 miles. An irregular trident of steep-walled bathyal troughs carried ocean waters to the interior of this intra-continental shelf-sea. Exuma Sound, Tongue of the Ocean, and similar channels perform like service for the Bahama Banks. Flat shallow shelves surrounded and separated the three deep Permian troughs. Steep slopes bordering these shelves plunged down to the nearby flat floors of the troughs. The shelves, slopes, and basin floors have been named undaform, clinoform, and fondoform segments, respectively (Rich, 1951). Shallow early Permian shelves were open, and fresh sea waters circulated freely across them. Irregularities on the bottoms, adverse winds, and fluid inertia undoubtedly combined to limit or check cross-shelf circulation. Wherever this happened, evaporitic controls

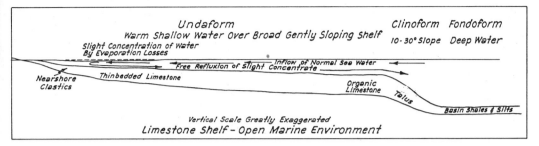

Fig. 1.—Diagram of open marine environment on limestone shelf.

directed movement of the waters.

Loss of water by evaporation lowered the water level on the shelf and increased the concentration and specific gravity of the remaining brine. This heavier brine sank and drained seaward down the sloping shelf. Seaward escape of dense brine is called reflux action (King, 1947). Water flowing in from the open ocean, replaced that lost by refluxion and by evaporation, and maintained circulation across the shallow shelves. With free refluxion, concentrations on even the broadest shelves seldom became great enough to inhibit organic activity. Increased temperatures, lack of food, and the accumulation of waste products were probably more important than increased salinity in limiting carbonate-secreting animal growth. Limestones on early epicontinental flats accumulated more slowly and were denser than those forming on the rapidly growing modern oceanic banks. Most ancient shelf carbonates were torn up and reworked time after time. During this reworking they were almost continuously bathed in sea water. The fact that they are limestones probably indicates that normal sea water has never been an effective dolomitizing agent. The thick Early Permian marine section indicates that a balance between rate of subsidence and rate of deposition prevented restriction and maintained free circulation throughout the Wolfcamp epoch.

RESTRICTED SHELF ENVIRONMENT

The onset of post-Wolfcamp tectonic stabilization slowed subsidence of the shelves to perhaps 0.007 foot per year. Cool, food- and mineral-charged sea water, pushed onto the shallow sunlit shelves, produced lush plant and animal growths. Aggressive shelf-edge organisms combed a disproportionate share of food and minerals from the inflowing currents. Shell- and skeleton-forming inhabitants in this favored marginal zone built the shelf upward much faster than normal subsidence lowered it. Sediment-binding, agitation-resisting organisms congregated here and built reef mounds. When the shelf-edge buildups reached wave base the lateral extension of these mounds into continuous ridges resulted in shelf-edge barrier reefs. Once established, these barriers were largely self-perpetuating. Storm waves breaking against the reefwalls swept detritus into the protected backreef zones. The waves also piled rubble into the forereef deeps. This rubble built shallow foundations of talus across which the reefs forestepped seaward.

Eventually the barrier reef zones became wide and high enough to check free surface refluxion. Development of circulation-restricting barriers and elimination of free refluxion led to the formation of shelf-wide evaporite lagoons (Fig. 2). The appearance of dolomites in the Permian section practically coincided with the formation of these lagoons.

Several times during the Permian, sporadic increases in rate of subsidence allowed fresh sea water to flood the lagoons. During these open-shelf intermissions normal marine limestones spread shoreward across the lagoonal evaporites. In intervening periods of quiescence the reef barriers re-established themselves and lagoonal deposits buried the widespread limestone ledges. Alternating subsidences, stillstands, and even minor upwarps produced, overwhelmed, and recreated a series of shifting environmental provinces. The surprisingly uniform sedimentary profiles resulting from each recurrent fluctuation show that the environmental patterns developed were not fortuitous.

EVAPORITE LAGOON EVIRONMENT

In the Permian evaporite lagoons stream inflow and rainfall were insufficient to offset evaporation losses. Resulting deficiencies were balanced by

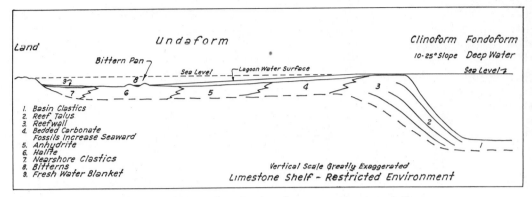

Fig. 2.—Diagram of restricted environment on limestone shelf.

ocean waters drawn inward across the barrier reefs. Influxing sea water flowed toward the depressions caused by evaporation. During this extended journey across the wide Permian lagoons the inflowing waters became more and more concentrated. Permian lagoonal geometry prevented this dense brine from refluxing prematurely. Lagoons 200 miles wide and 10 feet deep had floor slopes averaging about 1 foot in 20 miles. Hydrodynamic currents pushing down these almost imperceptible slopes were persistent rather than precipitous. With free reflux checked, concentrations continued increasing until the brines became dense enough to force a passage through the lagoon floor. Effective dolomitization apparently occurred as these hypersaline brines seeped downward through the underlying carbonates. For this reason a discussion of the origin and composition of the brines is essential.

Most sediments deposited in the shelf-wide Permian lagoons behind the circulation-restricting barriers were derived from sea water (Fig. 3). Modern sea waters are characterized by salinities of approximately 35,000 parts per million. For simplification this value is commonly expressed in parts per thousand and written 35°/oo. Having already undergone several billion years of pre-Permian development, the Permian ocean waters probably approached this same 35°/oo. Relative proportions of solutes making up the 35°/oo also probably resembled those present in modern oceans.

Brine Stratigraphy
Reef and Vitasaline Environments

Waters crossing the lagoonal barriers came from the upper oceanic layers where salinities, temperatures, and other factors favored organic activity. Rocks of the reef and near-backreef areas reflect this vitasalinity (Lang, 1937). Reef and vitasaline limestones include skeletons of framework-builders, shells and tests of mobile or attached forms, and bioclastic debris. Primary textures were relatively coarse. As temperatures and concentrations increased farther within the lagoons, habitability of the inflowing waters declined. Inner limits of the vitasaline zones are poorly defined biologically because different groups of organisms have widely varying toler-

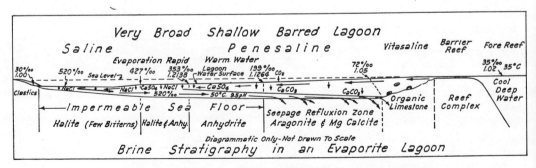

Fig. 3.—Idealized section of brine stratigraphy in evaporite lagoon.

ances to salt. Few forms, however, thrive and reproduce in salinities above 72°/oo, at which concentration chemical precipitation of $CaCO_3$ begins (Scruton, 1953, after Clarke, 1924). The intersection of the 72°/oo isosalinity contour and the lagoon surface thus supplies a practical boundary between the vitasaline and penesaline parts of the inflowing current. Hypersaline brines refluxing across the lagoon floor had concentrations far greater than those of the inflowing surface currents. Where the influxing water rested on hypersaline brine at the base of the lower backreef slope, bottom salinities increased from less than 72°/oo to more than 199°/oo within a few inches.

PENESALINE ENVIRONMENT

Chemically extracted, penesaline limestones precipitated from the inflowing waters until salinities of 199°/oo and densities of 1.126 were reached. A rise in temperature and increase in the concentration from 72°/oo to 199°/oo caused expulsion of carbon dioxide from the influxing sea water. This loss of CO_2 from solution led to the cessation of limestone deposition. With expulsion of CO_2, the pH of the brine probably rose to 9.0 or more. Continued increases in temperature and concentration in later evaporation stages prevented any resorption of CO_2 from the atmosphere.

Limestones of the outer penesaline zone contain tests and skeletons of swimming and floating organisms killed by increasing concentrations of the inflowing currents. These shells increased the textural range and permeability of the normally fine-grained penesaline carbonates. Because of the high temperature, aragonite was the commonly precipitated carbonate mineral. Calcite, other than that in the shells and skeletons, was apparently magnesium-free. Presence of calcite laminae in the Castile anhydrite (Adams, 1944) indicates that some, and probably all, chemically precipitated carbonates were originally non-dolomitic.

After the practical disappearance of CO_2 and the concentration of brine to 199°/oo, precipitation of calcium sulphate began. Gypsum, rather than anhydrite, was deposited at temperatures below 34°C. (McDonald, 1953) but Permian brines were seldom that cool. The gypsum crystals that did form later dehydrated after settling into the dense brines refluxing across the lagoon floors (Conley and Bundy, 1958). Anhydrite precipitated until salinities of 427°/oo and densities of 1.257 were reached. Deposition of anhydrite reduced the quantity of sulphate ions in solution and this decrease further raised the pH of the remaining brine. Salinity increases expelled free oxygen and rendered the brine essentially Eh neutral. In some Permian lagoons anhydrite was the highest-rank evaporite precipitated. In such lagoons brine densities never rose above 1.213, but in most lagoons more soluble salts were also deposited. Small amounts of magnesite were frequently precipitated with the anhydrites but most of the magnesium originally in the influxing water remained in solution. $CaCO_3$ introduced into brines with concentrations above 200°/oo might have combined with magnesium to form dolomite, but natural methods for introducing the $CaCO_3$ were lacking.

SALINE ENVIRONMENT

Halite, or rock salt, characterized evaporite deposits in the 353°/oo to 523°/oo salinity range. Brine density in the halite environment varied between 1.213 and 1.306, or roughly 18 to 28 per cent greater than that of normal sea water. Halite and anhydrite both precipitate through the range from 353°/oo to 427°/oo, but crystallize only at the brine-atmosphere interface. The anhydrite zone was thus limited by the intersections of the 199°/oo to 427°/oo isosalinity contours with the surface of the lagoon. Halite was deposited beyond the 353°/oo isosalinity contour.

Halite deposition, however, was not limited to submerged areas in the Permian evaporite lagoons. Highly concentrated brines yielded large amounts of salt with only slight fluid losses. Excessive precipitation beneath the centers of evaporation raised the lagoon floors to sea-level. Brine films spreading over the exposed surfaces then built atoll-like salt ramparts rising well above the water level of the lagoon. In the locally arid climates hypersaline brines migrated over these ramparts and deposited their bittern loads of hygroscopic potash and magnesium within the desiccating pans. Bitterns were preserved only where deposited far enough from shore to escape solution during infrequent stream water flooding. This salt rampart explanation for bittern pans is separate from, and entirely unrelated to, the dolomitization hypothesis here advocated.

Bitterns, being the final products of complete evaporation, should include all the sea water solutes not previously deposited. Magnesium and potassium salts, deposited from a unit volume of sea water evaporated to dryness, should be several times as abundant as the anhydrite. Instead, the Permian bittern deposits are very minor compared with other evaporites. Bitterns are only one of many shorted evaporite clan members in the Permian lagoonal sequences. Sea water evaporated to dryness should yield 19 times as much halite as anhydrite and 14 times as much anhydrite as limestone (Delwig, 1955, p. 104). Instead of this ideal 1:14:266 ratio, the Permian average is closer to 1:8:40. It was this disproportionate lack of high rank evaporites that led King (1947) to develop the refluxion theory.

Dolomitization by Seepage Refluxion

The scarcity of high-rank salts in the Permian evaporite deposits indicated that highly concentrated brines escaped before complete desiccation. Refluxion was a simple and generally effective process. The heaviest brines migrated to the lowest closed depressions on the lagoon floor. These heavy brines were covered by a blanket of lighter water which protected them from further desiccation. As a result they accumulated, filled the depressions, and flowed on, seeking a natural spillway. Those parts of the floor of the lagoon which were covered by tightly recrystallized anhydrite or halite were effectively impermeable. The slightly greater permeability of the penesaline carbonates provided the spillway for the heavy brines. Pores in these chemically precipitated limestones were of capillary dimensions. Connate water in these pores prevented the entrance of light fluids. Even heavy brines seeped so slowly through them that the lagoonal circulation was uninterruptedly maintained. Slight variations in rates of seepage refluxion determined the maximum concentrations attainable in the lagoons.

The highly concentrated brines were chemically potent solutions. They were hot, with temperatures of 35°C. or above; highly alkaline, with a pH of 9.0 or higher; and had nearly neutral redox potentials. The carbon dioxide, calcium, oxygen, and sulphide contents were abnormally low. Concentrations of magnesium, potassium, sodium, bromine, fluorine, and some other elements were excessively high. The Mg-Ca ratio was many times greater than that in normal sea water. While the evaporite lagoon persisted, the supply of this chemically unbalanced brine was assured.

The refluxing hypersaline brine seeping through the lagoon floor was much denser than the connate water it displaced. The brine, therefore, tended to seep directly downward and only followed bedding planes when vertical migration paths were unavailable. Rocks through which the brine seeped were composed of metastable aragonite and high-magnesian calcite. These chemically and physically unstable carbonates were readily dolomitized by the potent hypersaline brines. In rocks with varying permeabilities, the seeping brines migrated mainly through the more porous zones and bypassed the denser limestone lenses. Permeability, though low in the penesaline carbonates, was uniformly effective; therefore, these sections are almost completely dolomitized. In vitasaline carbonate sections, however, relatively permeable dolomites are commonly interbedded with unaltered lithographic limestones. Coarse-grained porous Permian dolomites are limited to beds previously composed of coarse-grained porous limestones. Fine-grained dense dolomites occupy those zones in which lithographic limestones normally form on open shelves. Dolomite textures, thus, are caused by primary permeability and crystallinity rather than by dolomitization.

Constantly renewed supplies of brine continued to pour through the Permian carbonates for appreciable spans of geologic time. Spillways through which the brines entered the rock column shifted seaward as the shelves regressed. The lagoonal sedimentary profile followed the progressively forestepping reefwall. Old escape zones, overlapped and sealed off by the advancing evaporites, were replaced by similar outlets farther seaward. With each forestep, previously uninvaded vitasaline and reefwall limestones were exposed to the dolomitizing brines. Regressions were normally so slow that all limestones susceptible to alteration were dolomitized before the supply of brine was cut off.

Changes in the composition and character of the refluxing brine and of the invaded rocks limited the downward extent of dolomitization. Magnesium losses and calcium additions, during dolomitization, reduced the overwhelming Mg-Ca ratio of the primary brine. Loss of heat to the cooler limestones and to the displaced connate waters lowered the temperature of the refluxing brine. Carbon dioxide from the displaced connate waters tended to neutralize the alkalinity of the brine. Calcitization of the thick foreset talus was largely completed before brine invasion. Aragonites in these heavily loaded zones which escaped early calcitization were stabilized by increasing overburden pressures (Jamieson, 1953). Dolomitization of calcite and stabilized aragonite has not been recognized in Permian rocks.

Refluxion Footprints

If the hypersaline brines followed the paths through the Permian limestones here suggested, they should have left footprints behind them. Even a cursory examination of the Permian carbonates associated with evaporite lagoons shows many of these footprints. Among these are inclusions of anhydrite that are hundreds or thousands of feet vertically and miles horizontally from deposits of bedded anhydrite (Adams, 1932). These inclusions fill worm borings, fractures, fossil casts, solution cavities, and primary interstitial openings. Many shells are completely replaced. Others, especially the fusulinids, retain parts of the original wall structures suspended in the anhydrite. Benthonic animals whose shells are replaced by anhydrite, and worms that dug the anhydrite-filled borings did not live in an anhydrite-depositing environment. The anhydrite is clearly secondary. The increasing content of anhydrite, as the evaporite boundary is approached, points to the source. The decrease in pressure resulting from the escape of hypersaline brines through capillary passages into open cavities probably explains the deposition of the inclusions. Loss of heat may have aided the process. The inclusions of relatively insoluble anhydrite are preserved in cuttings and cores from wells, but have been dissolved from outcrops. The total volume of the inclusions within the Permian dolomites may well run into cubic miles.

The second set of footprints is similar to the first. Some dolomite cores cut with oil-base muds or highly saturated brines, instead of fresh water, show inclusions of halite with the anhydrite. The halite is washed out of most cuttings and its presence was unsuspected in the early drilling and in examination of outcrops. Halite inclusions normally take the shape of the vugs, fossil casts, and other openings that they occupy. Increased oil production, obtained when porous reservoir dolomites are washed with fresh water, suggests an abundance of crystalline halite in these rocks. Inclusions of anhydrite could form from brines with concentrations of $200°/oo$, but crystalline halite presumably required concentrations above $353°/oo$. The escape of concentrated saline brines demonstrated by these inclusions explains the common lack of high rank evaporites in many Permian lagoons. Anhydrite and halite inclusions within many Permian dolomite reservoirs are regarded as proof that emplacement occurred before entrapment of oil. The abundance of hypersaline brines trapped in pre-evaporite and evaporite-equivalent carbonates furnishes a third set of footprints. Salinities of these brines range up to $360°/oo$ or slightly above the lower limits for precipitation of halite. Still higher concentrations probably exist in favorably located traps. Salinities between $199°/oo$ and $300°/oo$ are common in off-structure wells. These brines trapped high in the Permian section thus have concentrations similar to those in evaporite lagoons.

The thick sections of sub-lagoon dolomites supply a fourth group of seepage refluxion footprints. Adequate magnesium for dolomitizing these 100- or 1,000-cubic-mile masses was available only in the evaporite lagoons. Heavy brines seeping downward through unstable carbonates were readily capable of emplacing the amounts of magnesium required within the time available. In their downward passage, they carried the displaced calcium with them. Part of this calcium was redeposited as vein fillings and other secondary growths. The remainder, still in solution, gives the trapped hypersaline brines an otherwise unexplainable non-lagoonal character. Dolomitization, as evidence of hypersaline brine refluxion, gains significance when supported by inclusions of halite and anhydrite, and by trapped supersaline brines. These four footprints and associated data furnish the basis for this approach to dolomitization.

Stratal Geometry of Dolomitization

The Permian dolomites (Fig. 4) were formed on and around shallow inundated shelves whose interiors were occupied by restricted evaporite lagoons. These deposits developed during the regressive phase of a major sedimentary cycle. Aggressive organic activity along the edges of the shelves maintained the crests of the barrier reefs at or near sea-level. Talus, broken from the reef crests by wave action, built shallow forereef platforms across which the reefwalls progressively forestepped during periods of slow subsidence of

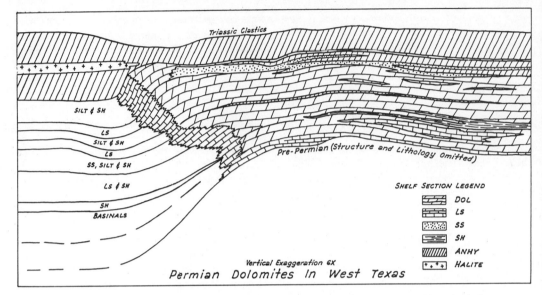

Fig. 4.—Basin-shelf cross section showing generalized lithologic and structural relations of Permian dolomites and associated sediments.

the basin. Concurrent subsidence, upbuilding, and forestepping allowed thick evaporite sections to accumulate in the continuously shallow lagoons. During the migration of environmental provinces seaward, evaporites overlapped penesaline carbonates, which in turn submerged reefwall deposits while new reefwalls formed over the freshly deposited talus. The refluxing brines recrystallized large parts of these topographically lower sections to produce the extensive dolomite complex. Carbonates deposited above refluxing brine levels remain undolomitized unless subsequently invaded by brines from higher lagoons.

The geological settings for other dolomites are more diversified than those associated with the reef-bordered shelf-wide Permian Basin lagoons. Dolomites apparently form in both the transgressive and regressive phases of sedimentary cycles. Environmental variations, however, are largely differences in proportion rather than in kind. Restrictions of circulation are indicated for most dolomites. Dolomite-linked evaporites may form behind longshore bars, within atoll rings, on alternately flooded and emergent mud flats, and in many other environments. Truncation in many places removed the bedded evaporites, leaving inclusions of anhydrite as the only evidence of their former existence.

Ordovician, Devonian, and Cretaceous dolomites of Texas which have varying amounts of evaporites associated with them show evidence of deposition on partially restricted shelves. Similar evaporite-dolomite-limestone sequential relationships are known or suspected for deposits of dolomite the world over.

Summary

Dolomitization was a normal process in carbonate rocks on barred shelves. The reef barrier prevented free refluxion of waters from the lagoon. Following this restriction of circulation the salinity of the brines increased and evaporites formed. The hypersaline brines eventually became heavy enough to displace the connate waters and seep slowly downward through the slightly permeable carbonates of the lagoon floor. During this seepage, the magnesium replaced part of the calcium and the metastable aragonites and high magnesian calcites recrystallized as dolomites. Chemical and physical changes, in the brines and in the rocks through which they flowed, limited the extent of dolomitization.

References

Adams, J. E., 1932, "Anhydrite and Associated Inclusions in the Permian Limestones of West Texas," *Jour. Geol.*, Vol. 40, pp. 30–45.

———, 1944, "Upper Permian Ochoa Series of Delaware Basin, West Texas and Southeastern New

Mexico," *Bull. Amer. Assoc. Petrol. Geol.*, Vol. 29, pp. 1596–1625.

CLARKE, F. W., 1924, "Data of Geochemistry," 5 ed., *U. S. Geol. Survey Bull. 770*.

CONLEY, R. F., AND BUNDY, W. M., 1958, "Mechanism of Gypsification," *Geoch. Cosmoch. Acta.*, Vol. 15, pp. 57–72.

DELLWIG, L. F., 1955, "Origin of the Salina Salt of Michigan," *Jour. Sed. Petrology*, Vol. 25, pp. 83–110.

EPSTEIN, S., AND LOWENSTAM, H. A., 1957, "On the Origin of Sedimentary Aragonite Needles of the Great Bahama Banks," *Jour. Geol.*, Vol. 65, pp. 364–75.

FAIRBRIDGE, R. W., 1957, "The Dolomite Problem," in "Regional Aspects of Carbonate Deposition," *Soc. Econ. Paleon. and Mineral. Spec. Pub. 5*, pp. 125–78.

ILLING, L. V., 1954, "Bahaman Calcareous Sands," *Bull. Amer. Assoc. Petrol. Geol.*, Vol. 38, pp. 1–95.

JAMIESON, J. C., 1953, "Phase Equilibrium in the System Calcite-Aragonite," *Jour. Chem. Phys.*, Vol. 21, pp. 1385–90.

KING, R. H., 1947, "Sedimentation in Permian Castile Sea," *Bull. Amer. Assoc. Petrol. Geol.*, Vol. 31, pp. 470–77.

LANG, W. B., 1937, "The Permian Formations of the Pecos Valley of New Mexico and Texas," *ibid.*, Vol. 21, pp. 833–98.

MCDONALD, G., 1953, "Anhydrite and Gypsum Relations," *Amer. Jour. Sci.*, Vol. 252, p. 884.

NEWELL, N. D., 1955, "Bahamian Platforms," in "Crust of the Earth," *Geol. Soc. America Spec. Paper 62*, pp. 303–19.

RICH, J. L., 1951, "Three Critical Environments of Deposition and Criteria for Recognition of Rocks Deposited in Each of Them," *Bull. Geol. Soc. America*, Vol. 62, pp. 1–20.

SCRUTON, P. S., 1953, "Deposition of Evaporites," *Bull. Amer. Assoc. Petrol. Geol.*, Vol. 37, pp. 2498–2512.

DISCUSSION

DOLOMITIZATION BY SEEPAGE REFLUXION[1]

GEORGE V. CHILINGAR[2] AND HAROLD J. BISSELL[3]
Los Angeles, California, and Provo, Utah

The reviewers would like to commend J. E. Adams and M. L. Rhodes for an interesting paper titled, "Dolomitization by Seepage Refluxion," published in the December, 1960, issue of the *Bulletin*, which furthers our knowledge of the complex problem of dolomite formation. The reviewers suggest that pooling of experience, resources, and data may be necessary to solve this complex problem. The following thoughts and discussion are presented here with this objective in mind.

1. The reviewers tend to disagree with the latter part of the following statement: "Close associations of evaporites, dolomites, and marine limestones are common enough to suggest that restricted lagoon-open ocean couples are essential for the formation of extensive dolomites. This idea is not new, but a workable, theoretical basis for applying it has been lacking" (p. 1912). Teodorovich (1942, 1946, 1950, 1958) and Strakhov (1956a, 1956b) emphasized the importance of dolomite formation in saline lagoons and worked out elaborate chemical schemes which, however, are contradictory in some instances. Teodorovich (1946, 1950, 1958) recognizes the following important types of dolomite deposits: (1) normal-marine calcareous dolomites and replacement dolomitic limestones; (2) primary pelitomorphic dolomites and sulphate-dolomite rocks; (3) calcareous dolomite deposits of salinified seas, lagoon-like bays on reefs, and periodically drying calcareous muds; and (4) primary-chemical, pelitomorphic, calcareous-dolomitic muds, which form in continental lakes having dry and hot (sometimes periodical) climate. Type 2 dolomites formed in salinified lagoons and are interbedded with beds and lenses of anhydrite-gypsum; or they are primary, spotted, microspotted, or more or less homogeneous sulphate-dolomite rocks.

[1] Manuscript received, January 17, 1961.

[2] Petroleum Engineering Department, University of Southern California.

[3] Department of Geology, Brigham Young University.

Strakhov (1956a) distinguishes the following four facies types of dolomite deposits: (1) dolomites of peripheral, salinified parts of large, mainly platform-type, marine basins; (2) dolomites which formed in the central parts of epicontinental seas; here Strakhov includes replacement dolomites (metasomatic) and primary dolomites of highly mineralized lagoons with interbedded anhydrite; (3) normal dolomites of geosynclinal seas; and (4) dolomites of marine bays of arid zones, with brackish water due to influx of river waters.

Teodorovich (1946; 1950; 1955; 1958, p. 302–311) explains in detail the chemistry of dolomite formation in saline lagoons.

2. The reviewers also would like to question the following statement on page 1912: "Geologists generally agree that most bedded dolomites were produced by post-depositional alteration of limestones." This would not hold true for those geologists familiar with enormous primary dolomites of Precambrian and Early Paleozoic times. These dolomites precipitated directly out of sea water largely due to the high pCO_2 of the ancient atmosphere. Evidence such as the association of primary dolomites with other primary sediments, lack of primary porosity, fineness of grains (≤ 0.01 mm.), and preferred orientation of C-axes of grains (crystals), all point toward their primary origin (Chilingar, 1956c). Possibility of direct precipitation of dolomite on slow evaporation out of sea water has been demonstrated by several authors: Kazakov, Tikhomirova, and Plotnikova, 1957; Chilingar, 1956b; and Baron and Favre, 1959. Teodorovich (1946, p. 819) explains chemistry of the precipitation in saline lagoons (also see Chilingar, 1956a).

3. Adams and Rhodes' theory of dolomitization by seepage refluxion is similar to that of Teodorovich (1955, p. 77), in which it was envisioned that slow, warm bottom currents play an important role in the formation of marine replacement dolomites. Rivière (1939a, 1939b) showed experimentally that slow movement of warm sea water

above pure powdery $CaCO_3$ resulted in a marked enrichment of the latter in $MgCO_3$ content. According to Revière, this process is an extremely slow one and is controlled by the fineness of the grains and porosity of the calcareous mud being enriched in Mg content. The extent of seepage of hypersaline brines into penesaline carbonates having *low* permeability (Adams and Rhodes, 1960, p. 1917) possibly could be checked by some laboratory experimental work.

The temperature of 35° C. or higher quoted by Adams and Rhodes (p. 1917) is a favorable one for the formation of dolomite, because Yanat'eva (1957) showed that the region of crystallization of dolomite at a given pCO_2 reaches maximum proportions at temperatures between 30° and 45° C.

4. The reviewers believe that the fine-grained, dense dolomites described by Adams and Rhodes are probably a result of direct precipitation out of sea water; whereas dolomitization gave rise to the porosity of coarse-grained porous Permian dolomites. The only time dolomitization does not give rise to secondary porosity is possibly during the very early diagenesis when there is no solid framework. Obviously, subsequent cementation could obscure the fact that dolomitization gives rise to secondary porosity (Chilingar and Terry, 1954; Chilingar, 1956c).

5. The pH of 9.0 or higher assigned to hypersaline brines by Adams and Rhodes (p. 1917) raises some questions. In salinified lagoons the pH should be lower than 8.0–8.1, which is confirmed by the experimental work of A. V. Nikolaev (*in:* Teodorovich, 1946, p. 820), by theoretical considerations, and by the measurements of pH of waters from present-day saline lagoons. At the present time, reviewers are evaporating sea water in an electric vacuum oven having an hydraulic thermostat which controls the temperature; and are determining the precipitates formed, and the pH of the remaining brines. Results of this study probably will shed some light upon the problem.

6. Haidinger's reaction is often brought forward as a mechanism of dolomitization:

$$CaCO_3 + MgSO_4 \rightarrow MgCO_3 + CaSO_4 \text{ (in solution)}$$
$$CaCO_3 + MgCO_3 = CaCO_3 \cdot MgCO_3$$

Obviously, the reaction will proceed to the right only if waters are saturated with respect to $CaSO_4$, thus resulting in precipitation of $CaSO_4$ (Teodorovich, 1958, p. 305).

It is interesting to note that findings of Yanat'eva (1949, 1959, 1955) indicate that at high $MgSO_4$ concentration dolomite is unstable, whereas magnesite (or hydromagnesite) is stable. Thus, at high $MgSO_4$ concentration in solution the second part of Haidinger's reaction could not take place. For example, Teodorovich (1955, p. 99) attributes the absence of dolomite in the deposits of saline Kara-Bogaz-Gol gulf (Caspian Sea) to the high $MgSO_4$ content.

LITERATURE CITED

Adams, J. E., and Rhodes, M. L., 1960, Dolomitization by seepage refluxion: Am. Assoc. Petroleum Geologists Bull., v. 44, no. 12, p. 1912–20.

Baron, G. A., and Favre, J. H., 1959, Recherches experimentales sur le role des facteures physico-chimique dans la synthèse de la dolomie: World Petroleum Cong., sec. I, paper 3, p. 19–25.

Chilingar, G. V., 1956a, Note on direct precipitation of dolomite out of sea water: Compass of Sigma Gamma Epsilon, v. 34, no. 1, p. 29–34.

——— 1956b, Relationship between Ca/Mg ratio and geologic age: Am. Assoc. Petroleum Geologists Bull., v. 40, no. 9, p. 2256–66.

——— 1956c, Use of Ca/Mg ratio in porosity studies: Am. Assoc. Petroleum Geologists Bull., v. 40, no. 10, p. 2489–93.

——— and Terry, R. D., 1954, Relationship between porosity and chemical composition of carbonate rocks: Petroleum Engineer, v. 26, p. 341–43.

Kazakov, A. V., Tikhomirova, M. M., and Plotnikova, V. I., 1957, System of carbonate equilibriums (dolomite, magnesite): Trudy Inst. Geol. Nauk Akad. SSSR, geol. ser., no. 64, Vypusk 152, p. 13–58. (Chilingar failed to reproduce the results described by these writers. Instead of dolomite he obtained giobertite).

Revière, A., 1939a, Sur la dolomitisation des sédiments calcaire: Acad. Sci. Comptes Rendus, tome 209, no. 16.

——— 1939b, Observation nouvelles sur le mecanisme de dolomitisation des sédiments calcaire: Acad. Sci. Comptes Rendus, tome 209, no. 19

Strakhov, N. M., 1956a, Toward understanding of diagenesis, *in* Questions of mineralogy of sedimentary formations: Kniga 3 and 4, Izd. Lvov. Govt. Univ.

——— 1956b, Types and genesis of dolomite rocks, *in* Types of dolomite rocks and their genesis (symposium): Akad. Nauk SSSR Izd., Trudy Geol. Inst., Vypusk 4, p. 5–27.

Teodorovich, G. I., 1942, Dolomitization of reef formations of Ishimbay petroliferous area: Dokl. Akad. Nauk SSSR, v. 34, no. 6.

——— 1946, On the genesis of the dolomite of sedimentary deposits: Dokl. Akad. Nauk SSSR, v. LIII, no. 9, p. 817–20.

——— 1950, Lithology of Paleozoic carbonate rocks of Ural-Volga region: Izd. Akad. Nauk SSSR, Moscow-Leningrad, 213 p.

——— 1955, Toward the question of formation of sedimentary calcareous-dolomitic rocks: Trudy Inst. Nefti, Akad. Nauk SSSR Izd., tom 5, p. 75–107. Translated by Mark Burgunker, Inter. Geol. Review, v. 1, 1959, no. 3, p. 50–73.

——— 1958, Study of sedimentary rocks: Gostoptekhizdat, Leningrad, 572 p.

Yanat'eva, O. K., 1949, Solubility in system Ca, MgllCO$_3$, SO$_4$-H$_2$O: Dokl. Akad. Nauk SSSR, tom 67, no. 3, p. 479–81. Also see Chilingar, 1956, Am.

Assoc. Petroleum Geologists Bull., v. 40, no. 11, p. 2770–73.

——— 1950, Solubility of dolomite in aqueous solutions of salts: Izv. Sect. Physical-Chemical Analysis, Inst. General and Inorgan. Chem., Akad. Nauk SSSR, tom XX.

——— 1955, Effect of aqueous solutions of gypsum on dolomite in the presence of carbon dioxide: Dokl. Akad. Nauk SSSR, tom 101, no. 5, p. 911–12. Also see Chilingar, 1956, Am. Assoc. Petroleum Geologists Bull., v. 40, no. 4, p. 762–64.

——— 1957, About polytherm of solubility of ($CaCO_3$ +$MgSO_4 \rightleftharpoons CaSO_4 + MgCO_3$)—$H_2O$ system: Dokl. Akad. Nauk SSSR, tom 12, no. 6.

(Note: Most of the Russian references can be obtained from George V. Chilingar).

DEPOSITIONAL ENVIRONMENTS OF PHOSPHORIA FORMATION (PERMIAN) IN SOUTHEASTERN BIGHORN BASIN, WYOMING[1]

CHARLES V. CAMPBELL[2]

Tulsa, Oklahoma

ABSTRACT

The objective of this study is to reconstruct the depositional environments of Permian sediments that formed the Phosphoria Formation in the southeastern Bighorn Basin, Wyoming. Data of stratigraphy, petrology, and sedimentation are synthesized to achieve this objective.

Upper and lower members of the Phosphoria are defined. Each is characterized by distinctive facies patterns, separate sources of sedimentary materials, and differences in depositional environments. Extensive eastward transgressions and lesser regressions produced the facies change in the upper member from dolomite in the southwest to anhydrite and redbeds eastward. Sand grains derived from the underlying Tensleep Sandstone characterize the lower member; this detritus was supplied from topographic highs of the Tensleep erosion surface and was deposited in shallow bays bordering the transgressing lower Phosphoria sea.

Five types of dolomite are described: (1) dolomite mudstone, (2) detrital dolomite, (3) skeletal dolomite, (4) pelletal dolomite, and (5) oölitic dolomite. These types are distinguished on the basis of limestone megatextures that persisted through dolomitization. The dolomite is associated with claystone, anhydrite, intraformational breccia, sandstone, and conglomerate.

Dolomitization and redistribution of anhydrite are the most important diagenetic processes that have modified or destroyed porosity.

The study area lay in an embayment extending eastward across the stable shelf bordering the Permian miogeosyncline. This bay occupied the present western half of Wyoming, and the study area is a part of the eastern shoreline. The floor of this bay had a relief of somewhat more than 100 feet as Phosphoria deposition began, but it became a nearly plane surface dipping seaward at a few feet per mile by upper Phosphoria time. Landward topography during both phases was an extension of the submarine topography. Surface sea water flowed shoreward across this embayment from the area of upwelling along the margin of the miogeosyncline; and the temperature, salinity, and pH increased as the water approached shore. Highly saline water returned seaward along the bottom, and this water may have been the agent that dolomitized the calcium carbonate sediments. The climate was arid and either tropical or subtropical.

Depositional conditions varied within the embayment where the different rock types were deposited. The dolomites are dolomitized counterparts of the types of sediments currently forming on the Bahama Banks; they were deposited under conditions similar to those now existing in this area of present sedimentation, where diverse megatextural types form in response to currents of different energies. The claystone is a tidal flat deposit. Dynamic rather than physical restrictions during deposition of the upper member aided the concentration of sea water to the point where calcium sulfate precipitated; however, physical barriers isolated bays where calcium sulfate accumulated in the lower member. Intraformational breccias were dumped in channels crossing the tidal flats. Sandstones formed as nearshore deposits or as parts of deltas, and the conglomerates are intraformational deposits or residual gravels.

Most Phosphoria oil is produced from dolomite within 100 feet of the top of the formation. Eastward migration of oil is limited by the facies change from dolomite to anhydrite and claystone, and reservoir rocks are types of dolomite consisting predominantly of relic grains. Dolomitization modified the original porosity, and anhydrite further plugged porosity locally. Phosphoria rocks contain both source and reservoir beds.

INTRODUCTION

Depositional environments of Permian sediments that formed the Phosphoria Formation in the southeastern Bighorn Basin, Wyoming (Fig. 1) are reconstructed. Phosphorites and petroleum, associated with Phosphoria rocks, had distinct environments of deposition. Characteristic features of these environments are presented to facilitate search for commercial deposits of these materials in the Phosphoria and other formations.

Data of stratigraphy, petrology, and sedi-

---

[1] Manuscript received, April 20, 1961; revised, September 6, 1961. This paper is based on a Ph.D. dissertation submitted to Stanford University in 1956.

[2] Jersey Production Research Company.

The writer thanks all those persons who aided him in various ways. Special thanks are due G. K. Brasher and L. A. Murphy who assisted in running well samples, and J. J. Graham, Claude Minard, and Klaus Küpper who measured the surface sections. Discussions

with V. E. McKelvey, R. P. Sheldon, T. M. Cheney, and R. A. Gulbranson aided in understanding many phases of the geology of the Phosphoria Formation. In addition, R. A. Gulbranson made X-ray diffractometer equipment of the United States Geological Survey available to the writer, supervised its use, and assisted in the interpretation of results. G. A. Thompson supervised the original dissertation, and R. M. Mitchum read and suggested revisions to the revised manuscript. Interpretations and conclusions presented in this paper, however, are solely the writer's.

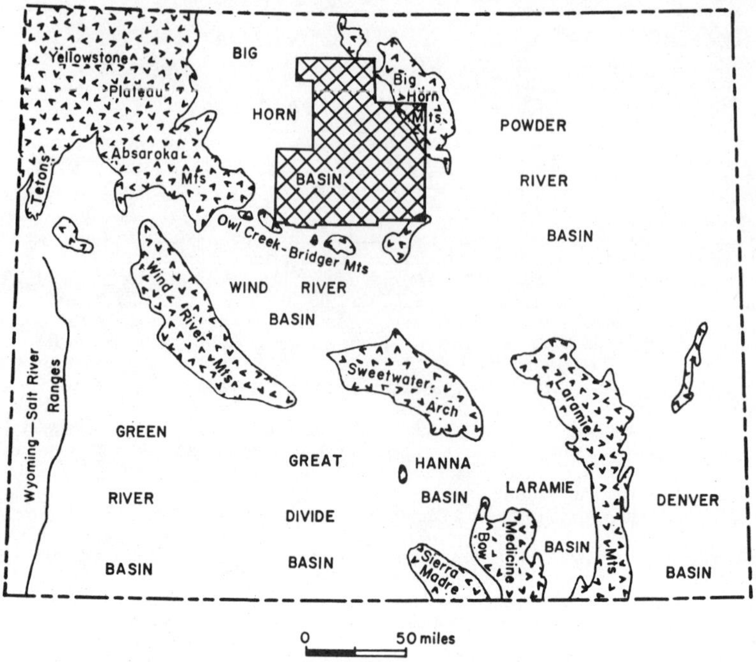

Fig. 1.—Index map of Wyoming locating study area.

mentation are synthesized. Rocks composing the Phosphoria are classified and correlated, and facies relationships are mapped. Petrology is described along with some effects of diagenesis. Tectonic, physical, and physico-chemical controls of depositional environments of the different rock types are interpreted. Finally, the relations between oil occurrence and depositional environments are observed.

Detailed logging of cuttings and cores from 59 wells and samples from 25 surface sections provided basic data; supplemental were the study of 91 thin sections, X-ray diffractometer analyses of 23 samples, and organic analyses of six samples. Sizes of grains and crystals are designated according to the Wentworth grade scale. The lithologic log for each well was adjusted to corresponding mechanical logs for accurate determination of the thickness of lithologic intervals used in facies mapping. Thicknesses of stratigraphic intervals were corrected for dip. Control points average one well or surface section for 6 square miles; only selected wells drilled prior to April, 1954, were studied.

STRATIGRAPHY

Phosphoria rocks of Permian age grade eastward from carbonate rocks to redbeds across the study area. The carbonate rock facies has recently been assigned to the Park City Formation (McKelvey et al., 1956, 1959), and the redbed facies has been named the Goose Egg Formation (Burk and Thomas, 1956), which also includes the equivalents of the overlying Dinwoody Formation of Triassic age. Use of Phosphoria Formation was restricted to the phosphatic mudstone and chert facies in southeastern Idaho and adjacent areas (McKelvey et al., 1956, 1959). Though this facies nomenclature is correct and useful in oil exploration, it has not been widely adopted by geologists in oil companies; they use Phosphoria for all three facies in the Bighorn Basin as does this report.

The Dinwoody Formation of early Triassic age overlies the Phosphoria (Fig. 2). An anhydrite marker bed with the texture of a microbreccia marks the top of the Dinwoody and the base of the Chugwater Formation, also of Triassic age. This marker corresponds with the Dinwoody "kick" on electrical resistivity logs. The remainder of the Dinwoody consists of 30–75 feet of grayish red or greenish gray claystone or dolomite with some beds of anhydrite.

The Tensleep Sandstone underlies the Phos-

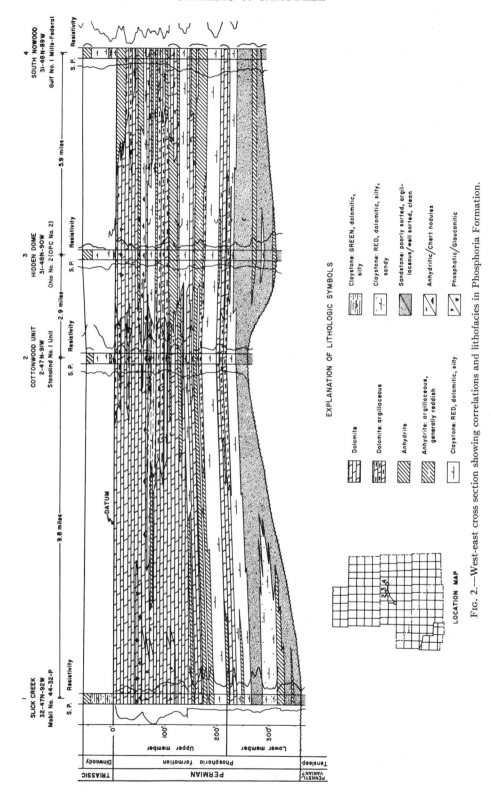

Fig. 2.—West-east cross section showing correlations and lithofacies in Phosphoria Formation.

phoria (Fig. 2) and is typically very fine-grained, very well sorted sandstone.

According to McKelvey et al. (1953a, p. 42), the Phosphoria Formation and its partial stratigraphic equivalents of marine origin cover an area greater than 225,000 square miles in southwestern Montana, western Wyoming, northeastern Colorado, eastern Idaho, northern Utah, and northeastern Nevada. Thickest stratigraphic sections are found in the miogeosyncline in northwestern Utah and northeastern Nevada, where Cheney et al. (1956, p. 1717) report 3,000–4,000 feet. In southwestern and south-central Montana the Phosphoria Formation wedges out northward because of non-deposition and truncation.

The Phosphoria Formation ranges from 192 to 361 feet thick in the southeastern Bighorn Basin (Fig. 3). The irregular isopachous pattern results from topographic differences on the Tensleep erosion surface when Phosphoria deposition began.

SUBDIVISIONS OF PHOSPHORIA

Upper and lower members of the Phosphoria Formation are defined in this report on the basis of lithologic and mechanical log characteristics (Fig. 2); they are consistently recognizable throughout their areal extents in the Bighorn Basin, in the Powder River Basin on the east, and the Wind River Basin on the south. The lower member wedges out north of the study area due to non-deposition, and the upper member wedges out due to non-deposition in south-central Montana.

Some of the members of the Park City and Phosphoria Formations (McKelvey et al., 1956, 1959) and the Goose Egg Formation (Burk and Thomas, 1956) were recognized and used as guides for correlation. The upper and lower members, however, were not subdivided because they are two separate transgressive depositional cycles; and units designated by others are only parts of these cycles. Each member exhibits a different facies distribution, a different source of component sedimentary particles, and different factors dominating its environment of deposition.

UPPER MEMBER OF PHOSPHORIA

Definition.—The upper member consists of strata between the base of the Dinwoody Formation and the top of the lower member (Fig. 2). At the top and base of this member, respectively, are eastward-extending dolomite tongues, except locally where the dolomites are replaced laterally by anhydrite. Characteristic abrupt downward increases in resistivity on electric logs correspond with these contacts. Where the claystone overlying these dolomite units becomes very dolomitic or where dolomite occurs, the resistivity "kick" is less pronounced. However, these contacts are readily recognized in cores and cuttings.

Distribution and thickness.—The upper member in the southeastern Bighorn Basin ranges from 154 to 238 feet thick (Fig. 4). It is irregular in thickness along the eastern margin of the study area, but gradually thickens westward from the central part.

Facies relationships.—Figure 4 shows that the upper member consists wholly of dolomite or dolomite facies in the southwest and mostly of grayish red claystone or redbed facies eastward. Locally, anhydrite constitutes more than half of the redbed facies.

The intertonguing and intergrading of the dolomite and redbed facies are illustrated in Figure 2. The facies change progresses eastward from the base to the top of the member; successively younger beds in the dolomite facies intergrade and interfinger with beds of the redbed facies farther and farther east. This change resulted from the extensive eastward transgressions and minor regressions of the sea.

Member boundaries.—The upper member and the overlying Dinwoody Formation in the study area appear conformable and gradational. The uppermost Phosphoria bed is a light-colored dolomite, except in the eastern part of the area where replaced by anhydrite. The basal Dinwoody beds are greenish gray claystone or argillaceous dolomite in the west and grayish red claystone in the east. As the Dinwoody grades downward, dolomitic claystone gives way successively to dolomitic claystone containing anhydrite nodules, then to dolomite containing anhydrite nodules and inclusions, and finally to the upper dolomite bed of the Phosphoria, which commonly contains anhydrite inclusions. This vertical interval of gradation is less than 10 feet.

Though physical evidence for a hiatus between the Phosphoria and the Dinwoody is lacking in the study area, Newell and Kummel (1942) and Kummel (1950) have presented faunal evidence demonstrating an unconformity in western Wyoming. Burk and Thomas (1956, p. 9) state

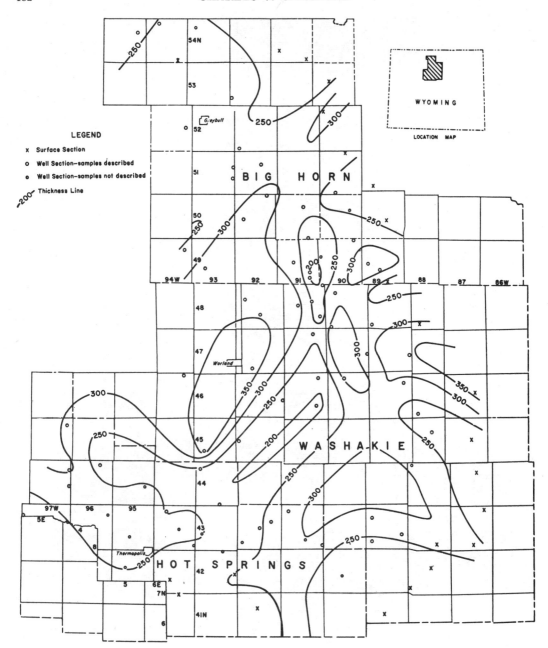

FIG. 3.—Thickness map of Phosphoria Formation, southeastern Bighorn Basin, Wyoming.

that Ochoan rocks are apparently absent in Wyoming, but that evidence of an unconformity between the Dinwoody and Phosphoria is insignificant, and that the lithologic change between the two is slight. Love (1948, p. 97) notes that the discordance between the Phosphoria and Dinwoody is apparent only in regional studies.

The upper and lower members of the Phosphoria are conformable. A grayish red claystone unit, averaging 25 feet thick, is found throughout the study area at the base of the upper member, except where the claystone becomes greenish gray and grades laterally to greenish gray dolomite in the southwestern part. This claystone rests on the

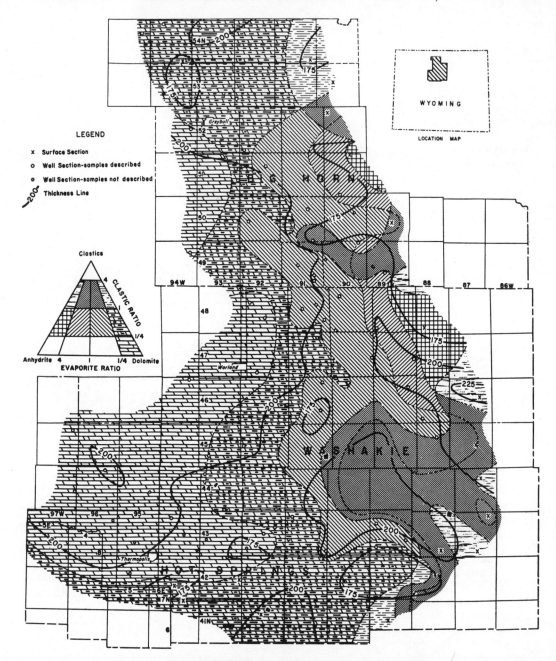

Fig. 4.—Thickness and lithofacies of upper member of the Phosphoria Formation, southeastern Bighorn Basin, Wyoming.

upper dolomite bed of the lower member, or locally on a thin (5 feet or less) anhydrite or sandstone bed that intervenes between the claystone and the dolomite.

Correlation and age.—Correlation lines for lithologic units within the upper member are drawn in one of two ways: (1) if the top and base of the member are approximately parallel, the lithologic correlations are drawn parallel with the top; and (2) if the top and base diverge, the lithologic

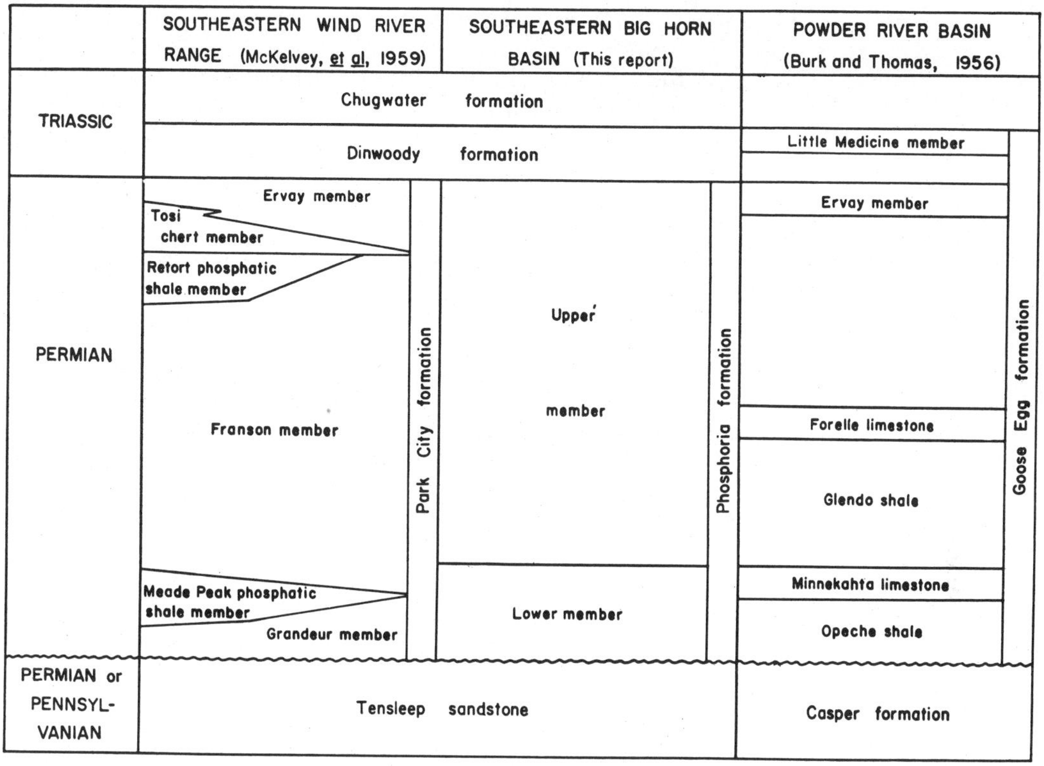

	SOUTHEASTERN WIND RIVER RANGE (McKelvey, et al, 1959)	SOUTHEASTERN BIG HORN BASIN (This report)	POWDER RIVER BASIN (Burk and Thomas, 1956)
TRIASSIC	Chugwater formation		
	Dinwoody formation		Little Medicine member
PERMIAN	Ervay member / Tosi chert member / Retort phosphatic shale member / Franson member / Meade Peak phosphatic shale member / Grandeur member (Park City formation)	Upper member / Lower member (Phosphoria formation)	Ervay member / Forelle limestone / Glendo shale / Minnekahta limestone / Opeche shale (Goose Egg formation)
PERMIAN or PENNSYL- VANIAN	Tensleep sandstone		Casper formation

Fig. 5.—Correlation chart.

correlation for any selected horizon diverges from the top in proportion to the increase in thickness of the upper member. Detailed correlations across the study area show that the facies change takes place without change in thickness. As no unconformities are found within or bounding this member, the dolomite facies and the redbed facies are lateral time equivalents; and the upper member is a time-rock unit. The top and base are then time planes, and the rate of deposition varied as the member thickened and thinned (Fig. 2).

Correlations of the upper member with the Goose Egg Formation east and southeast of the study area and the Park City Formation south and southwest are shown in Figure 5. The Ervay Member and Glendo Shale of the Goose Egg are recognized in the redbed facies of the upper member, but no equivalent of the Forelle Limestone was identified. The Ervay can also be identified in the dolomite facies in the southwestern corner of the area, where the Tosi Chert and Retort Phosphatic Shale Members separate the Ervay and Franson Members of the Park City. Elsewhere in the dolomite facies the Ervay and Franson are inseparable because chert beds, ranging from a few inches to 30 feet thick, appear at various stratigraphic horizons; these chert beds are of little use for correlation in areas much larger than a few townships.

The upper member of the Phosphoria is Guadalupian in age, which is assigned by correlation with the Phosphoria in the Wind River Mountains south of the study area. There the upper part of the Phosphoria contains a well preserved and diversified Guadalupian fauna (Burk and Thomas, 1956, p. 9).

LOWER MEMBER OF PHOSPHORIA

Definition.—The lower member of the Phosphoria Formation consists of the strata below the upper member and above the Tensleep Sandstone of Pennsylvanian and Permian age (Fig. 2). A dolomite unit, locally capped with an anhydrite or sandstone bed less than 5 feet thick, composes the upper unit of this member. An abrupt downward increase in resistivity on electric logs coincides with the top of the dolomite unit. Sand grains reworked from the Tensleep characterize the lithologic types in the lower part of the lower member, and the basal bed is usually poorly sorted sandstone or conglomerate. No characteristic electric log feature coincides with the base throughout the study area; examination of well samples is necessary to satisfactorily place the contact, especially in the northern part of the area.

Distribution and thickness.—The lower member ranges in thickness from 20 to 150 feet in the southeastern Bighorn Basin (Fig. 6). Comparison of Figure 6 with Figure 3, thickness of total Phosphoria, shows that thick and thin areas on the two maps coincide; the thickness variation for the total Phosphoria, therefore, results mostly from thickness changes of the lower member.

Variation of rock types.—A persistent dolmite unit, locally with an anhydrite or sandstone bed less than 5 feet thick at the top, is the upper unit of the lower member; and the lower part is composed of dolomite, claystone, and sandstone, with some conglomerate and anhydrite (Figs. 6 and 2).

Member boundaries.—The lower member of the Phosphoria conformably underlies the upper member as previously described.

An unconformity separates the lower member from the underlying Tensleep Sandstone. Physical evidence of this unconformity is local slight angular discordance, channeling, local presence of basal conglomerate, abundant sand grains in the lower member derived from the Tensleep, and the irregular topography of the contact.

The duration of erosion represented by the unconformity includes upper Wolfcampian and some of lower Leonardian time. Lower Phosphoria rocks of Leonardian age rest on Tensleep rocks of Wolfcampian age (Verville, 1957, p. 350) in the southeastern Bighorn Basin.

The stratigraphic hiatus extends from middle Pennsylvanian to middle Permian in the southeastern Bighorn Basin. Desmoinesian fusulinids (Henbest, 1954, p. 52-53) in the uppermost Tensleep beds in the northern part of the study area underly Phosphoria rocks of probable Leonardian age.

Correlation and age.—Correlation lines for lithologic units within the lower member are drawn parallel with the top of the member. The Phosphoria was deposited on the irregular surface of the Tensleep by an eastward transgressing sea. Until the local high areas were covered, they supplied sand to the low areas. Deposition ended as the upper dolomite unit spread over the study area, forming a nearly plane sea floor and a time plane.

Figure 5 shows the correlation of the lower member with formations in adjacent areas. The upper dolomite unit of the lower member corresponds with the Minnekahta Limestone of the Goose Egg Formation on the east, and the lower part of the lower member equals the Opeche Shale. Westward the upper dolomite correlates with the 23 feet of predominantly carbonate rocks overlying the "lower phosphate" of King (1947), which is now recognized as the Meade Peak Phosphatic Shale Member of the Park City Formation. The phosphatic shale was not identified in the study area, but the lower part of the lower member appears equivalent to the Grandeur Member of the Park City.

By correlation with fossiliferous sections, the lower member is Leonardian and possibly lowermost Guadalupian. Both the Meade Peak Phosphatic Shale of the Phosphoria and the Grandeur Member of the Phosphoria are Leonardian (McKelvey et al., 1959, p. 39), and the upper dolomite unit of the lower member, or Minnekahta equivalent, may be Guadalupian.

PETROLOGY

Of the six different rock types found in the Phosphoria Formation, dolomite, claystone, and anhydrite are constituents of both members; intraformational breccia is found only in the upper, and sandstone and conglomerate in the lower.

By definition, the texture of a rock is the size, shape, and arrangement of component elements. Megatextural elements of carbonate rocks are: (1) grains, such as pellets, oöliths, skeletal fragments, lumps (aggregates of grains formed by accretionary processes), and detrital grains (fragments of pre-existing rocks); and (2) the interstitial materials, such as carbonate mud matrix or sparry carbonate cement. Each megatextural element is composed of crystals which are the microtextural elements. An element of both mega- and microtextures is void space or pores between elements. Study of megatextures provides significant data about the depositional environment, and study of microtextures reveals features of the diagenetic environment.

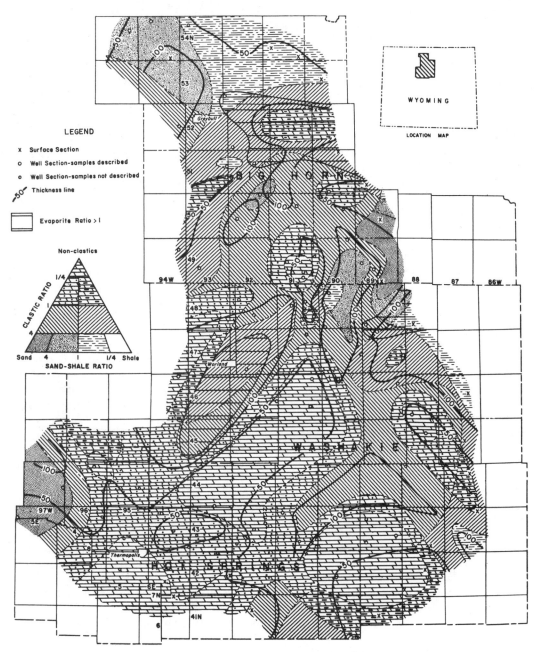

Fig. 6.—Thickness and lithofacies of lower member of Phosphoria Formation, southeastern Bighorn Basin, Wyoming.

TYPES OF DOLOMITE

Five types of dolomite are recognized in thin sections on the basis of different relic limestone megatextures.

1. Dolomite mudstone
2. Detrital dolomite
3. Skeletal dolomite
   a. Skeletal dolomite with mudstone matrix
   b. Very fine crystalline skeletal dolomite
4. Pelletal dolomite
5. Oölitic dolomite

All gradations between types were observed, and dolomite composed predominantly of relic lumps might be found if enough samples were examined.

Accessory silt and clay are included with carbonate particles of similar size in distinguishing the types of dolomite. Inclusion of quartz and clay detrital grains with carbonate grains of similar size is believed justified because mechanically deposited grains of similar size were deposited under essentially the same hydraulic conditions, regardless of whether they were quartz, carbonate, or clay.

Dolomite mudstone.—Dolomite mudstone is composed essentially of fine silt- and clay-size dolomite crystals; it contains less than 30 per cent relic limestone grain types (Fig. 7). This type of dolomite was observed in all sections of the Phosphoria and is the most abundant type, beds of which are less than 5 feet thick. Dolomite mudstone grades laterally into pelletal dolomite, anhydrite, claystone, or detrital dolomite.

Most dolomite mudstones are composed of subhedral, uniform-size crystals less than 4 microns in diameter, but some have average crystal sizes as large as 30 microns. Dolomite rhombs as large as 540 microns are more or less uniformly scattered throughout this rock type. Dolomite crystals as a whole appear dusty because of films and inclusions of organic matter.

Impurities in dolomite mudstone include carbonaceous material, clay, quartz silt and sand, authigenic and detrital glauconite, collophane skeletal grains, hematite or pyrite, mica, chert, and anhydrite.

Some mudstones are homogeneous with no laminations (Fig. 7a). Others have essentially parallel, commonly wavy laminations (Fig. 7b) or lensing laminae (Fig. 7c). A peculiar type of "churned" structure, caused by burrowing organisms, is shown in Figure 7d. This "churned" structure resembles that illustrated by Stratten (1959, Fig. 2) from Recent tidal flat deposits.

Detrital dolomite.—Thirty per cent or more sand or larger grains derived from pre-existing rocks (detrital grains) are essential components of detrital dolomite (Fig. 8). Detrital dolomite is found in beds less than 5 feet thick; it grades laterally into dolomite mudstone, dolomitic claystone, or dolomitic sandstone.

The essential detrital grains may be quartz from the underlying Tensleep in the lower member; quartz from a northwestern source in the upper member in the western part of the area; or dolomite mudstone, pelletal dolomite, or claystone in the eastern part of the area.

The matrix between the detrital grains is identical with the dolomite mudstone previously described (Fig. 8). In amount, the matrix in the samples described ranges from 50 to 70 per cent.

Impurities include carbonaceous material, clay, quartz silt and sand, pyrite or hematite, detrital glauconite, collophane skeletal grains, chert, and anhydrite.

Skeletal dolomite.—Skeletal dolomite contains 30 per cent or more relic skeletal grains as the dominant grain type (Fig. 9). This type of dolomite is most abundant in the western part of the study area and is rarely found in the eastern part. Beds are usually less than 5 feet thick. Skeletal dolomite grades laterally into pelletal dolomite or dolomite mudstone.

Two subtypes of skeletal dolomite are distinguished: (1) skeletal dolomite with a dolomite mudstone matrix (Fig. 9a, b, c), and (2) very fine crystalline skeletal dolomite (Fig. 9d).

Mudstone matrix is homogeneous and shows no evidence of laminations. The amount ranges from 70 to 25 per cent in the thin sections studied.

Very fine crystalline skeletal dolomite consists of anhedral crystals ranging from less than 4 microns to 5 millimeters maximum diameter; the average crystal size is in the very fine grade. The characteristic microtexture is similar to the "grain growth" in the limestones described by Bathurst (1958, p. 24–31). Grains, other than skeletal grains composed of collophane or francolite, are recognizable only as ghost forms marked by absence or concentration of carbonaceous impurities. Many relic skeletal grains now consist of crystals averaging two or more times the size of the enclosing crystals. Others consist of crystals similar in size to the enclosing crystals. These crystals commonly cross grain boundaries

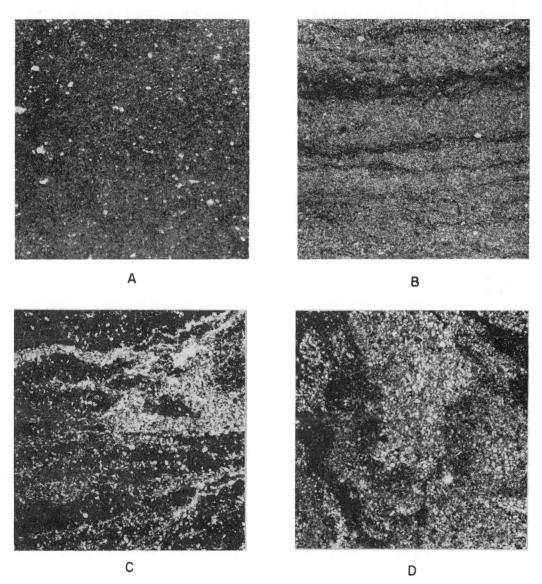

Fig. 7.—Variations in dolomite mudstone.

A.—Homogeneous dolomite mudstone. Clear grains are dolomite rhombs and detrital quartz. (Plane polarized light, 40×)

B.—Dolomite mudstone showing wavy laminae. Clear grains are detrital quartz and black cubic grains are pyrite. (Plane polarized light, 30×)

C.—Silty dolomite mudstone showing wavy lensing laminae. Clear angular grains are detrital quartz, and clear rhombic forms are dolomite. Irregular concentration of quartz silt probably is burrow filling. (Plane polarized light, 30×)

D.—Silty dolomite mudstone showing "churned" structure. Clear angular forms are detrital quartz, black squares are pyrite, and clear rhombic forms are dolomite. "Churned" structure is due to burrowing organisms, and irregular vertical form filled with quartz silt is probably a burrow. (Plane polarized light, 30×)

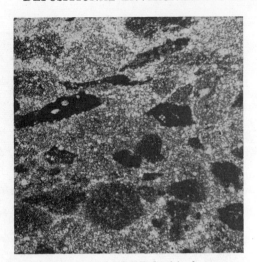

Fig. 8.—Detrital dolomite: Dolomitic claystone composed of black grains; and dolomite mudstone, gray. (Plane polarized light, 30×).

as clearly shown in the relic pellets in Figure 9d. The pellets are defined by concentrations of carbonaceous inclusions, and some of these ghosts are part of as many as three dolomite crystals; other pellets are confined to a single dolomite crystal, which has the "dusty" pellet form in its center and clear borders. Parts of some pellets consist of dolomite crystals smaller than 4 microns, like the pellets subsequently described.

Skeletal grains consist of dolomite, collophane, francolite, and rarely calcite. Collophane or francolite constitutes fragments derived from echinoid spines, bone tissue, fish scales, brachiopods, bryozoans, and rare fish teeth and algae. Dolomite composes fragments of brachiopods, pelecypods, gastropods, echinoid spines, bryozoans, and rare sponge spicules, foraminifera, and ostracods. Rare bryozoan fragments from the western part of the area studied consist of calcite.

The skeletal grains range from less than very fine sand sizes to more than 15 millimeters long. They are poorly sorted and rounded; their shape depends on the original shell form as well as the abrasion history of the fragments.

Minor constituents of skeletal dolomites include pyrite, glauconite, carbonaceous material, quartz, muscovite, clay, chert, and anhydrite.

The two types of skeletal dolomite probably formed under different conditions. Skeletal dolomite with a mudstone matrix accumulated where currents were too weak to winnow the mud from the skeletal grains. Very fine crystalline dolomite probably formed from partly cemented skeletal grains; absence of remnants of dolomite mudstone and argillaceous impurities between grains suggests that none was present. Present high porosity in spaces between relic grains in this dolomite indicates intergranular porosity in the original limestone. This porosity can not be explained solely by leaching of original rock material; solutions could not have entered and dissolved appreciable amounts of material or uniformly dolomitized the limestone if original porosity did not exist. This porosity is not related to fractures. Currents during depositon of this second type of skeletal dolomite were probably strong enough to winnow out the fine mud, leaving only loosely cemented grains.

Pelletal dolomite.—Pelletal dolomite contains thirty per cent or more relic pellets as the dominant grain type (Fig. 10). Relic pellets are discrete ovoid to irregular grains which are composed of silt- to clay-size dolomite crystals and show no regular internal structure. This type of dolomite is present in all except a few well sections logged, but it is most abundant in a band extending north to south through the central part of the area studied. Beds of pelletal dolomite are usually less than 5 feet thick and grade westward into skeletal dolomite and eastward into dolomite mudstone or in some places claystone.

Dolomite crystals composing the pellets appear subhedral and commonly measure less than 9 microns maximum dimensions, though some consist of crystals as large as 16 microns. Carbonaceous inclusions give pellets a dusty appearance and darker color in thin sections than the dolomite matrix. Some pellets have nearly clear centers and marginally show an increase in concentration of organic matter, which was expelled from the centers of the pellets during recrystallization and dolomitization.

Lumps or aggregates of grains resembling those described from the Bahama Banks (Illing, 1954, p. 29–35) are present in varying amounts in many pelletal dolomites (Fig. 10b). Pellets in the aggregates are identical with the pellets previously described and are bonded by carbonaceous dolomite similar to that constituting the pellets. Some aggregates have a rim darker than the interior of the lump because of a greater concentration of organic matter in the rim.

Most pelletal dolomite is cemented by colorless dolomite crystals that contain less organic matter

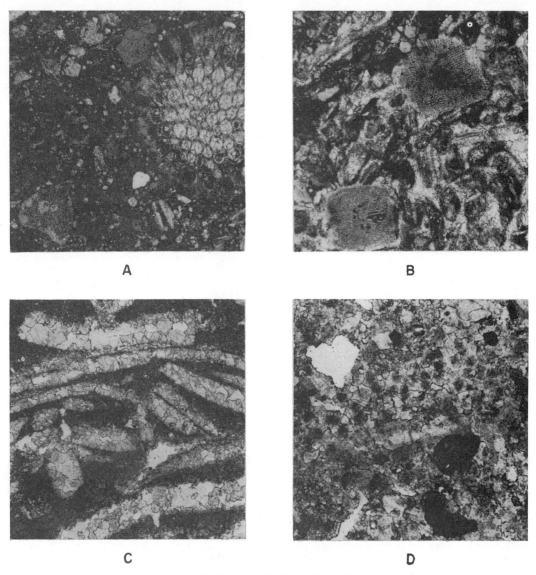

Fig. 9.—Variations in skeletal dolomite.

A.—Mudstone matrix. Dolomitized bryozoans and echinoid fragments are in upper right and lower left, respectively. (Plane polarized light, 30×)
B.—Mudstone matrix. Relic skeletal grains are derived from echinoids and either pelecypods or brachiopods. (Plane polarized light, 30×)
C.—Mudstone matrix. Relic skeletal grains are fragments of pelecypod shells. (Plane polarized light, 30×)
D.—Very fine crystalline. Collophane composes black skeletal grains. Clear colorless linear areas are relic skeletal grains, and gray ovoid grains are relic pellets. (Plane polarized light, 30×)

than the pellets (Fig. 10). Crystals of dolomite cement increase in size from the walls toward the centers of the interstices between pellets.

Some dolomites contain mudstone matrix, ranging in amount from 10 to 70 per cent, which is usually difficult to recognize because the pellets merge into the matrix.

The average size of pellets differs from bed to bed; the smallest are very fine and the largest are medium. Ovoid shapes are most common with the ratio of length to width consistently approximating 3 to 2. Sorting varies from good to poor; those

 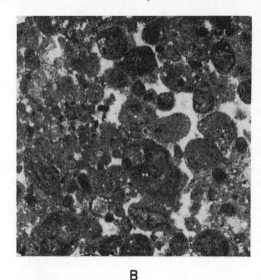

Fig. 10.—Variations in pelletal dolomite.
A.—Pelletal dolomite. Dark-colored ovoid masses are relic pellets. (Plane polarized light, 30×)
B.—Pelletal dolomite. Colorless dolomite between relic pellets increases in crystal size from walls toward centers of interpelletal spaces. Some larger grains are lumps. (Plane polarized light, 30×)

pelletal dolomites containing the most lumps show the poorest sorting.

Carbonaceous material, clay, quartz silt, pyrite, glauconite, collophane, chert, and anhydrite are minor constituents of pelletal dolomite.

Among primary sedimentary structure in pelletal dolomites are wavy lensing laminae, usually less than 4 millimeters thick, which result from slightly different average pellet sizes in different laminae. Discontinuous irregular lenses of dolomite mudstone, usually less than 4 millimeters thick, are common; these lenses are composed of even wavy laminae suggestive of algae. Lineations are expressed in most samples of this dolomite by parallel orientation of the long dimension of constituent grains. Burrows, minute scour structures, and stylolites are also found in this type of dolomite.

In specimens larger than thin sections, pelletal dolomite containing mudstone matrix commonly shows a particular texture termed "birdseye" in limestones. This texture is characterized by small, sharply defined augen which are usually irregularly lenticular. Augen appear between grains and may be either voids or filled with clear dolomite; adjoining augen are commonly connected by a film of clear dolomite, and their long axes tend to be oriented parallel with the bedding. This "birdseye" texture in limestones has been classified as "dismicrite" by Folk (1959, p. 28), "dismicrite" or "dispellet" by Wolf (1960, p. 1415–1416), "pelleted lime mud" by Illing (1959, p. 15), and "merged" calcarenites or bahamites by Beales (1958, p. 1872–1873). Both Pedry (1957) and Boyd (1958) illustrate Phosphoria reservoir rocks having "birdseye" texture.

Oölitic dolomite.—Oölitic dolomite contains 30 per cent or more relic oöliths as the dominant grain type (Fig. 11). This type is apparently restricted to local areas along the eastern margin of the study area; it commonly grades laterally into pelletal dolomite, but may also grade to dolomite mudstone and claystone.

Relic oöliths tend toward spherical shapes, are predominantly of fine and very fine sizes, and appear well sorted in any one laminae (Fig. 11). They are recrystallized to dolomite, and films of carbonaceous material show their concentric structure. Nuclei are commonly dolomitized pellets having diameters half that of the oöliths. Both the nucleus and concentric rings are composed of fine silt- or clay-size dolomite identical with that previously described for pellets.

The material between oöliths may be dolomite mudstone, clear dolomite, or anhydrite cement. Dolomite mudstone is the same as previously

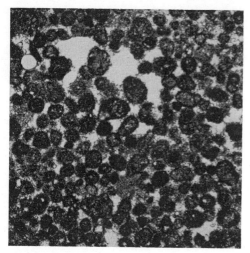

FIG. 11.—Oölitic dolomite. Relic oöliths are mostly of superficial type, and clear areas are pores. (Plane polarized light, 30×).

described and may be present in amounts ranging from 30 to 70 per cent. Clear dolomite increases in grain size toward the centers of the interstices, similarly to that in pelletal dolomites. Anhydrite may fill interstices, not previously filled by dolomite, or may replace clear dolomite or dolomite mudstone.

CLAYSTONE

A lithologic type consisting of approximately equal proportions of dolomite, clay minerals, and silt-size quartz is designated claystone, the name commonly applied by geologists working with the Phosphoria.

The number and thickness of claystone beds in the upper Phosphoria increase eastward until the upper member is all claystone. Claystone usually grades westward into dolomite mudstone, or anhydrite, but locally into pelletal dolomite or oölitic dolomite. In the lower member claystone commonly grades into sandstone.

Claystone is predominantly grayish red, but greenish gray claystone is usually a westward gradation, which in turn grades to greenish gray dolomite mudstone. In thin sections, ferric iron in hematite is seen to color the grayish red type, and ferrous iron in pyrite is present in the greenish gray.

Other minor constituents in claystone include magnetite, orthoclase, zircon, tourmaline, a bright green mineral having the optical properties of glauconite, chert, and anhydrite.

Grayish red claystone often contains greenish gray "reduction spots." The centers of some spots contain a carbonaceous film showing parallel striae suggestive of plant imprints. Others contain a pyrite cube or the pseudomorph of a pyrite cube in their centers.

Thin sections show curved lenses and irregular masses of argillaceous dolomite enclosed in dolomitic, argillaceous siltstone (Fig. 12a and b). The result is a "churned" structure similar to that in dolomite mudstones. Some definite burrows are recognizable, indicating that churning is due to reworking of sediments by burrowing organisms. Stratten (1959, Fig. 2) illustrates a similar structure from Recent tidal flat deposits of the Dutch Wadden Sea.

ANHYDRITE

Anhydrite beds are most abundant in the area of facies change, but they occur both eastward and westward. Beds are usually less than 10 feet thick but may exceed 20 feet locally. In outcrops anhydrite is partially altered to gypsum. All gradations between anhydrite and dolomite mudstone and anhydrite and claystone were observed.

Anhydrite forms massive or nodular beds and rarely shows laminations. Nodules range in size from 2 millimeters to greater than 5 centimeters. Red or green argillaceous dolomite outlines nodules and fills interstices (Fig. 13).

Minor constituents in the anhydrite also include pyrite, chert nodules, euhedral authigenic quartz crystals, and hematite in pale red varieties.

INTRAFORMATIONAL BRECCIAS

Breccias were observed only in outcrops of the upper Phosphoria along the eastern margin of the area; they fill channels 2–15 feet deep and spread out from the channel as thin beds, which wedge out into grayish red claystone or anhydrite. Usually the breccias weather to a "box-work" structure because the fragments leach out, leaving their molds in the more resistant matrix. Collapse structures are absent in rocks overlying the breccia.

Breccia consists of randomly oriented, predominantly tabular fragments of grayish red claystone in a somewhat more dolomitic matrix. The matrix is usually grayish red claystone, but may be very light gray or yellowish gray dolomite. Some fragments of dolomite mudstone, chert, and greenish gray claystone are also present. The

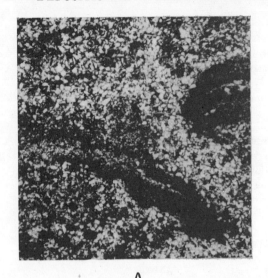 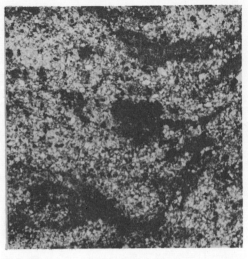

Fig. 12.—Variations in claystone.

A.—Grayish red, dolomitic, silty claystone. Angular clear grains are detrital quartz silt, and black areas are hematitic argillaceous dolomite. "Churned" structure is due to burrowing organisms. Form marked by concentration of quartz silt and transverse to laminations is probably burrow. (Plane polarized light, 30×)

B.—Greenish gray, dolomitic, silty claystone. Black cubic grains are pyrite, and angular clear grains are detrital quartz silt. "Churned" structure is due to burrowing organisms. (Plane polarized light, 30×)

tabular fragments are poorly sorted and are as large as 5 centimeters long and usually less than one centimeter thick. Their corners show little rounding.

SANDSTONE

Sandstone composes a major part of the lower member in the central, in the northwestern, and in the southwestern parts of the area of this study. As a rule, the lower member is predominantly sandstone where it is thicker than 75 feet; individual beds are rarely thicker than 10 feet. All gradations between sandstone, dolomite, and claystone were observed; and lenses of any one of these lithologic types may be found in any of the others.

Typical sandstone consists of 50 per cent very fine quartz grains, 25 per cent scattered larger grains, and 25 per cent interstitial material (Fig. 14); the texture is microconglomeratic. Chert, dolomite, feldspar, and collophae comprise some larger grains, but most are quartz. Pyrite is a common accessory mineral.

A mixture of dolomite and anhydrite (Fig. 14) bonds this sandstone along with minor amounts of clay minerals and hematite in the northern part of the area. Dolomite is usually dominant in the southern part of the area and anhydrite in the north. Most thin sections show anhydrite definitely replacing dolomite, and chert locally replaces these cementing materials.

Sedimentary structures present in Phosphoria sandstone are minute scour features, cross laminations, and distorted argillaceous laminae. Ripple

Fig. 13.—Nodular anhydrite. Nodules are outlined by black, irregular streaks of argillaceous dolomite. (Crossed nicols, 30×).

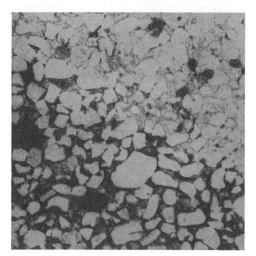

Fig. 14.—Sandstone. Dark material between clear quartz grains consists of silt-size dolomite crystals, and lighter material is anhydrite. (Plane polarized light, 30×)

marks were noted in a few surface outcrops.

Fine and medium quartz grains are more abundant in the lower member than in the typically very fine-grained Tensleep Sandstone and were probably concentrated during reworking of the Tensleep detritus.

CONGLOMERATES

Conglomerates, usually less than 2 feet thick, appear at different horizons throughout the basal 50 feet of the lower member. As many as four separate conglomerates are found in a single well, and they are erratic in distribution and vertical position.

Conglomerate fragments are predominantly dolomite, except at the base of the lower member where quartz and chert are dominant. Sizes range from granules to pebbles 10 centimeters maximum diameter. Fragments in the southern part of the area include dolomite mudstone, pelletal dolomite, and skeletal dolomite; in the northern part they are typically grayish red, purple, argillaceous, silty dolomite identical to that from the Tensleep in this area. Fragments of chert show colloform banding or are jasper that contains relic organic structures, apparently fusulinids.

The matrix of these conglomerates is usually sandstone, but a few contain dolomite mudstone that carries abundant subangular to rounded quartz sand and silt grains. The sandstone matrix is similar to the sandstone in the lower member, except it contains a larger amount of coarse sand

grains (Fig. 15). The larger rounded grains in the southwestern part of the area may be collophane fossil fragments, probably bone tissue, or different types of dolomite. Fossil fragments, apparently derived from brachiopods or pelecypods, are partly or wholly replaced by anhydrite in one sample (Fig. 15).

Conglomerates containing quartz and dolomite grains derived from the Tensleep are basal conglomerates, and those with Phosphoria type dolomite fragment in a dolomite matrix are intraformational and properly belong in the category of detrital dolomite.

DIAGENESIS

Secondary alterations indicate that Phosphoria rocks are products of diagenesis. This reducing and alkaline environment differed from the oxidizing and alkaline environment prevailing during deposition. Consequently, effects of diagenesis must be evaluated in order to reconstruct the depositional environments.

Diagenesis is defined as the changes, not caused by weathering, affecting a sediment between deposition and true metamorphism (Correns, 1950, p. 49).

Combinations of physical, biological, and chemical processes affected diagenesis of the sediments that formed Phosphoria rocks. The chemical processes of dolomitization and the redistribution of anhydrite most affected reservoir

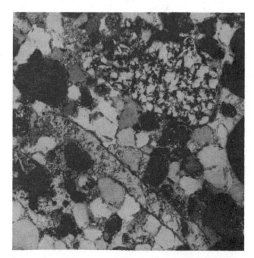

Fig. 15.—Matrix of conglomerate. Grains are quartz, dolomite, chert, and anhydrite. Anhydrite has replaced original shell material of elongate skeletal grain in center of photograph (Crossed nicols, 30×)

properties and are further discussed. Other chemical reactions including phosphatization, pyritization, silicification, and glauconitization took place under reducing and alkaline conditions. Still other chemical reactions between the sediments and fluids (as cementation and leaching) occurred but are difficult to assess because they have been obscured by dolomitization. Such processes probably were concurrent, except that silicification and redistribution of anhydrite post-date the others.

Dolomitization.—X-ray diffraction studies of 21 representative carbonate rocks from the Phosphoria in the study area show that the carbonate mineral is dolomite, except one sample from the western part which contains some calcite skeletal grains.

The Phosphoria dolomites are dolomitized limestones. The skeletal grains were derived from organisms that do not build their hard parts of dolomite (Vinogradov, 1953), and oöliths and pellets currently forming in marine waters are nowhere known to consist of dolomite. The calcite and aragonite originally forming these shells and grains have been replaced by dolomite. Skeletal, oölitic, and pelletal dolomite grade into dolomite mudstone throughout the area. Many geologists consider that fine grain size and association with anhydrite indicate primary dolomite. However, the writer believes that the grains and original calcium carbonate mud were converted to dolomite by the same process, approximately at the same time.

Dolomitization must have been nearly contemporaneous with deposition, because the dense nature of many Phosphoria rocks, especially those containing fine silt- and clay-size dolomite crystals, precludes free circulation of any magnesium-bearing waters after the initial stages of compaction. The complete and uniform dolomitization must have occurred while the sediments were still in contact with the overlying sea water.

Replacement of calcium carbonate by dolomite proceeds according to the reaction:

$$2CaCO_3 + Mg'' = CaMg(CO_3)_2 + Ca''$$

Equilibrium will be determined by the relative activities of the calcium and magnesium ions in solution. If concentrations of these two ions in sea water are used as criteria for their activities, dolomite should be stable with respect to calcite in sea water; calcite and aragonite on the sea floor should react with sea water to form dolomite (Garrels *et al.*, 1960). Studies of the thermodynamics of this reaction also imply that dolomite should be forming on the floor of our modern oceans (Garrels *et al.*, 1960; Kramer, 1959; also Halla and Ritter, 1935; Faust and Callaghan, 1948).

Only traces of dolomite have been found in Recent marine sediments, contrary to the predictions from the chemical reaction for dolomitization. Garrels *et al.* (1960) suggest two possibilities for this difference: (1) the concentration of the magnesium ion in sea water may differ vastly from its activity, or (2) the calculated value for the free energy of formation of dolomite may be in error, though Garrels *et al.* think their value is approximately correct. Physical chemistry of dolomitization is thus not known, but relative concentrations of calcium and magnesium in sea water are probably governing factors.

Redistribution of anhydrite.—Widespread replacement of all other rock types by anhydrite was observed througout the Phosphoria; it is associated with fractures or otherwise originally porous and permeable rocks. Original pores may be partially to completely plugged by anhydrite, probably precipitated from solutions that carried the replacement anhydrite. Anhydrite, dissolved from anhydrite beds, was distributed along paths of permeability. When running sets of well samples, the writer observed that the abundance of irregular masses of replacement anhydrite increased as an anhydrite bed is approached.

DEPOSITIONAL ENVIRONMENTS

A depositional environment is a dynamic complex of tectonic, physical, chemical, and biologic conditions that prevail at the depositional site of sedimentary particles. This environment is characterized by physico-chemical properties of the depositing fluid, and tectonic and physical conditions along with biologic processes that influenced this fluid. These factors are complexly interrelated.

Biologic factors were important in the depositional environment of the Phosphoria but are difficult to evaluate. The fragmental condition and diagenetic alteration of fossils rarely permits generic identification, and rounding of skeletal grains indicates deposition away from their place of growth. Some of the pellets in the pelletal dolomites are probably of fecal origin, which suggests one way organisms influenced sedimentation. Activity of burrowing organisms, shown by "churned" structures in dolomite mudstones and claystones, is another.

TECTONIC SETTING

The study area lies at the eastern margin of a shallow embayment on the stable shelf bordering the Permian miogeosyncline. This embayment occupied approximately the western half of Wyoming and northeastern Utah. Dolomite predominates in the eastern part of this embayment and grades westward into cherts, phosphorites, and mudstones in western Wyoming and northern Utah along the margin of the miogeosyncline (McKelvey et al., 1959). Phosphoria rocks are sandstone northwest of the study area on the seaward margin of the shelf. Dolomite intergrades and intertongues with anhydrite and redbeds northward, eastward, and southward; and the study area lies across this gradational belt.

ORIGIN OF PHOSPHORITES

Kazakov's (1937) hypothesis for the formation of phosphorites has been modified and applied to the Phosphoria Formation by McKelvey et al. (1953b, p. 54–62), who state:

"The Phosphoria formation accumulated in a large shelving embayment bordered by lands of low relief that contributed little detritus to the sea. Cold, phosphate-rich waters upwelled into this basin from the ocean reservoir to the south or southwest. Phosphorite was deposited from these ascending waters, probably in depths of 1000 to 200 meters, as their pH increased along with increase in temperature and decrease in partial pressure of CO_2. Carbonates were precipitated from these waters when they reached more shallow depths, at a somewhat higher pH. The phosphate-rich waters nurtured a luxuriant growth of phytoplankton, as well as higher forms of plant and animal life, some remains of which were concentrated with fine-grained materials in deeper waters away from shore. Part of the phosphate and probably some of the fine-grained silica in the formation were concentrated by these organisms. Finally, these conditions persisted over much of Permian time."

Krumbein and Garrels (1952, p. 8–9) state that the ions constituting phosphorites and carbonates are independent of Eh, but pH of solutons saturated with ions constituting these minerals is an important control. Phosphorites precipitate at pH values above 7.1 (Kazakov, 1950), and abundant calcium carbonate begins to form when the pH rises to 7.8 (Krumbein and Garrels, 1952, p. 25). Phosphate could precipitate along with calcium carbonate from solutions having pH values above 7.8, but the ratio is much in favor of calcium carbonate because the absolute solubility of calcium phosphate is so much less than for calcium carbonate and because the solubility-pH curves of these two materials are parallel (Krumbein and Garrels, 1952, p. 9). Phosphorites then will be the predominant precipitate in the pH range from 7.1 to 7.8; and calcium carbonate, when the pH is above 7.8.

PHYSICAL FACTORS

Climate.—Climatic conditions were probably semi-arid and subtropical or tropical. An abundance of evaporites indicates that evaporation exceeded the influx of fresh water, so the annual rainfall was probably low. Borderlands supported some flora, for scattered plant remains are found in the claystones along with rare unidentified spores. The red iron oxide coloring matter in claystones and dolomites has doubtful climatic significance because the red iron oxide may be a product of post-depositional changes of originally greenish gray sediments (Krynine, 1949). Interpretations of the Permian climatic zones based on paleontology (Newell et al., 1953, p. 184–185; Stelhi, 1957) and paleomagnetism (Opdyke and Runcorn, 1960, Fig. 4, p. 969) conflict, but both place the region including the study area in a subtropical or tropical zone. The presence of upwelling currents also implies an arid climate in the adjacent coastal area based on conditions where upwelling is occurring today (Brongersma-Sanders, 1948, p. 19).

Submarine topography.—Phosphoria deposition began in shallow water of the northward and eastward transgressing Permian sea that worked landward following the valleys on the Tensleep erosion surface. Between valley floors and uplands, relief was greater than 100 feet; this relief is indicated on Figure 5 by thickness lines, which also show the topography of the Tensleep erosion surface at the end of lower Phosphoria time.

By the end of lower Phosphoria time the valleys were filled, and the sea bottom was essentially a plane surface. The eastward extending dolomite tongue at the top of the lower member marks a time plane and a nearly level sea floor surface of deposition. Tongues of upper Phosphoria dolomite, uniform in thickness, extend many miles eastward into the claystone facies, and a slight change in sea-level apparently moved the strand line many miles landward or seaward. Shallow water extended many miles from shore during upper Phosphoria time. Recent counterparts of dolomitized limestones, known to form only in water less than 30 feet deep, also suggest deposition in shallow water.

No physical barriers that restricted shoreward circulation of ocean waters were found in the upper member. However, lateral restrictions to circulation along the north-south coast existed; they are indicated by eastward extending tongues of anhydrite that are bordered on their northern and southern sides by westward extending claystone tongues.

Terrestrial conditions.—In lower Phosphoria time the coastal belt probably had relief of a few hundred feet. The low hills and broad swales sloped southward and eastward toward the sea. Land persisted throughout lower Phosphoria deposition about 6 miles north of the study area, and the lower member lies progressively on older Tensleep rocks from south to north.

Sandstone fringes the topographic highs on the Tensleep erosion surface and was derived from the highs (Fig. 6). Clay and silt detritus must have been supplied largely from areas may miles north and east, and the predominance of claystone in the northeastern corner of the area suggests an estuary was present.

In upper Phosphoria time the coastal area was essentially an extension of the gently westward sloping sea floor. No well defined streams would be apt to flow over such a plane surface, and most land waters would probably reach the sea by seepage. Approximately equal thicknesses of the time-equivalent dolomite and redbed facies indicate that all sediments accumulated at the same slow rate. Data are lacking to evaluate the importance of transportation by wind.

Current patterns.—Physical barriers restricting circulation in the lower Phosphoria were the islands and headlands between the arms of the transgressing sea. Evaporites were deposited in some of these restricted bays. Elsewhere, lower Phosphoria rocks were probably deposited under conditions of open circulation of marine waters.

The dominant current pattern prevailing during upper Phosphoria time is inferred from the eastward gradation of dolomite to anhydrite and claystone. For evaporites to form, evaporation must have exceeded the combined influx of rainfall and terrestrial waters. This condition existed in the absence of physical barriers over a wide area of shallow water marginal to the shore. The evaporation required to precipitate calcium sulfate would lower the level of nearshore waters and cause a net shoreward flow of oceanic waters (Fig. 16). Water concentrated by evaporation would be denser than the inflowing sea water and would sink to the bottom. These denser waters would flow back toward the ocean over the bottom sediments because they usually were not concentrated to the point where salts more soluble than calcium sulfate were precipitated; to the writer's knowledge, only casts of salt crystals have been reported from Phosphoria equivalents east of the study area (Privrasky *et al.*, 1958). These bottom waters would be enriched in magnesium and might penecontemporaneously dolomitize the bottom sediments, if a high concentration of magnesium is a principal factor in the alteration of calcium carbonate to dolomite.

The dominant current pattern would be modified by wind and tide currents as well as friction between opposing currents and with the bottom.

PHYSICO-CHEMICAL FACTORS

Temperature.—The temperature of the water of the Phosphoria sea increased eastward across the study area. If the phosphorite deposits west of the area of study originated in association with upwelling oceanic currents, surface water temperatures may have been colder than 15°C. (Sverdrup *et al.*, 1942, p. 132). This cold water would be warmed gradually as it passed shoreward across the wide shallow shelf, and temperatures may have exceeded 30°C. along the shoreline.

Absence of appreciable growths of reef corals is puzzling in light of this inferred temperature range. Hermatypic corals grow in sea water warmer than 18°C. and are best developed where the mean annual water temperatures are between 23°C. and 25°C. (Wells, 1957, p. 609). On the other hand, sediment binding algae, indicative of shallow warm water, flourished in areas of pelletal dolomite formation. Possibly, where Permian waters were warm enough to support profuse growth of colonial corals, the salinity was too high.

Salinity.—A lateral salinity gradient is suggested by the lateral gradation of dolomite to anhydrite, assuming the composition and initial salinity of Phosphoria sea water was the same as for present-day sea water. This salinity increased from that of normal sea water where phosphorites were depositing west of the study area to the point where calcium sulfate precipitated along the eastern margin. Experiments indicate that sea water must be concentrated to a salinity of about 120 parts per thousand before anhydrite or gypsum precipitate (MacDonald, 1953; Posnjak, 1940). Gypsum has been observed depositing in lagoons or estuaries where the salinity is 200 parts per thousand or more (Bramkamp and Powers, 1955; Morris and Dickey, 1957). Where streams entered the Phosphoria sea, these high salinities would be decreased locally by the influx of fresh water.

Skeletal dolomites, common only in the western part of the area, grade eastward to sparsely fossiliferous dolomites, which supports the inferred salinity distributions. Marine faunas only flourished in areas some distance seaward from the shore of the Phosphoria sea.

A vertical salinity gradient, corresponding to the inferred current pattern, probably existed in the Phosphoria sea near the shore.

pH.—The pH of the ocean waters flowing eastward across the area studied probably increased. Phosphorites to the west precipitated from water having a pH less than 7.8; the calcite and aragonite, which altered to dolomite in the western and central parts of the area, were precipitated from waters having pH greater than 7.8. Though the precipitation of calcium sulfate is independent of pH, limited data suggest an increase in pH corresponding with increased salinities (Krumbein and Garrels, 1952); and the pH where calcium sulfate was precipitating may have been as high as 9.

Eh.—The Eh of Phosphoria sea water was probably positive throughout the study area. The shallow water was periodically agitated and oxygenated by wind-generated waves. In the western part of the area enough hydrogen sulfide may have evolved from decaying organisms to temporarily produce negative Eh values in the bottom water.

DEPOSITIONAL ENVIRONMENT OF SPECIFIC ROCK TYPES

Dolomites.—The depositional environment of the sediments which formed Phosphoria dolomites is similar to that found on the Bahama Banks today. The Phosphoria environment, however, differs from the Recent one in two respects: (1) appreciable land detritus was supplied to the Phosphoria sea but is not contributed to the Bahama Banks, and (2) Phosphoria dolomites are associated with anhydrite but no anhydrite is now depositing in the Recent environment.

Various workers (Illing, 1954, p. 17–45; Newell *et al.*, 1959) have described Recent counterparts of the dolomite mudstone and skeletal, pelletal, and oölitic dolomite from the Bahama Banks. Similar Recent carbonate sediments are found in the Gulf of Batabano, Cuba, where detrital carbonate sediments are present in addition to the other types (Daetwyler and Kidwell, 1959).

Physical barriers to water circulation and influx of large amounts of terrestrial debris restricted carbonate deposition in lower Phosphoria time, but other conditions were favorable for carbonate accumulation.

Stratigraphic relations of upper Phosphoria rocks suggest deposition on a shelf covered by a few feet of water and uninterrupted by physical barriers. Sea water increased in temperature and salinity as it flowed landward across this shelf.

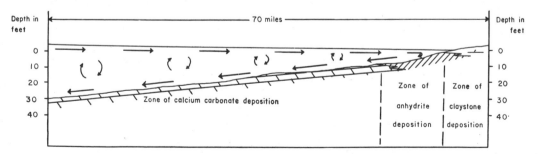

Fig. 16.—Cross section showing inferred current pattern perpendicular to shoreline of Phosphoria sea. Arrows indicate current directions. Due to evaporation of nearshore water, net flow of surface water is shoreward. Calcium carbonate and calcium sulfate would precipitate in zones indicated. Seaward-flowing bottom waters would be enriched in magnesium and might penecontemporaneously dolomitize calcium carbonate sediments.

These conditions characterize the Bahama Banks today (Smith, 1940), but they differ from the Gulf of Batabano where seaward physical barriers prevent direct shoreward flow of ocean waters (Daetwyler and Kidwell, 1959).

The general pattern of sediment types found on the Bahama Banks (Newell et al., 1959) appears in upper Phosphoria dolomites. On the margins of these Banks skeletal sands predominate, and they grade into non-skeletal sands which in turn grade into carbonate mud along the leeward side of Andros Island. Within the non-skeletal sands, oöliths lie seaward and grade landward into pellets or lumps, or lumps appear seaward from oöliths, or oöliths are absent and lumps are the only non-skeletal grain type. Dolomite mudstone is the most abundant dolomite type in the Phosphoria; however, relative amounts of associated types suggest a general shoreward change like that found on the Bahama Banks. Thus, skeletal dolomite is found in the western part of the study area, pelletal dolomite in the center, and dolomite mudstone in the east. In contrast to the Recent environment, oölitic dolomite in the Phosphoria is found along the shoreline.

The sedimentary pattern on the Bahama Banks is a reflection of submarine topography, and currents flowing over the Bank control both sedimentary facies and configuration of the bottom (Illing, 1954, p. 81–82). These currents are predominantly tidal and decrease rapidly in strength in shallow water toward the center of the Banks, where wind-generated currents, especially during storms, produce the only currents effective in reworking sedimentary particles.

Differences in current intensities probably controlled the distribution of calcareous sediments which formed Phosphoria dolomites. Very fine crystalline skeletal dolomite accumulated where currents were strong, and pelletal dolomite formed closer to shore where currents were weaker. Skeletal dolomite with a mudstone matrix deposited where currents were too weak to winnow mud from skeletal grains. Intermittent periods of agitation followed by periods of calm probably prevailed where dolomite mudstone was deposited, and mudstone having a "churned" structure probably accumulated on tidal flats. Partly consolidated sediments along the shoreline were broken up during storms and moved offshore to form detrital dolomites. Oölitic dolomite accumulated in agitated water along the shore.

Claystone.—Claystone in the upper member and much of that in the lower member was probably deposited on tidal flats. Clay minerals and detrital quartz silt predominate over dolomite, suggesting proximity to land and the mixing of land- and marine-derived components. The "churned" sedimentary structure, characteristic of many claystones, is essentially the same as found in Recent tidal flat deposits (Stratten, 1959, Fig. 2). Intertonguing and intergrading of dolomite and redbeds suggest deposition of the claystone along the shoreline of a shallow sea adjacent to a plane borderland rising a few feet above sea-level. Sandy claystone associated with sandstone probably accumulated in a nearshore or deltaic environment.

The abundance of burrows of organisms in the grayish red claystone suggests that the original sediments were greenish gray. Original red sediments, which must be lithified under oxidizing conditions to maintain their red color, probably would not contain enough organic matter to support a thriving population of burrowing organisms. Stratigraphic evidence suggests that the shoreward part of seaward advancing tongues of greenish gray claystone were subjected to desiccation and strong oxidizing conditions; thus originally greenish gray sediments would become grayish red by oxidation of the contained iron after deposition.

Anhydrite.—Anhydrite was deposited in bays along the eastern margin of the Phosphoria sea. Physical barriers probably isolated these bays during lower Phosphoria time. However, restrictions to shoreward movement of water in the upper Phosphoria were dynamic rather than static; and sea water, warmed and concentrated in its long passage shoreward across the shallow shelf, was further concentrated in bays to the point where calcium sulfate precipitated (Fig. 16). Scruton (1953) has suggested that these conditions might lead to the deposition of evaporites.

Intraformational breccia.—Intraformational breccia was probably deposited by torrential rains that swept across tidal flats, more or less following channels. The tabular shape of fragments suggests origin from mud-cracked beds, and the composition, texture, and stratigraphic relations suggest debris deposited by flood waters.

Sandstone.—Sandstone in the lower Phosphoria was deposited as nearshore sands and deltaic deposits. Where sandstone fringes topographic highs on the Tensleep erosion surface, the sand grains were eroded from the highs and distributed by marine currents in nearshore areas. Sandstone mixed with claystone in the northern and eastern parts of the area is probably deltaic; the sand grains of local origin were mixed with silt and clay, which had been transported great distances by streams.

Conglomerates.—Both intraformational and basal types of conglomerates contain marine fossils and were deposited in shallow water. Intraformational conglomerate formed after disruption of previously formed dolomite and deposition of the derived fragments in a mudstone matrix offshore. Nearby land contributed fragments and matrix in basal conglomerates. Some basal conglomerates may be residual gravels; others are associated with probable deltaic sediments.

Oil Occurrence

Oil is produced generally from Phosphoria dolomites within the top 100 feet of the upper member, and insignificant amounts of oil come from sandstone in the lower member.

Phosphoria oil accumulations are controlled by various conditions of deposition and diagenesis, and by structure which is not described in this paper.

The facies change from dolomite to anhydrite and claystone forms the principal eastward limit of Phosphoria accumulations with a lesser control where porous dolomite changes to impervious dolomite mudstone. Limiting proportions of clastics and evaporites are mapped in Figure 17. Phosphoria oil production is west of the evaporite ratio line of $\frac{1}{2}$, and prolific production is west of both the evaporite ratio line of $\frac{1}{4}$ and the clastic ratio line of $\frac{1}{2}$.

Reservoir rocks are generally types of dolomite consisting of relic grains. Very fine crystalline skeletal dolomite predominates in the western part of the study area, and pelletal dolomite in the central part. Dolomite mudstone usually has too low permeability for commercial production. Oölitic dolomite could provide reservoirs for oil, but was not the reservoir rock in any of the wells logged. Fractures increase permeability, but they do not reservoir commercial amounts of oil where other types of porosity are absent. Further division of the types of dolomite using one of the recently proposed limestone classifications (Folk, 1959; Wolf, 1960; Leighton, 1962) would be advantageous to further define and locate reservoirs in Phosphoria rocks.

Diagenesis has modified porosity in Phosphoria reservoir rocks. Original calcite and aragonite sediments had to be porous to permit passage of the solutions that uniformly dolomitized them. Dolomitization apparently destroyed porosity in muds converted to dolomite mudstone, but not in sediments composed mostly of grains. Subsequent to dolomitization, plugging of intergranular spaces by anhydrite occurred locally, particularly in areas marginal to maximum developments of anhydrite in the east-central part of Figures 4 and 17.

Phosphoria rocks in the Bighorn Basin are both sources and reservoirs for petroleum. Neither the overlying Dinwoody nor the underlying Tensleep Sandstone accumulated in environments favoring growth of abundant organisms or preservation of organic matter. However, organisms flourished in the nutrient-rich waters of the Phosphoria sea, and alkaline and reducing conditions during diagenesis favored both preservation of organic matter and conversion to petroleum.

Conclusions

Depositional environments have been interpreted from stratigraphic, petrologic, and sedimentational data and without detailed faunal information. Fossils are poorly preserved after dolomitization and are mostly fragments not in growth position; furthermore, Permian paleoecology is poorly understood. The other types of data, however, provide the same information that paleoecology might furnish.

Oil accumulations in Phosphoria rocks in the southeastern Bighorn Basin are restricted in the facies pattern, due to variations in depositional environments. Types of dolomite consisting predominantly of relic grains are the common reservoir rocks. Positions of these types is predictable from knowledge of depositional environments, and predictions could be refined in exploration work by applying recently proposed limestone classifications to Phosphoria dolomites.

The diagenetic environment is distinguished from the depositional environment, not only to determine nature of original sediments but also to evaluate effects on potential reservoir rocks. The most important diagenetic change in Phosphoria rocks is dolomitization, probably accomplished by highly saline waters returning seaward over cal-

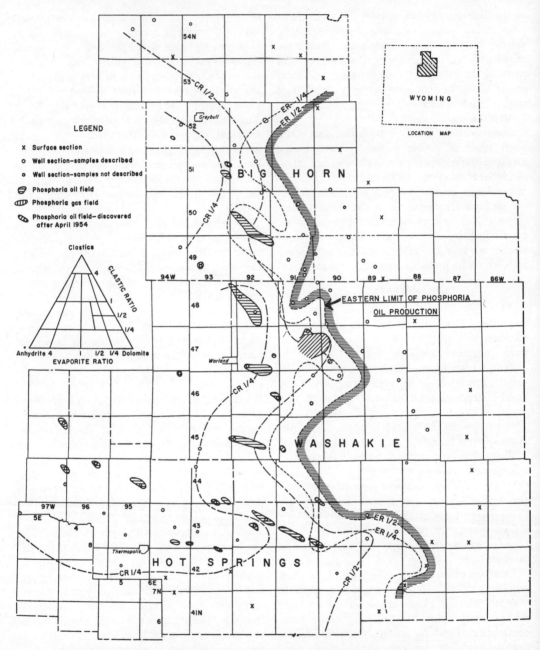

Fig. 17.—Relationship of oil occurrences and lithofacies in upper member of Phosphoria Formation, southeastern Bighorn Basin, Wyoming.

cium carbonate bottom sediments. Dolomitization commonly destroyed original porosity, but in many places only modified it. Anhydrite was redistributed and locally plugged porosity in subsequent diagenesis.

Distinction of depositional environments indicates probable areas for oil accumulation; but diagenetic changes determine the ultimate porosity distribution within the depositional pattern.

BIBLIOGRAPHY

Bathurst, R. G. C., 1958, Diagenetic fabrics in some

British Dinantian limestones: Liverpool and Manchester Geol. Jour., v. 2, pt. 1, p. 11–36.

Beales, F. W., 1958, Ancient sediments of Bahaman type: Am. Assoc. Petroleum Geologists Bull., v. 42, no. 8.

Boyd, D. W., 1958, Observations on the Phosphoria reservoir rock, Cottonwood field, Washakie County, Wyoming: 1958 Geol. Record, Rocky Mtn. Sec., Am. Assoc. Petroleum Geologists, pub. by Petroleum Information, Denver.

Bramkamp, R. A., and Powers, R. W., 1955, Two Persian Gulf lagoons (abs.): Jour. Sed. Petrology, v. 25, no. 2, p. 139–140.

Brongersma-Sanders, Margaretha, 1948, The importance of upwelling water to vertebrate paleontology and oil geology: Kon. Ned. Ak. Wet., Verh., Afd. Nat. (Tweede Sectie), Dl. XLV, no. 4, p. 1–112.

Burk, C. A., and Thomas, H. D., 1956, The Goose Egg Formation (Permo-Triassic) of eastern Wyoming: Geol. Survey Wyo., Rept. Inv. 6, 21 p.

Cheney, T. M., McKelvey, V. E., and Gere, W. C., 1956, Fusulinid-bearing rocks in Sublette Range, southern Idaho (Discussion with reply by Walter Youngquist): Am. Assoc. Petroleum Geologists Bull., v. 40, no. 7, p. 1716–1722.

Correns, C. W., 1950, Zur Geochemie der Diagenese. I. Das Verhalten von $CaCO_3$ und SiO_2: Geochim. Cosmochim. Acta, v. 1, no. 1, p. 49–54.

Daetwyler, C. C., and Kidwell, A. L., 1959, The Gulf of Batabano, a modern carbonate basin: 5th World Petroleum Cong. Proc., sec. 1, p. 1–21.

Faust, G. T., and Callaghan, Eugene, 1948, Mineralogy and petrology of the Currant Creek magnesite deposits and associated rocks of Nevada: Geol. Soc. America Bull., v. 59, no. 1, p. 11–74.

Folk, R. L., 1959, Practical petrographic classification of limestones: Am. Assoc. Petroleum Geologists Bull., v. 43, no. 1, p. 1–38.

Garrels, R. M., Thompson, M. E., and Siever, R., 1960, Stability of some carbonates at 25° C. and one atmosphere total pressure: Am. Jour. Sci., v. 258, no. 6, p. 402–418.

Halla, F., and Ritter, F., 1935, Eine Methode zur Bestimmung der Anderung freien Energie bei Reaktionen des Typus A(s)+B(s)=AB(s) und ihre Anwendung auf das Dolomit-Problem: Zeitschr. Physicalische Chemie, v. 175, p. 63–82.

Henbest, L. G., 1954, Pennsylvanian Foraminifera in Amsden Formation and Tensleep Sandstone, Montana and Wyoming: Billings Geol. Soc. Guidebook, 5th Ann. Field Conf., p. 50–53.

Illing, L. V., 1959, Deposition and diagenesis of some Upper Palaeozoic carbonate sediments in western Canada: 5th World Petroleum Cong. Proc., sec. 1, p. 23–52.

Illing, L. V., 1954, Bahaman calcareous sands: Am. Assoc. Petroleum Geologists Bull., v. 38, no. 1, p. 1–95.

Kazakov, A. V., 1950, Ftorapatitovaya Sistema Ravnovesii v. Usloviyakh Obrazovaniya Osadochnykh Porod (The Fluorapatite equilibrium system in the conditions of formation of sedimentary rocks): Akad. Nauk SSSR, Trudy Inst. Geol. Nauk, Geol. ser. Vyp. 114, no. 40, p. 1–21.

Kazakov, A. V., 1937, The phosphorite facies and the genesis of phosphorites, in Geological investigations of agricultural ores: Trans. Sci. Inst. Fertilizers and Insecto-Fungicides, no. 142 (pub. for 17th sess. Internat. Geol. Cong.), Leningrad, p. 95–113.

King, R. H., 1947, Phosphate deposits near Lander, Wyoming: Geol. Survey Wyo. Bull. 39, 84 p.

Kramer, J. R., 1959, Correction of some earlier data on calcite and dolomite in sea water (note): Jour. Sed. Petrology, v. 29, no. 3, p. 465–467.

Krumbein, W. C., and Garrels, R. M., 1952, Origin and classification of chemical sediments in terms of pH and oxidation-reduction potentials: Jour. Geol., v. 60, no. 1, p. 1–33.

Krynine, P. D., 1949, The origin of red beds: N. Y. Acad. Sci. Trans., v. 2, no. 3, p. 60–68.

Kummel, Bernhard, 1950, Triassic stratigraphy of the area around the Green River Basin, Wyoming: Wyo. Geol. Assoc. Guidebook, 5th Ann. Field Conf., SW. Wyo., p. 28–36.

Leighton, M. W., and Pendexter, C., 1962, Carbonate rock types, in Classification of carbonate rods: Am. Assoc. Petroleum Geologists.

Love, J. D., 1948, Mesozoic stratigraphy of the Wind River Basin, central Wyoming: Wyo. Geol. Assoc., Guidebook, 3d Ann. Field Conf., Wind River Basin, p. 96–111.

MacDonald, G. J. F., 1953, Anhydrite-gypsum equilibrium relations: Am. Jour. Sci., v. 251, no. 12, p. 884–898.

McKelvey, V. E., and others, 1959, The Phosphoria, Park City, and Shedhorn Formations in the western Phosphate field: U. S. Geol. Survey Prof. Paper 313-A, p. 1–47.

McKelvey, V. E., Swanson, R. W., and Sheldon, R. P., 1953a, Phosphoria Formation in southeastern Idaho and western Wyoming: Intermtn. Assoc. Petroleum Geologists, Guide to the Geology of Northern Utah and Southeastern Idaho, Fourth Annual Field Conference, p. 41–47.

——, ——, ——, 1953b, The Permian phosphorite deposits of western United States: Comptes Rendus 19th Internat. Geol. Cong., sec. XI, p. 45–64.

McKelvey, V. E., Williams, J. Steele, Sheldon, R. P., Cressman, E. R., Cheney, T. M., and Swanson, R. W., 1956, Summary description of Phosphoria, Park City, and Shedhorn Formations in western Phosphate field: Am. Assoc. Petroleum Geologists Bull., v. 40, no. 12, p. 2826–2863.

Morris, R. C., and Dickey, P. A., 1957, Modern evaporite deposition in Peru: Am. Assoc. Petroleum Geologists Bull., v. 41, no. 11, p. 2467–2474.

Newell, N. D., Imbrie, John, Purdy, E. G., and Thurber, D. L., 1959, Organism communities and bottom facies, Great Bahama Bank: Am. Mus. Nat. Hist., v. 117, art. 4, p. 181–228.

——, and Kummel, Bernhard, 1942, Lower Eo-Triassic stratigraphy, western Wyoming and southeast Idaho: Geol. Soc. America Bull., v. 53, no. 6, p. 937–995.

——, Rigby, J. K., Fischer, A. G., Whiteman, A. J., Hickox, J. E., and Bradley, J. S., 1953, The Permian reef complex of the Guadalupe Mountains region, Texas and New Mexico: 236 p., San Francisco, W. H. Freeman & Co.

Opdyke, N. D., and Runcorn, S. K., 1960, Wind directions in the western United States in the Late Paleozoic: Geol. Soc. America Bull., v. 71, no. 7, p. 959–972.

Pedry, J. J., 1957, Cottonwood Creek field, Washakie County, Wyoming, carbonate stratigraphic trap: Am. Assoc. Petroleum Geologists Bull., v. 41, no. 5, p. 823–838.

Posnjak, E., 1940, Deposition of calcium sulfate from sea water: Am. Jour. Sci., v. 238, no. 8, p. 559–568.

Privrasky, N. C., Strecker, J. R., Grieshaber, C. E., and Bryne, Frank, 1958, Preliminary report on the

Goose Egg and Chugwater Formations in the Powder River Basin, Wyoming: Wyo. Geol. Assoc. Guidebook, 13th Ann. Field Conf., Powder River Basin, p. 48–55.

Scruton, P. C., 1953, Deposition of evaporites: Am. Assoc. Petroleum Geologists Bull., v. 37, no. 11, p. 2498–2512.

Smith, C. L., 1940, The Great Bahama Bank: Jour. Marine Research, v. 3, no. 2, p. 147–189.

Stehli, F. G., 1957, Possible Permian climatic zonation and its implications: Am. Jour. Sci., v. 255, p. 607–618.

Stratten, L. M. J. U. Van, 1959, Minor structures of some recent littoral and neritic sediments: Geologie en Mijnbouw (N.W. Ser.), v. 21, no. 7, p. 197–216.

Sverdrup, H. U., Johnson, M. W., and Fleming, R. H., 1942, The Oceans: 1087 p., New York, Prentice-Hall, Inc.

Verville, C. J., 1957, Wolfcamp fusulinids from the Tensleep Sandstone in the Big Horn Mountains, Wyoming: Jour. Paleontology, v. 31, no. 2, p. 349–353.

Vinogradov, A. P., 1953, The elementary chemical composition of marine organisms: Sears Foundation for Marine Research, Mem. no. II, 647 p. (Translated from original Russian by Julia Efron and Jane E. Setlow.)

Wells, J. W., 1957, Coral reefs: Geol. Soc. America Mem. 67, v. 1, p. 609–631.

Wolf, K. H., 1960, Simplified limestone classification (geol. note): Am. Assoc. Petroleum Geologists Bull., v. 44, no. 8, p. 1414–1416.

PRE-TRENTON SEDIMENTATION AND DOLOMITIZATION, CINCINNATI ARCH PROVINCE: THEORETICAL CONSIDERATIONS[1]

WARREN L. CALVERT[2]
Columbus, Ohio

ABSTRACT

The pre-Trenton sedimentary history of the Cincinnati Arch geologic province involves the paleogeology and paleogeography of eastern North America from earliest geologic time to Late Ordovician. The early sedimentary record indicates that the North American continent was built up through geologic time by the addition of successive, broad, sedimentary continental shelves around an original shield area composed of igneous and metamorphic rocks. The shield area furnished the original clastic sediments for the shelf areas throughout many erosional cycles. Lack of terrestrial vegetation had a profound effect upon pre-Chazyan sedimentation; eolian activity was extremely important on a broad scale, and erosion, transportation, and deposition of clastic sediments were much more rapid than in subsequent geologic time in eastern United States.

A theory for the origin of "original" dolomite by the diagenetic decomposition of authigenic biotite mica accounts for production of dolomites of widespread formational extent. This theory provides a source of magnesium for the conversion of marine limestones to dolomites, and accounts for the presence of glauconite in sub-Trenton rocks. It explains formation of chert, kaolinite, and siliceous cement by the diagenetic decomposition of authigenic feldspar. "Redeposited" or precipitated dolomite is thought to have resulted from solution of "original" dolomite and subsequent precipitation. "Replacement" or secondary dolomite is attributed to the circulation of magnesium-rich waters through fractures and porous zones in pre-existing limestones.

The transgressive nature of Cambrian deposition involved the movement of three distinct zones of sedimentation northwestward across the Cincinnati Arch geologic province. This resulted in the deposition of three distinct but transitional and time-transgressive bodies of sediment prior to the deposition of the widespread Knox Dolomite Supergroup. The unconformity at the top of the Knox Dolomite Supergroup represents an important time break between two distinctly different periods of sedimentation and two distinctly different lithologic types and fauna.

The interval between the end of Grenville deposition (approximately 1,000 million years ago) and the beginning of Cambrian deposition (about 550 million years ago), represented by the Ocoee, most of the Chilhowee, the Grand Canyon, the Belt, and the Windermere Groups of rocks, and by the extensive period of erosion preceding Cambrian time, is equivalent to a geologic era. The name Lipozoic Era is proposed for this part of geologic time.

INTRODUCTION

This study is the result of the analysis and correlation of the sub-Trenton rocks of the Cincinnati Arch geologic province (Calvert, 1962a) by geophysical logs and detailed sample descriptions from more than 100 wells located in Ohio, Kentucky, Indiana, Pennsylvania, and New York. Two detailed cross sections have been published (Calvert, 1962b, 1963); their location is shown in Figure 1. These cross sections were first publicly displayed in connection with a talk given before the Indiana-Kentucky Geological Society, May 22, 1962, at Evansville, Indiana.

To describe and measure rock sections and to place them in cross sections is no doubt a contribution to any geological problem. A geologist, however, should go further; he should discuss the implications of such a contribution. It is important to know the character of the original material; the conditions under which the material accumulated; the source and direction from which the material came; the changes which occurred since deposition; and, in the case of sedimentary rocks, the agency by which the material was transported. Briefly, no report on the geology of an area is complete without a discussion of the geologic history of the area as revealed by an interpretation of the rocks.

The initial considerations in this study of pre-Trenton sedimentation in the Cincinnati Arch geologic province are to orient the province paleogeographically, paleogeologically, paleoclimatically, and paleontologically, starting as far back as possible. By a study of the rocks, a sequence of events and environments are reconstructed which are here referred to as pre-Trenton sedimentary history. The study is confined prin-

---

[1] Manuscript received, June 18, 1963.

[2] Head, Subsurface Geology Section, Ohio Division of Geological Survey. The helpful suggestions of Daniel A. Busch and Donald L. Norling, consulting geologists, and of George H. Davis, III, and Donald A. Zimmerman, Sun Oil Company, are gratefully acknowledged. Harold J. Flint drafted the illustrations.

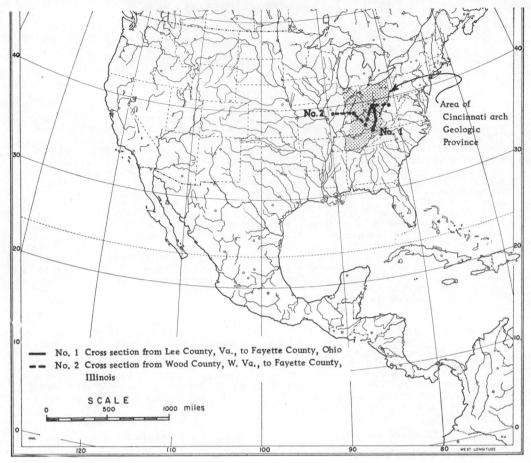

FIG. 1.—Location of Cincinnati Arch geologic province and relevant published cross sections.

cipally to the eastern half of North America, in which the Cincinnati Arch geologic province is located.

The Cincinnati Arch geologic province has been defined only vaguely. Darsie Green (1957, p. 627) states: "The Cincinnati Arch geologic province includes all the structure that separates the Appalachian, Michigan, and Illinois basins." This definition includes only the northern part of a structurally positive area, which extends from northern Alabama to northwestern Indiana (shown on the "Tectonic Map of the United States", 1944, published by the American Association of Petroleum Geologists). Most geologists think of the Cincinnati Arch geologic province as including the region surrounding the entire positive area which extends from Alabama to Indiana, and include the Findlay arch in northwestern Ohio. It is recommended that Green's definition be expanded to read "The Cincinnati Arch geologic province includes the entire positive structural area which separates the Mississippi embayment and Illinois basin on the west from the Michigan basin on the north and the Appalachian basin on the southeast." By this definition, the Cincinnati Arch geologic province includes the Nashville dome, the Cincinnati arch proper, the Ohio-Indiana platform, the Kankakee arch, and the Findlay arch. It seems desirable to put definite boundaries to such an area, even if they must be somewhat arbitrary. In the present report, the Cincinnati Arch geologic province is defined as that area covered by Ohio (except Williams, Fulton, Defiance, and Henry Counties in the northwest), Indiana (except Steuben, LaGrange, Noble, and DeKalb Counties in the

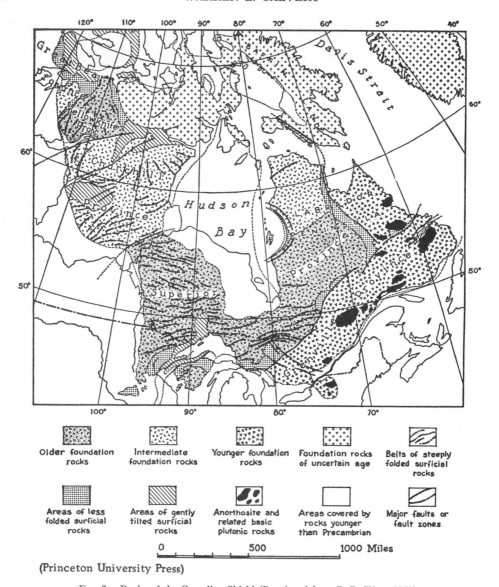

Fig. 2.—Rocks of the Canadian Shield (Reprinted from P. B. King, 1959).

northeast, and Posey, Vanderburgh, Warrick, Gibson, and Knox Counties in the southwest), Kentucky (except that part west of a line from Owensboro to Elkton), and those parts of Tennessee and Alabama enclosed by the Tennessee River (Fig. 1).

PRECAMBRIAN HISTORY OF EASTERN NORTH AMERICA

CANADIAN SHIELD

The main part of the Canadian Shield is an area of igneous and metamorphic rocks exposed north of the Great Lakes, extending northeastward to Newfoundland and northwestward to Great Bear Lake; it is a rock complex mostly of granites of deep-seated origin and gneisses and schists which have been intruded by these granites (Fig. 2). The rocks in large part are highly metamorphosed and complexly folded; parts of the complex are ancient lava flows. The oldest rocks are in the Superior Province (King, 1959), known by radiometric dating to be more than 2,000 million years old. On the north, rocks of the Churchill Province have been dated radiometrically at about 1,900 million years of age. Rocks of the Yellowknife and Great Bear Provinces, farther northwest, are probably somewhat younger. These four geologic provinces form the central Canadian Shield.

The original shield area of North America extended at least as far south as Texas; this is shown by well cuttings of basement rocks at relatively shallow depths beneath a thin cover of sedimentary rocks. The original North American shield might be called the "embryonic" North American continent (Fig. 3). Basement rocks similar to those of the central Canadian Shield are known in the subsurface of the Cincinnati Arch geologic province as far south as Lincoln County, Kentucky (Calvert, 1963, Pl. 1). The most ancient rocks of the shield have been traditionally called Archeozoic or Archean rocks; it is assumed that they rest on crystalline rocks of the original crust of the earth, known as Azoic rocks. Archeozoic rocks have been referred to recently by most geologists as "lower Precambrian." Superimposed on these rocks, in some areas, are younger sediments (Huronian) which have been dated radiometrically as about 1,400 million years old, classed by some geologists as "middle Precambrian." These were possibly the first shelf deposits of the "embryonic" continent. Still younger rocks (Keweenawan), dated as about 1,100 million years old, are classed as "upper Precambrian." These rocks were also possibly shelf deposits. The Huronian and Keweenawan rocks traditionally have been known as Proterozoic rocks; Proterozoic is preferred in this report. It is beyond the scope of this study to attempt to detail Azoic, Archeozoic, and Proterozoic geologic history. The "embryonic" North American continent is used as the starting point for the pre-Trenton geologic history of the Cincinnati Arch geologic province.

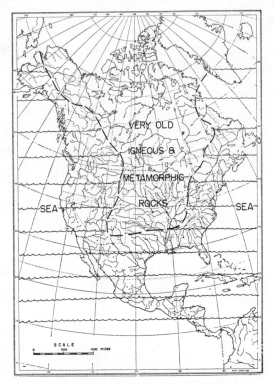

Fig. 3.—"Embryonic" (pre-Grenville) North American continent.

GRENVILLE HISTORY

Adjacent to the central Canadian Shield on the east is a wide belt of highly folded, faulted, intruded, and metamorphosed sediments of late Proterozoic age, containing carbonates with intrusions known to be (by radiometric dating) about 900–1,000 million years old. Much of this metamorphosed sedimentary section was originally limestone and dolomite, now altered to marble and calc-silicate hornfels. The associated gneisses, schists, and quartzites were originally shales, siltstones, and sandstones. These rocks are known as the Grenville Group (Fig. 2), commonly considered to be part of the Canadian Shield. Along the present outcrop, the Grenville province is as much as 250 miles wide, and the total thickness of the sediments is estimated to be more than 20,000 feet. It has been demonstrated by Bass and McCormick (McCormick, 1961, p. 56) that the Grenville Group underlies Ohio east of a line extending roughly from Cincinnati, Ohio, north-northeastward across western Ohio to Sandusky Bay, on Lake Erie. Grenville-type rocks occur in the subsurface as far south as Leslie County, Kentucky (Calvert, 1962b, p. 14). Inasmuch as the basement rocks west of the Grenville line are central shield-type rocks, either Grenville rocks were not deposited in the western part of the Cincinnati Arch geologic province or they were later removed by pre-Paleozoic erosion. It is not known to the writer whether or not Grenville-type rocks were deposited along the southern and western margins of the continental shield.

The Grenville rocks are evidently the first known carbonate shelf deposits to be formed along the eastern edge of the "embryonic" North American continent. If the Grenville limestone (at least 10,000 feet thick) is of biogenic origin,

Fig. 4.—"Infant" (post-Grenville) North American continent.

and was deposited in shallow, open marine waters on a continental shelf, as the writer believes, this would indicate a temperate climate, a continent of low relief, and the existence of an abundance of marine carbonate-secreting organisms during Grenville time. However, the intense deformation to which the Grenville rocks were later subjected has erased their original features and makes any interpretation of their origin and history highly speculative.

PRECAMBRIAN POST-GRENVILLE HISTORY

Subsequent to deposition and lithification, the Grenville rocks were highly deformed, intruded, and thrust toward the central Canadian Shield (King, 1959). The period of orogeny which followed Grenville deposition raised and extended the boundaries of the "embryonic" North American continent. This expanded continent might be called the "infant" North American continent (Fig. 4); it consisted of the central Canadian Shield proper, plus the Grenville belt. While the Grenville Mountains were being formed, the "infant" continent was being eroded severely and sediments were piling up in great clastic wedges around its borders to form a new and expanded continental shelf. The non-fossiliferous, commonly undisturbed, clastic sediments that underlie fossiliferous Cambrian rocks in many places were deposited during this interval of time. The rate of erosion and deposition was probably very rapid, compared with that of later sedimentary intervals, for reasons which are discussed on subsequent pages. Much of the sedimentary section deposited during this time in eastern North America was later highly folded, faulted, intruded, and metamorphosed to form the rocks of the crystalline Appalachians, where the earliest intrusions have been dated radiometrically as being about 440 million years old (Late Ordovician). The remainder of the post-Grenville clastic sediments of eastern North America are represented in the Great Smoky Mountains and the southern Appalachian Valley by the Ocoee Group and the lower part of the Chilhowee Group of rocks. That similar sedimentation was occurring in western North America is evidenced by the Grand Canyon, Belt, and Windermere Groups of rocks, which occur unconformably between the basement complex and Cambrian rocks. The post-Grenville clastic sediments apparently were deposited in an environment unfavorable to the existence of carbonate-secreting organisms.

By the time the continent was nearly base leveled, a new continental shelf had been formed, sloping eastward toward the sea. On the shallower part of the shelf, adjacent to the (by this time) ancestral Grenville Mountains, marine life probably continued to evolve, and the presence of carbonate-secreting organisms might have resulted in limestone deposition. Subsequent events were probably (1) the re-elevation of the continent, (2) the retreat of the sea to about the vicinity of the middle of the present Appalachian Valley, and (3) another long period of erosion, during which the post-Grenville carbonates were entirely eroded away, and the western post-Grenville clastics were removed and redeposited eastward. In this way, by successive elevation, erosion, and deposition, the continent was successively worn down and the continental shelf extended. The interval of time between the Proterozoic and the Cambrian was called by Walcott the Lipalian Interval (Schuchert, 1924, p. 179). East of the shore line, deposition was apparently continuous into the Paleozoic.

By the beginning of the Paleozoic Era, the North American continent was approaching maturity or the "adolescent" stage (Fig. 5). There is no evidence for the so-called borderland of "Appalachia" at the east. The crystalline Appalachian Mountains (Blue Ridge) did not appear until Late Ordovician or Early Silurian time, and it was then that the Appalachian geosyncline was initiated. The old Canadian Shield was ringed by a fringe of sediments which formed a long, seaward sloping continental shelf at the beginning of Cambrian time. Marine geophysical work (Drake et al., 1959) apparently does not support the concepts of an ancient borderland or a chain of volcanic islands (island arcs) off the eastern coast of North America during the Cambrian Period. These hypothetical landmasses were conceived mainly to (1) account for a nearby source of sediments, and (2) to support the idea of a geosyncline. There is an alternative explanation for sedimentation in eastern North America prior to the Blue Ridge and related orogeny. Before considering another theory, however, a recapitulation of Precambrian history appears warranted.

GEOLOGIC TIME SCALE

Thus far, the following events have been postulated: (1) the existence of the "embryonic" North American continent, (2) the deposition of thick Grenville shelf sediments along the continental border on the east and their lithification, (3) a long period of orogeny, which resulted in the intense folding, faulting, intrusion, metamorphism, and westward thrusting of the Grenville rocks, (4) a long period of erosion and peneplanation with concurrent deposition at the east, and (5) re-elevation, erosion, and re-deposition of sediments to form a broad continental shelf beyond the present Atlantic Coastal Plain. From this a geologic time scale for pre-Paleozoic time can be constructed, and the known time units since the beginning of the Cambrian Period may be added. The oldest rocks, the Archeozoic of the Canadian Shield, are known to be more than 2,000 million (2 billion) years old. The next group of rocks, the Proterozoic, including the Grenville Group, is known to be about 1,400–1,000 million years old. The beginning of the Paleozoic Era (Cambrian Period) is estimated to have occurred about 550 million years ago. The Mesozoic Era is thought to have begun about 225 million years ago, and the Cenozoic Era about 70 million years ago. It is

Fig. 5.—"Adolescent" (pre-Paleozoic) North American continent.

apparent that there is an unaccounted time gap of approximately 450 million years between the Proterozoic and Paleozoic Eras in the presently accepted time scale. It was during this time that the Ocoee, most of the Chilhowee, the Grand Canyon, the Belt, and the Windermere Groups of rocks were deposited (they are overlain by Cambrian rocks and underlain by the basement complex). This time interval represents a period of time longer than the entire Paleozoic Era; it is here proposed that it be called the Lipozoic Era (meaning missing life), because the clastic rocks which remain to represent this era were deposited in an environment unsuitable for the preservation of fossils.

The Grand Canyon, Belt, Windermere, Ocoee, and most of the Chilhowee Groups of rocks, here suggested as representing the Lipozoic Era, were deposited during the interval of time between about 550 and 1,000 million years ago. Recent investigations (Gulbrandsen et al., 1963) have dated glauconite from the Belt Group at about 1,070 million years of age, but this age is con-

TABLE I. PROPOSED REVISED GEOLOGIC TIME SCALE (MODIFIED FROM A. HOLMES, 1959)

ERA	PERIOD	TIME (Millions of years ago)		SPAN (Millions of years)
		Beginning	End	
Cenozoic		70	present	70
	Recent	0.03	present	0.03
	Pleistocene	1	0.03	0.97
	Tertiary	70	1	69
Mesozoic		225	70	155
	Cretaceous	135	70	65
	Jurassic	180	135	45
	Triassic	225	180	45
Paleozoic		550	225	325
	Permian	270	225	45
	Pennsylvanian	310	270	40
	Mississippian	350	310	40
	Devonian	400	350	50
	Silurian	440	400	40
	Ordovician	490	440	50
	Cambrian	550	490	60
Lipozoic		1000	550	450
Proterozoic		1500	1000	500
Archeozoic		2500	1500	1000
Azoic		4000	2500	1500

siderably older than previous ages given for these rocks. Illite studies (Goldich et al., 1959) gave ages ranging from 740 to 780 million years, and mica studies (Hunt, 1962) indicated ages of less than 844 million years for the Belt Group. All of the ages so far determined are within or close to the range here given for the Lipozoic Era. It will be of interest to note whether future radiometric determinations for the rocks here classed as Lipozoic fall within the postulated time interval.

Based on the preceding dates it is possible to construct a revised geologic time scale (Table I). The dates are only approximate; those for the Paleozoic and later are taken from Arthur Holmes (1959).

CAMBRIAN HISTORY OF EASTERN NORTH AMERICA

SOURCE OF CAMBRIAN SEDIMENTS

By the beginning of Cambrian time, it appears that the eastern part of the North American continent had been built eastward and southward from the Canadian Shield to beyond the present Atlantic coast, except possibly in the region of Florida (Fig. 5). Then a slow submergence of the eastern continental border began, marking the beginning of Cambrian time. The transgression of the sea was from the east and the source of the clastic sediments was at the west and north. The Cincinnati Arch geologic province was emergent and undergoing denudation. The surface of the continent probably was covered with a thick mantle of coarse, detrital material. The rivers were contributing considerable amounts of material at various points along the coast, and this material was being reworked by wave action and shore currents. The sedimentary material was not of the same character as that which would be deposited in a transgressing sea today. To understand the reason for this, the probable climate and vegetation of Cambrian time must be taken into account.

CAMBRIAN CLIMATE AND VEGETATION

In all probability, the climate of Cambrian time was not much different from that of the present. If it is assumed that the earth's axis during Cambrian time was inclined to the solar plane, climatic belts such as the present torrid, temperate, and frigid zones must have prevailed much as is the case today. The degree of difference in climatic extremes would have been determined by the degree of inclination of the earth's axis. Belts of prevailing westerly winds no doubt existed in the north temperate zone, wandering poles notwithstanding. High and low pressure areas, produced by temperature and humidity variations, must also have occurred. This would have caused cyclonic and anticyclonic circulation of the winds, just as at present, causing stormy periods and periods of clear, dry weather. Seasonal changes, such as those which occur now, also could be expected to have occurred in Cambrian time. Precipitation no doubt was dependent on latitude, ocean currents, mountain ranges, and other factors which affect precipitation today. The average rainfall of the earth was probably much the same in Cambrian time as now, although it probably was distributed differently.

The vegetation of Cambrian time is another story. If land plants existed, almost certainly they were of a very low order, without roots to bind a soil cover. Although marine plants are known in abundance from the Precambrian to the present, according to Seward (1931, p. 110), no traces of undoubted terrestrial plants of any kind have been found in rocks deposited prior to Late Silurian time. It is possible that some forms of marine algae may have migrated to a land habitat by Cambrian time, but there is no evidence to support this supposition. At any rate, it seems safe to assume that practically no land flora afforded cover to the North American continent in Cambrian time (Haught, 1956, p. 8). On the basis of recent work being done in several research laboratories (as yet unpublished), the earliest record of spores from possible land plants is in shales of Middle Ordovician age.

SOURCE AND TRANSPORTATION OF SEDIMENTS

Figure 5 shows that there were three classes of rocks furnishing sediments to the sea at the beginning of Cambrian time. One class was composed of the basaltic lavas, intrusive granites, and ancient metamorphics of the Central Shield; a second class consisted of the gneisses, schists, quartzites, marbles, anorthosites, and syenites of the Grenville belt; the third class of rocks was composed of the coarse clastic rocks of the Lipozoic belt nearest the coast.

It has been postulated that there probably were annual periods of plentiful rainfall during the Cambrian Period, so a considerable amount of material probably was transported to the coast by streams and rivers. Because there was no vegetation capable of binding loose material, the movement of sediment by water must have been very great during rainy periods. During dry periods, however, and in arid and semi-arid areas of the continent, eolian transportation surely replaced water transportation as the dominant force which transported detritus from the land areas to the sea. Dust and small particles of detritus undoubtedly were picked up by the wind in quantity, held in suspension, and transported many hundreds of miles. The larger particles were picked up by storms and gusts of wind, moved by leaps and bounds, or rolled along by traction to form dunes. Since the prevailing winds, then as now, were most likely from the west, the general direction of eolian transportation in eastern North America during Cambrian time was probably from west to east; or from the continent toward the area of deposition, the encroaching sea.

In addition to the work of the prevailing westerly winds, the daily action of land and sea breezes probably was very effective in Cambrian time. Such breezes are due to the differential in the rate of heating and cooling between the land and the sea (Grabau, 1924, p. 44); they are common along the coasts of present continents. During the day, the land is heated more quickly than the sea, causing the air above the land to become less dense and rise, reducing the barometric pressure over the land. This causes a sea breeze, a flow of air from the sea to the land. Along the eastern coast of North America, such a breeze would tend to counteract the influence of the prevailing westerly winds, and material in suspension would be blown back toward its source. At night, however, the land cools more rapidly than the water and the process is reversed, causing a land breeze to flow from the land toward the sea. Along the eastern coast of North America, this would accelerate the force of the prevailing westerly winds, and assist in transporting suspended material out to sea. In Cambrian time, the land and sea breezes were probably much stronger than those of today, because of the more rapid and intense heating and cooling of the land surface, due to the lack of soil cover.

Because of the complete lack of soil cover, the importance of the wind as an agent in the transportation of sediments, at least prior to Chazy deposition, can not be overemphasized. Never, since the advent of higher-order land vegetation, has eolian influence been so great in the sedimentary process on such a vast scale. Present-day examples of the transporting power of the wind are given in most textbooks on sedimentation. Grabau (1924, p. 60) states:

"The great dust storm of March 9–12, 1901, however, brought to Europe—1,960,420 tons of dust, while North Africa received—1,650,000 tons. This dust deposit thus covered at least 300,000 square miles of land surface and 170,000 square miles of ocean, and, according to Walther, traveled in part at least a distance of—2,500 miles." This means that 3,610,000 tons of sediment was moved up to 2,500 miles in three days. Twenhofel (1932, p. 65–85) published an excellent analysis of the work of the wind and gives the following example (p. 71, and p. 68).

"The dust fall on March 9, 1918, affecting Wisconsin and adjacent states, transported to these states at least a million tons of material (Winchell and Miller), and that of March 19, 1920, brought fully twice as much (Winchell)." The material of the dust fall of March 9, 1918, in the Upper Mississippi Valley, transported by suspension in the air, indicated that the place of origin was the semi-arid region of New Mexico, Arizona, and adjacent states, an area more than 2,000 miles distant. As to material moved by traction, Twenhofel (1932, p. 80–81) points out:

"The sand transported by traction generally

moves in the form of dunes which range in height from mere ripples to more than 200 feet, reaching 300 feet marginal to the Bay of Biscay and about twice that height on the coast of North Africa (Grabau)—dunes advance slowly, the rate on the Bay of Biscay ranging from 15 to 105 feet per year—in other dune regions—it may rise to as much as several miles per year."

Again, Twenhofel (1932, p. 81) states:

"In the early ages, before vegetation developed to protect the surface, there must have been vast areas covered by drifting sands; sands of eolian production should have been abundant, and there should be extensive deposits of atmosphere-transported dust."

There is evidence in eastern North America that the clastic sediments deposited during Cambrian time were at least in part of eolian origin. First, there are the rounded, frosted quartz sand grains, found in beds and scattered throughout the entire section of pre-Chazy rocks, even in the carbonate deposits. These ubiquitous sand grains have all the characteristics of dune sands and they are considered to be of eolian origin. A second indication of eolian conditions during the Cambrian Period is detrital feldspar in many of the clastic beds, far from the source of the sediments. In addition, a third indication of eolian transportation is muscovite mica in part of the clastic section. Mica and feldspar can be transported hundreds of miles by the wind without suffering decomposition, whereas, these minerals would be disintegrated rapidly in the course of fluvial transportation. Flakes of mica and small particles of feldspar are easily picked up and carried by winds of ordinary velocity, leaving behind the larger quartz sand grains to be rolled eastward toward the sea by traction, as a migrating sheet sand or in the form of dunes.

The wind, depending upon its velocity, picks up and holds in suspension particles up to 1 mm in size, and it moves by traction particles up to 2 mm in size. Many eolian particles moved by traction would eventually be transported to the rivers or to the sea coast, and those held in suspension would be dropped on the land or in the sea, wherever the wind slackened in velocity or became calm. The Cambrian Period lasted about 60 million years and tremendous quantities of eolian material could have been transported to the sea. The total amount in cubic miles probably was enormous, and the amount of material moved by eolian transportation probably was equal to or greater than that moved by fluvial transportation during the Cambrian Period, just the opposite of conditions in later periods, when vegetation covered most of the continent.

From the paleogeologic map (Fig. 5) the principal constituents of most of the rocks being disintegrated on the central continental shield were feldspar, mica, and quartz. These minerals, in addition to marble and calc-silicate hornfels, also compose a large part of the Grenville rocks. During Cambrian time they were the principal constituents of an almost continual fallout over the marine depositional zones along the eastern coast of North America, where normal marine sedimentation was taking place.

DEPOSITIONAL ENVIRONMENTS

The geologic record indicates that the initial transgressing sea of the Paleozoic Era advanced over a gently sloping coastal plain such as exists along most of the present North American Atlantic Coast. In the littoral zone, the coastal waters were redistributing the dune-sand present on the old subsiding land surface, and mixing it with the sand being dumped in at the mouths of rivers. The material was typical dune-sand, composed of rounded, frosted, well sorted quartz grains, free of dust and small particles of mica and feldspar which previously had been blown away or disintegrated. As transgression progressed, beyond the strong currents and clastic sediments of the littoral zone, calcium-carbonate-secreting marine organisms were thriving in the warm, shallow waters. This shallow-shelf zone is here called the "near-shelf" zone (near-neritic). Here marine limestone was forming, composed of shell sand and calcareous oölite. Such a limestone is forming today off the coast of Florida, between the present reefs and the Florida Keys (Grabau, 1924, p. 406). Beyond the zone of carbonate deposition, on the deeper side of the continental shelf and on the continental slope, few carbonate-secreting organisms could exist, but much clastic fallout was being deposited below wave-base. This deeper-shelf zone is here called the "far-shelf" zone (far-neritic), a zone of fine-clastic accumulation, which, due to its distance from shore (where land breezes died), also was receiving the bulk of the material transported in suspension by the winds. Seaward from the continental slope, a bathyal zone undoubtedly was receiving some very fine, wind-blown clastic sediment. The material carried by wind suspension which fell in the littoral zone generally was ground up and disintegrated by wave action, which constantly agitated the sediments in that zone. The suspended material which fell in the carbonate zone settled in the calcareous sediments, and was decomposed by chemical action.

In summary, in Early Cambrian time there were essentially three zones of marine deposition along the eastern coast of North America: a littoral zone of coarse clastic deposition; a shallow, off-shore, near-shelf zone of carbonate deposition; and a deeper far-shelf and slope zone of fine clastic accumulation (Fig. 6).

CAMBRIAN SUBSIDENCE

The extent and composition of the pre-Paleozoic North American continent is shown in Figure 5. At the beginning of Cambrian time, the continent not only began to subside slowly, but the southeastern part began to subside somewhat more rapidly than the remainder of the continent at the northwest. This is shown by the nature and thickness of the sedimentary rocks. As this subsidence continued through Early and Middle Cambrian time, a hinge line developed between the more stable continental mass and the more rapidly subsiding eastern and southeastern coast (Fig. 7). It was anchored on the north by the Adirondack Highlands and on the southwest by the Ozark Highlands; it extended between them with a gentle southeastward convexity. This hinge line is Woodward's (1961) "coastal declivity which may be a fault scarp or a steep coastwise cliff". The Cambrian section of rocks west of this line, in eastern North America, is generally relatively thin (1,000–3,000 feet), but the section east of the Adirondack-Ozark line is comparatively thick (3,000–10,000 feet). Both sections thicken eastward to form a sedimentary wedge, but the

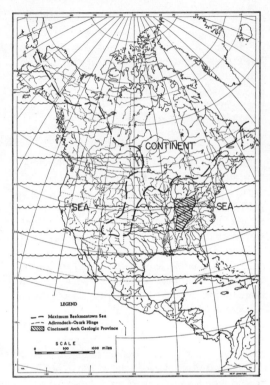

Fig. 7.—North America at end of Beekmantown advance.

thickening is much more rapid east of the Adirondack-Ozark hinge line. Although the slow subsidence of the continent continued during Cambrian time, causing a more or less continuous transgression of the sea, possibly marked by slight regressions, it appears that differential subsidence along the Adirondack-Ozark line ceased toward the end of Middle Cambrian time.

CAMBRIAN TRANSGRESSION

The evidence for continued transgression of the sea over the ancient Lipozoic surface during Cambrian time in eastern North America is in the stratigraphic succession of the rocks. The transitional sedimentary units suggest the following theory of transgressive sedimentation. Beginning with the slow eastward tilting of the continent, the advancing sea reworked and stratified the already clean dune sand in the littoral zone, depositing a basal sandstone, now known as the Weisner-Erwin-Antietam Sandstone (Table II). As transgression progressed, a shallow, off-shore depositional zone developed, the near-shelf zone,

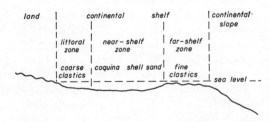

Fig. 6.—Cambrian depositional environments.

TABLE II. LITHOLOGIC CORRELATIONS OF CAMBRIAN FORMATIONS IN EASTERN UNITED STATES (NOT TIME CORRELATIONS)

	AREA OF LATE TRANSGRESSION			LITHIC CHARACTER			AREA OF EARLY TRANSGRESSION			
OUTCROP AGE	WISC.-MINN.	MISSOURI	NEW YORK	ZONE	DESCRIPTION	UNIT	PENNA.	MD.-VA.	TENN.-GA.-ALA.	OUTCROP AGE
EARLY ORDO-VICIAN?	Shakopee Dol. New Richmond Ss. Oneota Dol.	Powell Dol. Cotter Dol. Jeff. City Dol. Roubidoux Dol. Gasconade Dol.	Tribes Hill Dol.	EPICONTINENTAL	KNOX UNCONFORMITY Commonly sandy generally cherty dolomite.	CARBONATE	Bellefonte Dol. Axeman Ls. Nittany Dol. Stonehenge Fm.	Beekmantown Dol. Chepultepec Dol.	Mascot Dol. Kingsport Dol. Longview Dol. Chepultepec Dol.	EARLY ORDO-VICIAN?
	Jordon Ss. St. Lawrence Dol. Mazomanie Ss.	Eminence Dol. Potosi Dol.	Little Falls Dol.		Oolitic and pelletal dolomite, commonly sandy		Gatesburg Dol. Warrior Ls.	Conococheague Ls.	Copper Ridge Dol. Maynardville Dol.	LATE CAMBRIAN
LATE CAMBRIAN	Tomah-Birkmose Ss. Galesville Ss.	Doe Run Fm. Derby Dol. Davis Fm. (ELVINS GROUP)	U. Theresa Dol.	FAR-SHELF	TRANSITIONAL LITHIC BOUNDARY Fine to medium, micaceous, glauconitic, carbonate cemented sandstones and shales, with commonly sandy, glauconitic carbonate beds	FINE CLASTIC	Pleasant Hill Fm. Waynesboro Fm.	Nolichucky Sh. Maryville Ls. Rogersville Sh. Rutledge Ls. Pumpkin Valley Sh. ELBROOK FM. Rome Fm.	Conasauga Sh. Rome Fm.	MIDDLE CAMBRIAN
	EauClaire Ss.	Bonneterre Dol.	L. Theresa Dol.	NEAR-SHELF	TRANSITIONAL LITHIC BOUNDARY Partly oolitic dolomite. Sandy to north and west becoming dolomitic sandstone.	CARBONATE	Tomstown Dol.	Shady Dol.	Shady Dol.	EARLY
	Mt. Simon Ss.	Lamotte Ss.	Potsdam Ss.	LIT-TORAL	TRANSITIONAL LITHIC BOUNDARY Medium to coarse, clean, poorly cemented sandstone. Arkosic to NW.	COARSE CLASTIC	Antietam Ss.	Erwin Ss.	Weisner Qtz	CAMBRIAN

where shell and oölitic material accumulated to form a marine limestone. This limestone is now known as the Shady-Tomstown Limestone or Dolomite. The process by which it became a dolomite is discussed on subsequent pages. At each stage of transgression, the limestone was younger than the basal sandstone already present beneath it, but it was the same age as the westward extension of the basal sandstone contemporaneously deposited in the littoral zone. As tilting and subsidence continued, the littoral (coarse clastic) and near-shelf (carbonate) zones of deposition moved westward. A depositional zone farther east was then a little too deep for the existence of most carbonate-secreting forms of marine life, but it received vast quantities of fine-grained clastics by off-shore settling below wave-base, and from the prevailing winds and land breezes, whose velocities were commonly checked by distance from shore. This far-shelf zone, the zone of fine clastic deposition, produced the beds of the Rome-Waynesboro Formation and the Conasauga Shale (Table II). The fine clastic sediments, continually falling out of the atmosphere in a zone where the land breezes lost their velocity, gradually built up sand and mud bars, parallel with the shore, the tops of which were, in general, not far below the level of the sea. This accounts for the wave and ripple marks commonly observed in the fine-grained Rome Sandstone on the outcrop. Carbonate-secreting organisms did not thrive in this muddy and sandy environment. Minor regressions and fluctuations in subsidence varied the depth of the fine clastic zone, and occasionally favorable depth and non-turbulent conditions permitted the deposition of thin limestone beds as the result of an abundance of carbonate-secreting organisms. Of course, some fallout occurred in the littoral and near-shelf depositional zones, but not in the quantity dropped in the zone where the land breezes died.

As the westward transgression of the sea continued, the three zones of deposition, the littoral, the near-shelf, and the far-shelf, also moved westward. Contemporaneously, a basal sandstone, a limestone, and a shale (or fine-grained sandstone) were deposited. The final result is that a basal sandstone is overlain by a carbonate, which in turn is overlain by a fine clastic unit (Table II); all transgress time lines and all are the same age along a given time line.

By late Cambrian time, the Canadian Shield was reduced to low relief and the continental shelf, built up by aggradation, extended from Virginia to Wisconsin. The sea lapped well up on the Canadian Shield. A period of widespread epicontinental carbonate deposition ensued, except along the northern and western margins of the sea, beyond the Ozark, Wisconsin, and Adirondack regions, where mostly clastic littoral deposits still accumulated. This is the time when the Copper Ridge-Conococheague and Beekmantown carbonates were formed. It marks the culmination of Cambrian transgression in eastern North America (Fig. 7) and it was followed by almost complete emergence, prolonged denudation, and other events next to be considered.

KNOX UNCONFORMITY

Emergence.—At the end of Beekmantown deposition, the sea withdrew from the Cincinnati Arch geologic province, and in fact, it withdrew from most of North America. The rate of this marine retreat is not known, but it ended the first full depositional cycle of the Paleozoic Era. The Canadian Shield, which had been severely eroded and partly overlapped by the Cambrian deposits, was now raised (or sea-level was lowered) but not sufficiently to produce widespread clastic deposits (the St. Peter Sandstone is a special case) in eastern North America. A vast area of Cambrian carbonate rocks was exposed to the atmosphere, as well as that part of the Canadian Shield which had not been covered. Of all the forces of weathering and erosion brought to play upon this barren land surface, that of solution was probably the most effective against the carbonates. The relatively small amount of detrital insoluble material derived from the weathering of the carbonates was carried eastward by the wind or washed into sluggish rivers for transportation to the sea to form shales and siltstones. The relief of the Knox unconformity ranges from as much as 200 feet in Iowa to at least 400 feet in Virginia. The final relief of the land surface is by no means a measure of the total amount of carbonate rock removed during the post-Knox emergence, for many of the northern Cambrian outcrops of this time probably were removed entirely, and the general land surface lowered considerably.

Tilting and folding.—In the Cincinnati Arch geologic province, the Cambrian rocks are truncated northward beneath the Knox unconformity (Calvert, 1962b, Pl. 1). The Lambs Chapel Dolomite, of Lee County, Virginia, is truncated northward and it is completely absent north of the Ohio River region; only the lower part of the Chepultepec Dolomite remains in the Hopkins well of Fayette County, in south-central Ohio. In north-central Ohio the lower Chazy Dolomite rests unconformably on the lower part of the Copper Ridge Dolomite; in Sandusky and Erie Counties, Ohio (along the southwestern shore of Lake Erie), it rests on the Maynardville Dolomite; and in northern Ontario it finally rests on the Precambrian. This extensive truncation is evidence of the southward tilting of the rocks prior to the erosion shown by the Knox unconformity.

In addition, a southward plunging, truncated anticline is present below the Knox unconformity in central Ohio and eastern Kentucky. This anticline appears to have been along the axis of a Precambrian buried ridge, which was reactivated at the end of Beekmantown deposition. The Cambrian section is truncated to a greater degree over this anticline than on either side. Contrary to H. P. Woodward (1961, p. 1650), who shows Beekmantown rocks to be absent over this structure as far south as central Kentucky, Beekmantown rocks are present over the anticline almost as far north as Columbus, Ohio. During the post-Beekmantown emergence, moderate uplift began to occur west of the Cincinnati Arch geologic province in the Ozark region, on the north in the Wisconsin Highland region, and on the northeast in the Adirondack region of New York. These uplifts, however, did not furnish large quantities of sediment for subsequent deposition. The Cincinnati arch, a stable platform area between the Illinois, Michigan, and Appalachian basins, apparently was not a significant, positive feature, capable of affecting sedimentation, until post-Trentonian time.

Cambrian-Ordovician boundary.—The Knox unconformity is a major regional unconformity of vast time significance in the Cincinnati Arch geologic province and in most of North America. It has been called the post-Beekmantown, post-Arbuckle, post-Ellenburger, post-Bretonian, post-Pogonip, Post-Manitou, post-Prairie du Chein, post-Knox unconformity. It has the stratigraphic distinction of being one of the most widespread unconformities of the Paleozoic Era in North America, and its significance has been recognized more from recent subsurface studies than from earlier outcrop observations. The Beekmantown Group is considered to be approximately the time equivalent of the Tremadoc Slate in Wales. There is a notable stratigraphic break at the top of the Tremadoc Slate in Wales, represented by a major unconformity, which apparently has widespread eustatic significance. The Tremadoc Slate was placed at the top of the Cambrian System by Lapworth (1879, p. 1–15) in his original definition of the system. Other British geologists who consider the Tremadoc Slate to belong in the Cambrian System are Hicks, in 1881, Marr, in 1905 and 1913, Elles, in 1914, 1922, 1925, and 1937, Wood in 1914, Watts in 1917 and 1929, O. T. Jones, in 1936, and Stubblefield in 1956. The Geological Survey of Great Britain has always assigned the Tremadoc Slate to the Cambrian System of rocks. Although there is a general agreement that the Tremadoc Slate is equivalent to much of the Beekmantown, there is disagreement among American geologists whether the Tremadoc-Beekmantown belongs to the Cambrian or to the Ordovician chapter of earth history. This writer considers the Knox unconformity to mark the upper limit of the Cambrian depositional cycle. The transfer of the Beekmantown Group to the Lower Ordovician by many American geologists (about 1909?) does not seem justified, either on a stratigraphic or on a paleontological basis.

The stratigraphic reasons for placing the Beekmantown Group in the Cambrian are: (1) that it is genetically related to the rest of the Cambrian, —it was the culmination of the Cambrian depositional sequence in North America; (2) it is almost everywhere separated from the Ordovician by a great hiatus represented by the Knox unconformity; (3) it is lithologically similar to and transitional with the carbonates below, but it is lithologically totally unlike the succeeding clastics and carbonates of the Ordovician in North America and in other parts of the world.

The paleontologic reason for placing the Beekmantown Group in the Cambrian is that totally different faunas appear above the Tremadoc zone, even a different type of graptolite. Although Schuchert (1924, Pt. II, p. 235) placed the Beekmantown in the Champlainian (Ordovician), he gave a very cogent argument against such a classification, which is repeated here.

"In all the known areas of North America west of Appalachis and Acadis, there is a marked change in sedimentation between the Lower

Champlainian [Beekmantown]³ and the succeeding strata of Middle Champlainian [Ordovician] time. The older formations are dolomites, while the younger ones are thin-bedded limestones. Furthermore, the Middle Champlainian seas transgressed more widely and their faunas were totally different from those of the earlier epoch. This change is a striking illustration of the fact that the apparently insignificant break—the contact is everywhere a disconformable one—between the Lower [Beekmantown] and Middle Champlainian is of much time import in that greatly altered faunas have undergone a long evolutional change. In other words, the break in deposition represents a loss of record long enough for earlier faunas to have evolved into those so characteristic of Middle Champlainian time. This change is seen best in a completely different series of graptolites; a far greater prevalence of brachiopods, molluscs, and ostracods; the first crinids and fishes; and for the first time an abundance of bryozoans."

The Cambrian-Ordovician boundary has been discussed recently by several authors (Freeman, 1953; Patterson, 1961; Calvert, 1962b). There is additional, recent, world-wide evidence of the great faunal change above the Tremadoc zone. In discussing the Tremadoc graptolite faunas, William B. N. Berry (1960, p. 99) states:

"The local nature of the Early Tremadoc graptolite assemblages may be an indication that, at that time, graptolites were not truly planktonic. The Earliest Ordovician [Tremadoc] graptolites are dendroid in character, and although they lacked definite stems or basal connections such as are found in *Dendrograptus*, *Callograptus*, and other dendroid genera, perhaps they were, nevertheless, benthonic. Certainly, the Cambrian ancestors of *Dictyonema flabelliforme* and the anisograptids were sessile benthonic and *these Early Tremadoc forms were much like them*⁴ in numbers of branches and possession of three types of thecae. Structural changes had taken place in the direction of the graptoloids but most of the forms were still dendroids and probably had not attained the planktonic mode of life which was apparently characteristic of nearly all of the graptoloid graptolites."

Berry's conclusions indicate that there is no great difference in the graptolites of the Tremadoc (Beekmantown) and those of the underlying Cambrian rocks. The base of the Beekmantown Group is commonly designated in North America, on the basis of graptolites, as the present Cambrian-Ordovician boundary. Above the Tremadoc, however, Berry finds world-wide distribution of many new genera of graptoloid rather than dendroid graptolites.

Sokolov (1960, p. 50) in discussing the Ordovician of the U.S.S.R. states:

"In the lower series of the Ordovician with sufficient precision the Tremadoc stage with diverse faunas is well distinguished (Algain formation of Gornaya Shoria, Choi formation of Gorny Altay, deposits with Euloma, Niobe, and Ceratopyge of Southern Salair and northwestern part of Kuznetsky Alatau, *where they are closely bound with Upper Cambrian*)."⁵

Nevertheless, Sokolov "conforms with conceptions traditional for the Soviet stratigraphy," and he places the Tremadoc in the Ordovician.

Röömusoks (1960, p. 63) in discussing the Ordovician of Esthonia, states:

"In many localities the B_1 [Arenig] rests on the weathered surface of A_{2-3} [Tremadoc] sometimes on the Lower and Middle (?) Cambrian rocks (Mahnatch, 1958, Paskevicius 1958)."

Also

"The fauna of B_1 [Arenig, above Tremadoc] includes the first Ordovician trilobites and articulated brachiopods of Esthonia."

Lastly, Lindström (1960) in discussing conodont faunas in Sweden, cites numerous genera which appear for the first time above the Tremadoc (in the Arenig) and states (p. 92):

"In the strata below the Arenigian Lepidurus Limestone there are usually at least four times as many simple as compound conodonts."

There are, no doubt, many other examples in geologic literature which confirm the distinct faunal change above the Tremadoc (Beekmantown). Is not the Knox unconformity a more logical, recognizable horizon, both lithologically and paleontologically, at which to place a systemic boundary in North America, rather than at an indefinite horizon which commonly can not be found on the outcrop or in the subsurface of most areas?

---

³ Bracketed parts of quotations added.

⁴ Italics added.

⁵ Italics added.

Geologic systems are considered to be time-stratigraphic units based on paleontological similarities. Originally, in the early days of geology, large rock units related to a given depositional period of earth history were called systems of rocks; the fauna of these rocks was then studied and all other rocks having a similar fauna were said to be of the same age. Paleontological time boundaries became based on arbitrary criteria, independent of rock units. History has become confused with time. Should not rock units and depositional cycles be defined first, and then the fossils studied to determine what kind of organisms existed during various periods of earth history? It is hoped that this discussion will spur paleontologists to take another long, hard look at the *total* evolutionary change which occurred in North America during the long interval of time between Beekmantown and Chazy deposition. Possibly the present arbitrary criteria for designating the age of Beekmantown rocks as Lower Ordovician needs to be readjusted.

Ordovician History

Pre-Ordovician Surface

Cambrian history began in eastern North America with a westward advance of the sea, due to the slow subsidence of the continent; in like manner the next part of the history of the region began with a readvance of the sea. There is evidence that the Ordovician sea which advanced over the Cincinnati Arch geologic province covered a mature karst topography, characterized by carbonate hills 100–400 feet high. Old sink holes filled with sand, shale, and carbonate rubble are known from outcrop sections and from cores taken from wells in many areas. Joints filled with detrital material and lined with carbonate and quartz crystals to a depth of more than 100 feet below the Knox unconformity are well known in the carbonate formations of the Knox Supergroup. The sand which fills these joints and sink holes is the same rounded, frosted quartz of eolian origin which is found scattered throughout or bedded in many of the Cambrian rocks; much of it is probably local residue from the solution of previously deposited carbonates. Some of the sand could be fresh eolian sand from the Canadian Shield or from the erosion of the Cambrian clastics on the north and west.

Pockets and thin lenses of sand doubtlessly lay scattered over the old Cambrian surface east of central Indiana, but at the west a great cover of dune sand apparently had accumulated (now St. Peter Sandstone). This area of eolian accumulation now includes Illinois, Iowa, Missouri, Kansas, Oklahoma, northern Arkansas, western Kentucky, western Indiana, southern Minnesota, southern Wisconsin, and southwestern Michigan. It is possible that this sand came from the continental shield area on the north, transported by wind and possibly by an ancient river system which deposited the sand along flood plains, later to be scattered by the winds. It might have been wind-blown sand derived from that part of the shield area west (Rocky Mountain region) which had not suffered Cambrian submergence.

Ordovician Submergence

It appears that the Ordovician sea advanced over the eroded surface of the Cambrian rocks in eastern North America from the south and southwest. The marine advance seems to have been fairly rapid, caused by a general subsidence of the continent or a rise in sea-level. The first deposits were those of the Simpson Group exposed in southern Oklahoma. While these sediments were accumulating, the Everton Limestone of Arkansas and Missouri was being deposited. These formations are not present in the Cincinnati Arch geologic province. Next, the sea advanced farther north and covered the sand dune area in central United States, which resulted in the formation of the St. Peter Sandstone. Continued subsidence resulted in the submergence of most of northern and eastern United States, including the Cincinnati Arch geologic province. During this transgression, the Bromide deposits of Oklahoma, the Joachim Dolomite of Missouri and Illinois, and the Chazy Limestone of eastern United States were laid down (Table III). Additional subsidence resulted in the deposition of the Black River Group of limestones (including the lower part of the Viola Limestone of the Mid-Continent). This submergence covered much of the Canadian Shield and culminated in the greatest expansion of the Ordovician sea, the Trenton submergence, during which time the widespread Trenton Limestone was deposited. In many areas along the periphery of the Canadian Shield, the Simpson Group overlaps the eroded edges of the Beekmantown Group, the Black River Group overlaps the Simpson Group, and the Trenton Limestone overlaps the Black River Group. Most of the North

TABLE III. GENERALIZED SUB-TRENTON GEOLOGIC COLUMN FOR CINCINNATI ARCH GEOLOGIC PROVINCE

SEQUENCE	SUPERGROUP	GROUP	FORMATION	MEMBER
TIPPE-CANOE	OTTAWA LIMESTONE	BLACK RIVER	TRENTON LS.	
			EGGLESTON LS.	
			PLATTE-VILLE LS.	MOCCASIN LS.
				LOWVILLE LS.
~~KNOX~~ UNCONFORMITY~~		SIMPSON	CHAZY LS.	U. ARG. LIMESTONE
				M. LIMESTONE
			LAMBS CHAPEL DOL.	L. DOLOMITE
	KNOX DOLOMITE	BEEK-MAN-TOWN	ST. PETER SS.	
			CHEPULTEPEC DOL.	U. ARG. DOLOMITE
				L. SDY. DOLOMITE
		LEE VALLEY	COPPER RIDGE DOL.	BLOOMINGDALE DOL.
				MORRISTOWN DOL.
SAUK			MAYNARDVILLE DOL.	
	MONTE-VALLO	*KNOX CLASTIC	CONA-SAUGA SH.	NOLICHUCKY SH.
				MARYVILLE LS.
				ROGERSVILLE SH.
				RUTLEDGE LS.
				PUMPKIN VALLEY SH.
			ROME FM.	
			SHADY DOL.	
~~LIPALIAN~~ UNCONFORMITY~~			MT. SIMON SS.	
			BASAL ARKOSE	
			BASEMENT COMPLEX	

*Oostanaula (Calvert, 1963)

American continent appears to have been covered by the sea during Trenton (upper Viola) deposition, including much of the Arctic region. The progressive covering of source beds during Ordovician time had a direct effect upon the character of the sediments which accumulated in the Cincinnati Arch geologic province.

SOURCE AND NATURE OF SEDIMENTS

In the Cincinnati Arch geologic province little clastic material was received by the Simpson sea as it progressed over the eroded Beekmantown surface. The sea reworked the insoluble detrital material, mostly rounded frosted sand, clay, and chert, and redeposited it in poorly stratified beds. On the north and west, a part of the continental shield remained exposed, apparently near baselevel, and it was ringed by the tilted edges of Cambrian strata, but neither area was contributing much clastic material. True, both eolian and fluvially transported sediments were still being added to the sea, but the principal contribution was material in solution from Cambrian carbonates, which were by this time mostly dolomite. This resulted in the deposition of the very finely crystalline dolomites, greenish shales, and thin lenses of rounded, frosted sandstone which compose the lower Chazy Dolomite. As deposition continued, the interbedded very finely crystalline dolomites, dolomitic limestones, and lithographic limestones of the middle Chazy Limestone, and the silty limestones and shales of the upper Chazy Limestone were formed. The Simpson sea apparently advanced relatively rapidly; when it had covered most of the Cambrian dolomites, the deposition of lithographic limestone began. The continent apparently never was covered to any great depth by water; numerous intraformational conglomerates in the Simpson and Black River rocks are evidence of shallow-water conditions.

Black River deposition resulted mainly in the formation of thick calcareous deposits, later lithified into very finely crystalline to lithographic limestones containing little clastic material. Numerous thin beds of meta-bentonite in the Black River strata are the result of falls of volcanic ash over much of the eastern United States. The volcanism apparently occurred west of the Cincinnati Arch geologic province, probably in the Cordilleran region, and the prevailing winds scattered the resulting material as volcanic tuffs over a wide area. The meta-bentonite beds are particularly thick and prominent throughout the Eggleston Formation, the uppermost formation of the Black River Group. They appear to be excellent stratigraphic markers.

Black River deposition was followed by the widespread accumulation of coquina; the deposits now are known as the Trenton Limestone. The rock consists of the shells and shell fragments of numerous carbonate secreting organisms; the only clastic material present is in the form of thin shale partings and beds, indicating that very little clastic sediment was derived from rocks exposed on the north and west during Trenton deposition. In Trenton time, conditions were extremely favorable (water depth, temperature, and little clastic sediment) for the prolific existence of carbonate-secreting organisms over a wide area.

POST-DEPOSITIONAL CHANGES

GENERAL DISCUSSION

In previous reports by the writer (1962b, 1963), the sub-Trenton formations of the Cincinnati Arch geologic province are described, including a discussion of their lithologic character, thickness, and stratigraphic relations. Thus far, in the present report, the pre-Trenton sedimentary history of the region has been delineated, with emphasis on depositional environment, climate, vegetation, source of sediments, transportation of sediments, transgression, and erosion. The sediments are now limestones, dolomites, sandstones, and shales; they contain chert, glauconite, clay, pyrite, and other minerals. The lithologic and mineralogic characteristics now possessed by sub-Trenton rocks require explanation. The action and results of the processes involved in the lithification and alteration of the rocks are part of the geologic history of a region.

Diagenetic processes are those involving physical and chemical changes in sediment after deposition that converts it to consolidated rock. They include compaction, cementation, and recrystallization. Changes which occur after consolidation are known as post-diagenetic changes; these also may involve compaction, cementation, and recrystallization. Post-diagenetic changes are generally due to external influences, whereas diagenetic changes are mainly due to internal conditions. The diagenetic processes by which sediments are transformed, cemented, and lithified, are not well understood, particularly about the source and distribution of the material causing the transformation and cementation. The present theory of sub-Trenton sedimentation suggests some answers to problems involving the source of the material for some of the principal diagenetic processes.

DIAGENETIC PROBLEMS

Pettijohn (1957, p. 656) states "the problems of how and when sands become cemented and the source of the cementing material are still unresolved." He cites many explanations for the origin of cementing material, such as circulating meteoric waters, connate waters, solution at point of grain contact with redeposition in voids. solution of the fines, penecontemporaneous precipitation, and solid flow, all of which he rejects, and he concludes "the time and manner of origin of the silica cement in sandstone are not yet clearly established." He states further that "other cements, the carbonates for example, pose problems similar to those of silica." Because most of the sandstones in the sub-Trenton rocks of the Cincinnati Arch geologic province are more or less cemented, this problem is pertinent to this study.

A lengthy discussion by Pettijohn (1957, p. 439–444) is devoted to the origin of chert and he concludes that chert may be a polygenetic rock for which no single mode of origin exists. He cites proposals that nodular cherts form from direct precipitation from sea water, that cherts are epigenetic replacements by introduced silica, that cherts are reprecipitated from dissolved tests of organisms, and that volcanism may have supplied the source of the silica. He rejects volcanism as a source of silica and cites investigations which show that (1) concentrations of silica substantially greater than those of present-day stream waters (ocean waters have less silica concentration) can not be precipitated by any known agent, and (2) biochemical siliceous deposits of the present day are in waters too acid and too deep for the deposition of calcareous sediments. Pettijohn states "Introduction of silica seems inescapable" but no investigations or theories given show a satisfactory source for the silica necessary to form the tremendous amounts of chert present in sedimentary rocks as nodules, beds, and entire formations. Twenhofel (1932, p. 519–546) gives an exhaustive review of the various theories of chert formation; he accepts the theory that chert is derived from silica directly precipitated from sea water, but he advocates more research on the problem.

Dolomite is the principal constituent of most of the carbonates of the Cincinnati Arch geologic province below the Black River Group. Pettijohn (1957, p. 421–425) concludes that most dolomites are either replaced limestones or primary precipitates. He states "No organisms secrete dolomite." He gives no evidence for the source of the magnesium necessary to convert thousands of feet of limestone to dolomite, nor does he explain how the magnesium became evenly distributed. Fairbridge (1957, p. 169) states "there is very little evidence to support the idea of any primary precipitation of dolomite in the marine realm, either now or in the past." If dolomite is neither precipitated in marine waters nor secreted by organisms, how, then, can thousands of feet of formational dolomites be explained? Most geologists cite the magnesium in sea water as the source of the magnesium for dolomite, but nowhere is dolomite known to be precipitating in true marine environments today.

Glauconite is another constituent of the rocks of the region under discussion. Twenhofel (1932, p. 460) states:

"The existing state of knowledge with respect to the origin of glauconite supports the view that it is a product of diagenesis and that the glauconitic particles were originally pellets of mud containing finely divided and colloidal clay and iron oxide; that in some as yet unknown manner the aluminum of the clay was removed and its place taken by colloidal iron, and potash and colloidal silica were absorbed from the sea water or surrounding materials. The botryoidal shapes of the particles suggest additions of glauconite thereto from the surrounding waters, but nothing is known as to direct precipitation of glauconite from materials in solution."

He concludes "Although there may be some connection between shells, particularly those of foraminifera, and modern glauconites, it is difficult to find much evidence therefor in the glauconite of the geologic column." Pettijohn (1957, p. 467–469) cites several past explanations for the origin of glauconite, particularly that of Galliher, which indicates that glauconite is derived from biotite mica by a process of submarine weathering. On the following pages an alternative explanation for cementation, lithification, dolomitization, and the formation of chert and glauconite in the sub-Trenton rocks of the Cincinnati Arch geologic province is considered. Knowledge of the sedimentary conditions under which sub-Trenton rocks were deposited furnishes some very helpful clues in this study.

NATURE OF DOLOMITE, CHERT, AND GLAUCONITE

Types of dolomite.—Three distinctly different kinds of dolomite occur below the top of the Trenton Limestone in the Cincinnati Arch geologic province; original dolomite, redeposited dolomite, and replacement dolomite. Original dolomite is the type which is finely to coarsely crystalline; it contains vugs and vuggy porosity; and it is generally of formational extent. Such is the dolomite which occurs below the Knox unconformity. Redeposited dolomite is the type which is lithographic to very finely crystalline; it contains little original porosity and rare vugs; and it is also commonly of formational extent. This type of dolomite is found in sub-Trenton rocks above the Knox unconformity. Replacement dolomite is typically finely to coarsely crystalline, local in extent, and it is associated with fractures, faults, and solution channels. This type is apparently true secondary dolomite, formed by the circulation of magnesium rich waters through limestones, whereby some of the limestone is replaced by dolomite. Such is the dolomite found locally in northwestern Ohio and northeastern Indiana (in parts of the Trenton and Black River rocks) which acts as a reservoir for oil and gas.

Types of chert.—The chert in the sub-Trenton rocks of the Cincinnati Arch geologic province consists of two kinds: dense (cryptocrystalline) chert with a vitreous or waxy luster, which has a conchoidal fracture and is either opaque or translucent; and granular chert, with a dull luster and irregular fracture. The color of the cherts varies with the nature of the contained impurities, the most common colors are white or milky, gray, black, brown, blue, and pink. The dense cherts may be banded or mottled. They may contain inclusions of sand grains, oölites, pellets, fossil fragments, and pseudomorphs of other minerals such as calcite and dolomite. Some cherts are so oölitic as to be granular in texture. Other chert is white, granular, and with an earthy or chalky appearance (tripolitic), thought to be due to weathering. Many cherts contain irregular areas of clear, vitreous quartz, and these are called quartzose cherts. In the Beekmantown Group, chert forms the matrix of some thin sandstones. Chert occurs as scattered nodules, as zones of nodules, as lenses, and as distinct beds of wide areal extent. In connection with chert, it should be mentioned that clear, crystalline quartz also occurs as vug linings in the Cambrian dolomites.

Types of glauconite.—Glauconite occurs in the sub-Trenton rocks of the Cincinnati Arch geologic province as irregularly shaped particles, pellets, and as small fragments. It seems to be associated generally with sandy dolomites, sandy shales and siltstones, and micaceous shales. It has not been observed to be associated particularly with shell fragments; it generally occurs in random zones throughout a formation.

PECULIAR CHARACTER OF SUB-TRENTON SEDIMENTS

It has been postulated that the source of the clastic sediment deposited by wind and water in eastern North America, including the Cincinnati Arch geologic province, during Cambrian time, was a continental shield area extending from northern Canada to Oklahoma, composed of igneous and metamorphic rocks. Feldspar constitutes 59.5 per cent of the total material in all igneous rocks; mica occurs in abundance in many igneous and metamorphic rocks and biotite is by far the most common of the micas; quartz is the predominant mineral in many granites, porphyries, and quartzites. Since there was probably a more or less continuous fallout of these materials into a more or less continuously transgressing sea during Cambrian time, feldspar, quartz, and biotite mica should be present in considerable quantities in all of the rocks, carbonates and clastics alike. Quartz, in the form of rounded, frosted grains of various sizes and in varying amounts certainly is scattered or concentrated throughout the entire section of Cambrian rocks, and some feldspar and muscovite mica are still present in some of the clastic rocks. However, no biotite mica has been observed in any of the sediments, and feldspar and mica are both conspicuously absent in the carbonates. The absence of feldspar and biotite mica (which are thought to have fallen almost constantly in all the zones of deposition) in Cambrian carbonates is a condition which is explained by the following theory of diagenetic absorption.

DIAGENETIC ABSORPTION OF FELDSPAR AND BIOTITE

It is here proposed that the feldspar and mica particles which settled in the calcareous sediments of Cambrian time were absorbed in the diagenetic processes to form original dolomite, chert, glauconite, and kaolinite, and that the feldspar and biotite mica which fell in the clastic sediments decomposed to form mostly glauconite, kaolinite, and silica and dolomite cement. The writer believes that feldspar decomposed to form kaolin and silica in a carbonate environment; that in a clastic environment, feldspar was converted into kaolinite and silica cement. Biotite mica, upon decomposition, furnished the magnesium to convert calcium carbonate into dolomite, with such by-products as silica, glauconite, kaolinite, and pyrite, but in a clastic environment, biotite mica was converted into glauconite, kaolin, silica, and magnesite. It is probable that muscovite mica, being more stable than biotite, was decomposed in a carbonate environment, but little affected in a clastic area of deposition. Muscovite mica and feldspar contain little magnesium and therefore contributed little toward dolomitization. Phlogopite, a variety of biotite mica especially high in magnesium and low in iron, common to metamorphosed limestones and dolomites, is interestingly one of the principal accessory minerals of the Grenville carbonates, a belt of rocks subjected to erosion just west of the Early Cambrian depositional zones of eastern North America. Minor amounts of magnesium for dolomitization probably were furnished by the decomposition of other igneous and metamorphic minerals, such as pyroxenes, amphiboles, and serpentine, but these minerals are not so abundant as mica.

Feldspar decomposition.—The feldspars are complex aluminum silicates which contain variable amounts of potassium, sodium, calcium, and barium. Orthoclase, a potassium feldspar, is probably the most common of all the silicates. It decomposes readily under ordinary conditions of weathering to form kaolinite and silica, as follows.

$$\frac{KAlSi_3O_8 + HOH}{orthoclase} = HAlSi_3O_8 + KOH \quad \text{(hydrolysis)}$$

$$2 KOH + CO_2 = \frac{K_2CO_3 \text{ (soluble)} + H_2O}{potassium\ carbonate} \quad \text{(carbonation)}$$

$$HAlSi_3O_8 - 2SiO_2 = HAlSiO_4 + \frac{2\ SiO_2}{silica} \quad \text{(desilication)}$$

$$2\ HAlSiO_4 + H_2O = \frac{H_4Al_2Si_2O_9}{kaolinite} \quad \text{(hydration)}$$

Kaolinite may break down further into silica and gibbsite. All feldspars yield kaolinite upon decomposition, but not all produce free silica. Microcline and albite, which contain sodium, do not decompose as rapidly as the potassium and calcium-bearing feldspars.

It is probable that much of the argillaceous material in the sub-Trenton rocks of the Cincinnati Arch geologic province is derived from the decomposition of feldspar. The excess silica produced accounts for the presence of chert and quartz-lined vugs in the carbonates, and for siliceous cement and secondary grain enlargement in the sandstones.

Muscovite mica decomposition.—Muscovite micas are complex potassium-aluminum silicates which are stable under ordinary conditions of weathering, but which decompose slowly to form kaolinite when attacked by organic acids. The process is as follows.

$$\frac{KH_2Al_3Si_3O_{12} + HOH}{muscovite} = H_3Al_3Si_3O_{12} + KOH \quad \text{(hydrolysis)}$$

$$2 KOH + CO_2 = \frac{K_2CO_3 \text{ (soluble)} + H_2O}{potassium\ carbonate} \quad \text{(carbonation)}$$

$$2\ HAl\ SiO_4 + H_2O = \frac{H_4Al_2Si_2O_9}{kaolinite} \quad \text{(hydration)}$$

Magnesium biotite mica decomposition.—Biotite micas are complex magnesium-iron-potassium-aluminum silicates of variable composition. The general formula appears to be $HK(Mg,Fe)_2Al_2Si_3O_{12}$. A typical magnesium biotite mica will decompose into kaolinite, silica, and magnesite as follows.

$$\underset{\text{biotite}}{HKMg_2Al_2Si_3O_{12}} + 5\,HOH = H_6Al_2Si_3O_{12} + KOH + 2\,Mg(OH)_2 \quad \text{(hydrolysis)}$$

$$2\,KOH + CO_2 = \underset{\text{potassium carbonate}}{K_2CO_3\,(\text{soluble}) + H_2O} \quad \text{(carbonation)}$$

$$2\,Mg(OH)_2 + 2\,CO_2 = \underset{\text{magnesite}}{2\,MgCO_3 + 2\,H_2O} \quad \text{(carbonation)}$$

$$H_6Al_2Si_3O_{12} = \underset{\text{kaolinite}\quad\text{silica}}{H_4Al_2Si_2O_9 + SiO_2 + H_2O} \quad \text{(desilication)}$$

It is well known that $CaCO_3$ and $MgCO_3$ are unstable in association with each other and recrystallize as dolomite, $CaMg(CO_3)_2$. Thus a quantity of biotite mica containing magnesium, which decomposed in a calcium carbonate environment, would cause the calcium carbonate to recrystallize into dolomite, and would transform a lime deposit into a dolomite. The clay produced would form shaly partings or argillaceous material within the final dolomite, and the silica would form nodules and layers of chert. Naturally, any oölites, fossil fragments, sand grains, dolomite crystals, or other foreign material enclosed by the silica would be evident in the resulting chert.

Iron-rich biotite mica decomposition.—Glauconite is a hydrous silicate of iron and potassium which has a variable composition; the general chemical formula is $K_2(Mg,Fe)_2Al_6(Si_4O_{10})_3(OH)_{12}$. Galliher (1935a) has demonstrated that an iron-rich biotite mica will decompose into glauconite. He states that the alteration of biotite mica to glauconite involves the oxidation of iron, retention of potash, hydration, partial loss of alumina, and changes in structure. Galliher showed that present-day sediments containing glauconite are off-shore equivalents of those containing biotite mica. From dredging in Monterey Bay he found that micas seldom are found between the strand line and depths of 5–10 fathoms, due to turbulence; that micas are especially abundant between depths of 20–30 fathoms; that biotite mica decreases and glauconite increases in amount with distance from shore; that the final glauconite product contains about double the water contained in biotite mica and increases from 10 to 20 times in size, destroying all micaceous structure. Chemical analyses of biotite mica, and analyses of glauconite given by Galliher and others are shown in Table IV.

Galliher (1935a) found all stages of decomposition of biotite mica to pure glauconite in the sediments which he studied. He observed that recent occurrences of glauconite invariably lie adjacent to landmasses where plutonic or metamorphic complexes are exposed to erosion to furnish sources of biotite mica. However, Grabau (1924, p. 673) points out that glauconite may be derived from the erosion of older sediments and be redeposited, and he cites as examples the recent accumulations of glauconite from Cretaceous greensands along the Atlantic coastal plain, and the Cretaceous accumulations of greensands in northwestern Germany. It might be added here that the glauconite found in the deposits of the Mississippi Embayment and the Gulf of Mexico may be from the outcrops of Cambrian and Ordovician rocks within the drainage basin.

That glauconite forms from various materials in various environments has been demonstrated by Grim (1936, heavy minerals), Galliher (1935a, 1935b, 1939, biotite mica), Takahashi (1939, fecal pellets, clay, and silicates), Houbolt (1957, calcareous pellets), Burst (1958, degraded clay), and Wermund (1961, feldspar, quartz, and muscovite mica). Wermund (1961, p. 1689) found that minerals identified in the field as glauconite were commonly (1) illite-type, (2) montmorillonite-type, (3) kaolin-type, and (4) chlorite-type grains. Wermund concluded that most glauconite sediments are deposited in oxygenated, normally saline water but a few may be deposited in brackish water; the glauconite studied by Galliher formed in an alkaline, black mud environment which reeked with hydrogen sulphide.

Under special conditions, where decaying organic matter furnishes an abundance of hydrogen sulphide, the decomposition of iron-rich biotite mica would result in the formation of kaolinite, silica, and pyrite. The iron oxide from the biotite mica would combine with the hydrogen sulphide to form pyrite and water; the remaining material would decompose further into kaolinite and silica.

Because the chemical changes which take place in the transformation of biotite mica to glauconite are complex, and because the composition of both

TABLE IV. COMPARISON OF CHEMICAL ANALYSES OF GLAUCONITE WITH BIOTITE MICA (PER CENT)

	Biotite	Various Samples of Glauconite						
	1	2	3	4	5	6	7	8
SiO_2	36.25	55.95	51.90	53.61	54.84	49.47	48.12	48.18
Al_2O_3	18.25	11.56	1.52	9.56	3.52	5.59	9.60	6.97
Fe_2O_3	6.35	9.99	27.98	21.46	12.64	19.46	19.10	--
FeO	17.09	2.02	1.26	1.58	4.90	3.36	3.47	27.08
MgO	9.01	6.77	4.67	2.87	6.65	3.96	2.36	--
CaO	0.79	3.95	0.89	1.39	0.89	0.60	0.76	--
Na_2O	-	0.61	0.53	0.42	0.39	0.16	0.22	1.25
K_2O	8.68	4.12	4.90	3.49	7.00	8.04	7.08	7.40
H_2O	-	1.60	2.10	5.96	9.62	8.54	10.06	8.75
H_2O+	2.70	3.22	4.05					
P_2O_5	-	0.18	0.11	-	-	1.06	-	-
Organic	-	trace	trace	-	-	-	-	-
Total	99.12	99.97	99.91	100.34	100.45	100.24	100.77	99.63

1. Biotite, Monterey Bay, California (Recent) Galliher, 1935a
2. Firm glauconite, Monterey Bay, California (Recent) Galliher, 1935a
3. Spongy glauconite, Monterey Bay, California (Recent) Galliher, 1935a
4. Average, four analyses of glauconite (Recent) Twenhofel, 1932, p. 456 (modified)
5. Average, four analyses of celadonites (Recent) Twenhofel, 1932, p. 456 (modified)
6. Purified glauconite, New Jersey (Cretaceous) Twenhofel, 1932, p. 456 (modified)
7. Purified glauconite, England (Recent) Twenhofel, 1932, p. 456
8. Greensand grains, Minnesota (Lower Ordovician dolomites) Grabau, 1924, p. 672.

the original substance and the end product is variable, it is not practical to show by chemical formulae the decomposition reactions in the order in which they might occur. However, a comparison of the general formulae of the two minerals (Table IV) indicates that the processes involve the addition of 4 aluminum and 6 silicon atoms and the subtraction of 2 magnesium and 2 iron atoms from the original biotite mica. In addition, water is absorbed.

Cementation.—In a non-carbonate environment, such as that which results in the deposition of a sandstone or siltstone, the decomposition of magnesium-rich biotite mica would form kaolinite, magnesite, and silica, as previously demonstrated. The magnesite would coat the clastic grains or combine with calcium carbonate in the sea water to form a dolomite cement. The silica would be attracted to the quartz particles composing a sandstone or siltstone and cause secondary grain enlargement and silica cement. Chert could also be formed. Such chert could contain sand grains or form a chert-matrix sandstone. The kaolinite also would form a coating on the sand grains or result in interstitial clay. Many Cambrian sandstones contain grains coated with a white, earthy material which is probably either kaolinite or magnesite or both. The decomposition of feldspar and mica seems to be the principal source of the cementing material in the Cambrian sandstones of the Cincinnati Arch geologic province, as well as the source of the chert, glauconite, and dolomite found in the sub-Trenton rocks of the region.

RATE OF LITHIFICATION

The diagenetic processes of dolomitization, glauconitization, kaolinization, and silication apparently were very active in the Cambrian sediments of the Cincinnati Arch geologic province, therefore, lithification probably occurred soon after deposition. While not conclusive, one evidence for rapid lithification is the presence of intraformational conglomerates in dolomites stratigraphically short distances apart. The broken fragments of completely and uniformly dolomitized rock in these conglomerates, identical in composition with the underlying material, indicate that the underlying rock had become dolomitized and lithified soon after deposition. The absence of weathering products indicates that the conglomerates probably were formed by submarine wave action, prior to burial of the parent rock. Another evidence for rapid lithification is that carbonates forming today, off the coast of Florida and in other parts of the world, acquire a solid state in a relatively short time. The rate of present day glauconitization is also rapid and occurs before burial (Galliher, 1935a, p. 1362). Where deposition is slow, lithification of sediments may occur before appreciable burial.

REDEPOSITED DOLOMITE

Dolomites are slowly soluble; during the weathering process they are carried away in solution and leave behind only the insoluble accessory minerals contained in the original rock. Where original dolomites are exposed to erosion and weathering over wide areas of outcrop, extreme concentrations of dissolved $CaCO_3$ and $MgCO_3$ will be built up in restricted areas of adjacent seas. This will result in redeposition of dolomite by precipitation. Such redeposited dolomites will be very finely crystalline, and will commonly contain argillaceous material or be associated with thin beds of shale as a result of insoluble material derived from the weathering of the dolomite outcrops. Such an origin is here proposed for the dolomite which occurs in the lower part of the Chazy Limestone of the Cincinnati Arch geologic province and adjacent areas. Dolomites of this type occur in other sedimentary sequences throughout the world. It is highly probable that the source of the very finely crystalline Chazy dolomites was the vast expanse of original Cambrian dolomite over which the Simpson sea advanced, and from which the sea was receiving high concentrations of calcium and magnesium in solution. The presence of some chert, glauconite, and kaolinite indicates the addition of some wind-blown mica and feldspar to the Chazy sediments, but the volume of eolian material was apparently much less than in previous times, when more coarsely crystalline original dolomites were produced diagenetically.

The possibility also exists that magnesium for the transformation of a calcium carbonate sediment into a dolomite could be supplied to restricted areas of the sea by streams containing minerals directly derived from the decomposition of magnesium-rich igneous or metamorphic rocks. Such restricted areas of dolomite deposition might be evaporite basins, and the dolomites would be "precipitated." Dolomites of this type are not thought to be present in the sub-Trenton rocks of the Cincinnati Arch geologic province.

REPLACEMENT DOLOMITE

The dolomite which occurs in portions of the Trenton Limestone and in the upper part of the Black River Group in northwestern Ohio and northeastern Indiana is a coarsely crystalline, porous type of dolomite, very irregular in its stratigraphic occurrence. It appears to be the result of the invasion of an original limestone by magnesium-rich waters. The invasion apparently occurred vertically through deep-seated joints and faults, and then spread laterally along porous zones and bedding planes. The upward migration of the magnesian waters was arrested by a thick section of overlying shale. The source of the magnesium for the dolomitization was probably subsurface water containing excess magnesite in solution, obtained from the deep-seated Cambrian sediments. The type of dolomite produced by the replacement of solid limestone through the action of epigenetic magnesian waters is true, secondary dolomite.

COMPACTION

Sediments buried under subsequent deposits are subjected to pressure which results from the weight of the overlying material. This pressure causes the individual particles of a clastic rock to become more closely packed; it may cause shales and unconsolidated sandstones to readjust toward areas of less overburden. In partially cemented sandstones the pressure at the points of contact of the individual grains is increased and intrastratal solution may occur. Compaction from overburden will continue so long as sediments are unconsolidated and material is added above. If sandstones become well cemented and shales become fissile and indurated, compaction will practically cease. Compaction of clastic rocks results in compression of the contained fluids, causing the fluids to move through porous zones and fractures toward areas of less pressure. If such fluids are supersaturated, precipitation will occur where areas of lower pressure and temperature are reached, and result in secondary cementation, replacement dolomitization, and other phenomena. If the moving fluids are relatively unsaturated, solution will take place in relatively soluble materials, and result in solution channels and stylolites. The rate of compaction of clastic sediments is therefore variable and depends on the rate of addition of overburden, the quantity of overburden, and the degree of cementation.

Compaction apparently occurs early in the lithification of biogenic carbonate rocks and takes place prior to the cementation produced by diagenesis. The rapid burial of the accumulated shells and shell fragments of a carbonate beneath subsequent sediments no doubt produces some compaction in the carbonate prior to diagenesis. But the relative rapidity with which diagenetic processes accomplish the solidification of a biogenic carbonate precludes the occurrence of much compaction due to heavy overburden. On the other hand, precipitated carbonates are solid, homogeneous rocks upon deposition and suffer little post-diagenetic compaction. Compaction probably has not been an important factor in the lithification of the carbonates which occur in the sub-Trenton rocks of the Cincinnati Arch geologic province.

SUMMARY OF POST-DEPOSITIONAL CHANGES

The major part of the sub-Trenton rocks of the Cincinnati Arch geologic province apparently acquired their present chemical composition and character during the process of solidification. An exception is the dolomitization of the Trenton-Black River limestones of northwestern Ohio and northeastern Indiana. The great bulk of the material of which the rocks are now composed was authigenic, and it only has undergone the diagenetic changes necessary to establish chemical equilibrium within the formations and beds.

The pre-Chazyan carbonates appear to be the result of accumulations of biogenic calcarenite formed in relatively shallow, mildly agitated, marine shelf areas into which a fine, fairly constant atmospheric fallout composed principally of feldspar, mica, and quartz grains, was deposited. The feldspar and mica decomposed to form kaolinite, chert, glauconite, pyrite, and magnesite. The magnesite was absorbed by the calcarenite to form original dolomite by recrystallization while in a semi-solid state, causing vugular porosity.

The Chazy dolomites are thought to be principally chemical precipitates derived from the solution of extensive areas of original dolomite outcrops. Redeposition by means of solution and subsequent precipitation accounts for their generally dense to very finely crystalline nature.

The limestones of the Chazy Limestone and the Black River Group are thought to be chemical precipitates of calcium carbonate in very shallow water, which resulted from the attrition and solution of coquinas and calcarenite fines. The evidence that these rocks are precipitates rests in their dense, lithographic nature, and in the presence of numerous enclosed calcite crystals ("birdseyes"). Some beds which are coarsely crystalline may be the result of the recrystallization of less pulverized coquinas. Moderate amounts of windblown biotite mica and feldspar account for the local accumulations of silica, magnesite, kaolinite, and iron oxide necessary for the formation of the chert nodules, dolomitic zones, argillaceous zones, pyrite, and glauconite common in some beds. Glauconite is found only locally in the Chazy-Black River rocks, an indication that little iron-rich biotite mica was contained in the sediments. Wind-blown silt may have contributed to shale partings, and wind-blown volcanic ash certainly was deposited at various intervals to form metabentonites, but wind-blown sand grains are conspicuously absent.

The basal (Mt. Simon) sandstone appears to have been derived from a sheet or dune sand, reworked in the littoral zone of the advancing Cambrian sea. Prior to reworking, this sand had been winnowed by the wind of practically all mica and small feldspar particles, and similar particles subsequently received from fallout in the littoral zone probably were disintegrated or carried out into the near-shore zone by currents; this accounts for the absence of glauconite and chert in the Mt. Simon Sandstone; however, some feldspar decomposition locally produced kaolinite and a weak silica cement. Little diagenetic change apparently has occurred in the Mt. Simon Sandstone since deposition, except weak cementation; compaction seems to have been the only major epigenetic change.

The Rome and Conasauga Formations, composed of fine-grained, glauconitic, dolomitic sandstones and siltstones, glauconitic, argillaceous, partly oölitic dolomites and limestones, and red, green and brown, pyritic, dolomitic, and glauconitic shales, contain no biotite mica and generally little feldspar. Glauconite, dolomite, and kaolinite are accounted for by the diagenetic decomposition of biotite mica and feldspar originally present; the absence of chert indicates that the resultant silica was absorbed as cementing material and quartz grain enlargement. Chert appears to form more readily in a carbonate environment, where the silica is not attracted to existing quartz grains. The sandstones, siltstones, and shales contain large amounts of small, bronze-colored mica flakes, considered to be unaltered muscovite mica. In some areas west of the Cincinnati arch, however, a considerable amount of undecomposed feldspar is present in the Rome sandstones.

Conclusions

The following major conclusions are derived from the foregoing theoretical considerations regarding the sub-Trenton rocks of the Cincinnati Arch geologic province.

1. Pre-Chazy sedimentation was greatly affected by eolian transportation of sediments, which resulted from the lack of an effective soil cover on the North American continent.

2. Prevailing westerly winds are thought to have deposited erosional material, derived principally from the disintegration of igneous and metamorphic rocks of the continental shield, in a marine environment.

3. The diagenetic decomposition of wind-borne igneous and metamorphic material supplied the minerals for dolomitization, cementation, and the formation of chert and glauconite.

4. Sub-Chazy formations have transitional boundaries, and are mostly time-transgressive rock units.

5. The interval of time between Grenville deposition and Cambrian deposition is equivalent to a geologic era for which the term "Lipozoic Era" is proposed.

The theoretical considerations of this study are presented to stimulate interest in the Cambrian problems of eastern United States. It is hoped that future investigations will substantiate many of the concepts here presented.

Literature Cited

Berry, Wm. B. N., 1960, Correlation of Ordovician graptolite-bearing sequences: Rep. 21st Sess. Norden, Internatl. Geol. Cong., pt, VII, p. 97–108.

Burst, J. F., 1958a, "Glauconite" pellets: their mineral nature and applications for stratigraphic interpretations: Am. Assoc. Petroleum Geologists Bull., v. 42, p. 310–327.

―――― 1958b, Mineral heterogeneity in "glauconite" pellets: Am. Mineralogist, v. 43, p. 481–497.

Calvert, W. L., 1962a, Sub-Trenton rocks of the Cincinnati Arch geologic province (abs.), in Abstracts for 1961: Geol. Soc. America Special Paper 68, p. 144–145.

―――― 1962b, Sub-Trenton rocks from Lee County, Virginia, to Fayette County, Ohio: Ohio Geol. Survey Rept. Inv. 45, 57 p.

―――― 1963, Sub-Trenton rocks from Wood County, West Virginia, to Fayette County, Illinois: Ohio Geol. Survey Rept. Inv. 48.

Cloud, P. E., Jr., 1955, Physical limits of glauconite formation: Am. Assoc. Petroleum Geologists Bull., v. 39, p. 484–492.

Drake, C. L., Ewing, M., and Sutton, G. H., 1959, Continental margins and geosynclines: the east coast of North America north of Cape Hatteras, in Physics and chemistry of the earth: v. 3, p. 110-198, London, Pergamon Press.

Fairbridge, R. W., 1957, The dolomite question: Soc. Econ. Paleontologists and Mineralogists Special Pub. 5, p. 125–178.

Freeman, Louise B., 1953, Regional subsurface stratigraphy of the Cambrian and Ordovician in Kentucky and vicinity: Ky. Geol. Survey Bull. 12, 352 p.

Galliher, E. W., 1935a, Glauconite genesis: Geol. Soc. America Bull., v. 46, p. 1351–1356.

―――― 1935b, Geology of glauconite: Am. Assoc. Petroleum Geologists Bull., v. 19, no. 11, p. 1569–1601.

―――― 1939, Biotite-glauconite transformation and associated minerals, in Recent marine sediments: Am. Assoc. Petroleum Geologists, p. 513–515.

Goldich, S. S., Baadsgaard, H., Edwards, G., Weaver, C. E., 1959, Investigations in radioactivity-dating of sediments: Am. Asso. Petroleum Geologists Bull., v. 43, p. 654–662.

Grabau, A. W., 1924, Principles of stratigraphy: 2d ed., 1185 p., New York.

Green, D. A., 1957, Trenton structure in Ohio, Indiana, and northern Illinois: Am. Assoc. Petroleum Geologists Bull., v. 41, no. 4, p. 627–642.

Grim, R. E., 1936, The Eocene sediments of Mississippi: Miss. Geol. Survey Bull. 30, 240 p.

Gulbrandsen, R. A., Goldich, S. S., Thomas, H. H., 1963, Glauconite from the Precambrian Belt Series, Montana: Science, v. 140, p. 390–391.

Haught, O. L., 1956, Probabilities of the presence of reservoirs in the Cambrian and Ordovician of the Allegheny synclinorium, in Proceedings of the technical session, Kentucky Oil and Gas Association, May 25, 1956: Ky. Geol. Survey, ser. IX, special pub. 9, 7–16.

Holmes, A., 1959, A revised geological time scale: Edin. Geol. Soc. Trans., v. 17, pt. 3, p. 183–216.

Houbolt, J. J. H. C., 1957, Surface sediments of the Persian Gulf near the Qatar Peninsula: 113 p., The Hague, Netherlands.

Hunt, G., 1962, Time of Purcell eruption in southeastern British Columbia and southwestern Alberta: Jour. Alberta Soc. Petroleum Geologists, v. 10, p. 438–442.

King, Philip B., 1959, The evolution of North America: 190 p., Princeton, N. J.

Lindström, Maurits, 1960, A lower-middle Ordovician succession of conodont faunas: Internatl. Geol. Cong., 21st, Copenhagen, Rept., pt. 7, p. 88–96.

McCormick, Geo. R., 1961, Petrology of Precambrian rocks of Ohio: Ohio Geol. Survey, Rept. Inv. 41, 60 p.

Patterson, J. R., 1961, Ordovician stratigraphy and correlations in North America: Am. Assoc. Petroleum Geologists Bull., v. 45, no. 8, p. 1364–1377.

Pettijohn, F. J., 1957, Sedimentary rocks: 718 p., New York.

Rõõmusoks, A., 1960, Stratigraphy and paleogeography of the Ordovician of Esthonia: Rept. 21st Sess. Norden, Internatl. Geol. Cong., pt. VII, p. 58–69.

Schuchert, C., 1924, A textbook of geology: pt. 2, 724 p., New York.

Seward, A. C., 1931, Plant life through the ages: 601 p., Cambridge, England.

Sokolov, B. S., et al., 1960, Stratigraphy, correlation, and paleogeography of the Ordovician deposits of the U.S.S.R.: Rept. 21st Sess. Norden, Internatl. Geol. Cong., pt. VII, p. 44–57.

Takahashi, J. I., 1939, Synopsis of glauconitization, *in* Recent marine sediments, P. K. Trask, ed.: Am. Assoc. Petroleum Geologists, p. 503–515.

Twenhofel, W. H., 1932, Treatise on sedimentation: 2d ed., 926 p., Baltimore, Md.

Wermund, E. G., 1961, Glauconite in Early Tertiary sediments of Gulf Coastal province: Am. Assoc. Petroleum Geologists Bull., v. 45, no. 10, p. 1667–1696.

Woodward, H. P., 1961, Preliminary subsurface study of southeastern Appalachian interior plateau: Am. Assoc. Petroleum Geologists Bull., v. 45, no. 10, p. 1634–1655.

Edwards Formation (Lower Cretaceous), Texas: Dolomitization in a Carbonate Platform System[1]

W. L. FISHER[2] and PETER U. RODDA[2]
Austin, Texas 78712

Abstract The Edwards Formation is characterized by rudist bioherms, carbonate grainstone and mudstone, and evaporites which were deposited on an extensive, shallow-water, marine platform bounded by deeper water basins in which chiefly carbonate muds were deposited. Rudist bioherms were constructed principally along platform edges peripheral to an evaporitic lagoon. Main dolomite deposits are in a concentric belt marginal to the lagoonal facies.

Two main types of dolomite are present: (1) stratal dolomite—fine grained, tightly knit fabric, laminated to thin bedded, very slightly porous and permeable, associated with thin-bedded, mud-cracked, stromatolitic carbonate mudstone, ripple-marked carbonate grainstone, and thin evaporite-collapse layers; dolomite units are generally less than 2 ft (0.6 m) thick; magnesium carbonate content ranges irregularly from low to high; and (2) massive dolomite—fine- to coarse-grained loosely knit euhedral crystals, moderately to highly porous and permeable, replacing thick-bedded, fossiliferous carbonate grainstone; dolomite units are commonly more than 10 ft (3 m) thick and underlie prominent evaporite-solution units; magnesium carbonate content is low or high, with few intermediate values.

Both stratal and massive dolomites are judged to be products of metasomatic replacement of calcium carbonate which resulted from contact with magnesium-enriched brines. Features of stratal dolomite indicate prelithification replacement of carbonate muds and grains in extensive, low-relief, supratidal and intertidal zones along the southern margin of the lagoon. Massive dolomite was created by postlithification replacement of reef-trend carbonate grainstone along the northern margin of the lagoon as a result of seepage refluxion of lagoon brines. Both dolomitization processes were part of the original depositional system, and type of dolomitization was controlled by specific depositional facies of the system.

Introduction

In the past decade knowledge concerning the formation of dolomite has been augmented by (1) field evidence of dolomite or protodolomite formed on modern carbonate tidal flats by metasomatic replacement of carbonate muds in Florida and the Bahamas (Shinn et al., 1965), the Persian Gulf (Illing et al., 1965), the Netherlands Antilles (Deffeyes et al., 1964, 1965), and South Australia (Skinner, 1963, and others); (2) laboratory precipitation of protodolomite at normal pressure and temperature (Siegel, 1961); and (3) development of the seepage-refluxion model (Adams and Rhodes, 1960), based on regional distribution of dolomite and related carbonate-evaporite facies. Models developed from these studies applied within the framework of regional stratigraphy aid in recognition and genetic interpretation of at least certain types of dolomite, as well as time of dolomitization. This approach demonstrates that dolomitization is commonly, if not generally, a basic element of the original depositional system. The writers present results obtained by applying current models of dolomitization to the Edwards Formation of Texas, a widespread Lower Cretaceous carbonate unit which includes prominent dolomite facies.

Regional Stratigraphy

The Edwards Formation is characterized by carbonate grainstone and rudist bioherms and biostromes which were deposited or grew on an extensive, shallow-water, medium- to high-energy, marine platform. This feature, designated the Comanche platform (Fisher and Rodda, 1967), was constructed on the tectonically positive Llano and Devils River uplifts in Texas, the Coahuila Peninsula of northern Mexico, and associated smaller positive areas (Fig. 1). The regional platform includes smaller scale, locally designated units—the San Marcos platform, the Devils River platform, and the Coahuila platform. The Comanche platform was bounded on the east and south by a relatively deep-water oceanic basin—the Ancestral Gulf of Mexico—and on the north and west by an extensive shallow-water, open-marine basin—the North Texas-Tyler basin. Contemporaneous with grainstone deposition on the Comanche

---

[1] Manuscript received, October 5, 1967; accepted, December 19, 1967.
[2] Bureau of Economic Geology, The University of Texas at Austin.
Publication authorized by the Director, Bureau of Economic Geology.
Acknowledgement is made of critical reading of the manuscript by Peter T. Flawn, Director, L. F. Brown, Jr., and G. K. Eifler, Jr., Bureau of Economic Geology; Keith P. Young and Peter R. Rose, Department of Geology, The University of Texas at Austin; and O. T. Hayward, Department of Geology, Baylor University. Parts of this paper were presented by W. L. Fisher to the 3rd Annual Forum on Geology of Industrial Minerals, Lawrence, Kansas, on April 6, 1967.

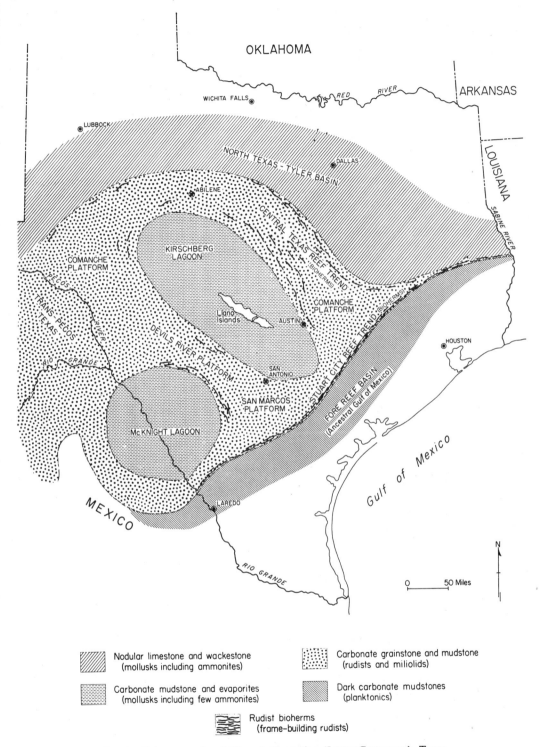

FIG. 1.—Paleogeography of Edwards Formation (Lower Cretaceous), Texas.

platform, low-energy basinal deposition on the north and west was chiefly of carbonate mud and clay; they include shelly and marly clay of the Walnut Formation and nodular carbonate mudstone and wackestone of the Goodland and Comanche Peak Formations. Deposition in the Ancestral Gulf of Mexico, on the basis of outcrops in northern Mexico (Smith, 1966) and the subsurface of South Texas, was mainly of dark carbonate mud and clay.

Two areas of periodically restricted deposition developed on the Comanche platform—the Kirschberg lagoon (designated by Fisher and Rodda, 1967, after Kirschberg Evaporite, Barnes, 1944), a broad, very shallow, shelf-type lagoon in Central Texas, and the McKnight lagoon (designated by Fisher and Rodda, 1967, after McKnight Formation of Winter, 1962), known from the subsurface of South Texas and outcrop of northern Mexico (Maverick basin). Alternating evaporites and shallow-water carbonate mudstone and grainstone were deposited in the Kirschberg lagoon; thin-bedded, ammonite-bearing black shale, carbonate mudstone, and evaporites were deposited in the McKnight lagoon (Lozo and Smith, 1964; Smith, 1966, and in press).

The name Edwards Formation as used herein includes rocks known as Edwards north of the Colorado, rocks commonly designated as "lower Edwards" or "Edwards B" in outcrop and subsurface south of the Colorado River, and the Edwards Plateau "unnamed lower unit" of Lozo and Smith (1964). Moore (1967) has defined similarly the Edwards Formation in west-central Texas. It is the approximate stratigraphic equivalent of the West Nueces and McKnight Formations and lower part of the Devils River Formation of southwestern Texas and northern Mexico (Lozo and Smith, 1964), the lower part of the Stuart City reef trend (Winter, 1962), and the Del Carmen Formation of Trans-Pecos Texas and northern Mexico (Maxwell and Dietrich, 1965; Maxwell et al., 1967).

Outcrop of the Edwards Formation generally parallels the Balcones escarpment and extends from the Parker-Hood County line on the north to northeastern Medina County on the southwest. It forms extensive outliers along the divide of the Colorado and Brazos Rivers in north-central and west-central Texas and is a prominent outcrop in the northeastern part of the Edwards Plateau and in local outliers on the Llano uplift.

The Edwards is a featheredge along the

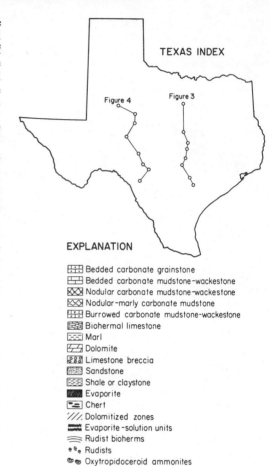

FIG. 2.—Index and explanation of symbols for stratigraphic cross sections, Figures 3–4.

northern margin of the Comanche platform and increases in thickness southeastward; sections up to 450 ft (123 m) thick are present in the subsurface of South Texas a few miles west of the eastern platform edge (Figs. 2, 3). Equivalent units farther south in Texas and northern Mexico (lower Devils River, West Nueces, McKnight, and Del Carmen) are as much as 600 ft (183 m) thick (Figs. 2, 4).

Throughout its extent the Edwards Formation (as here defined) is overlain by a thin sequence of approximately time-equivalent, open-marine marl and nodular carbonate mudstone containing oxytropidoceroid ammonites. This facies, characteristically basinal, is best developed in the North Texas-Tyler basin, where it constitutes the Kiamichi Formation. In the Edwards Plateau this facies is referred to informally as the "Dr. Burt ammonite beds" (Lozo

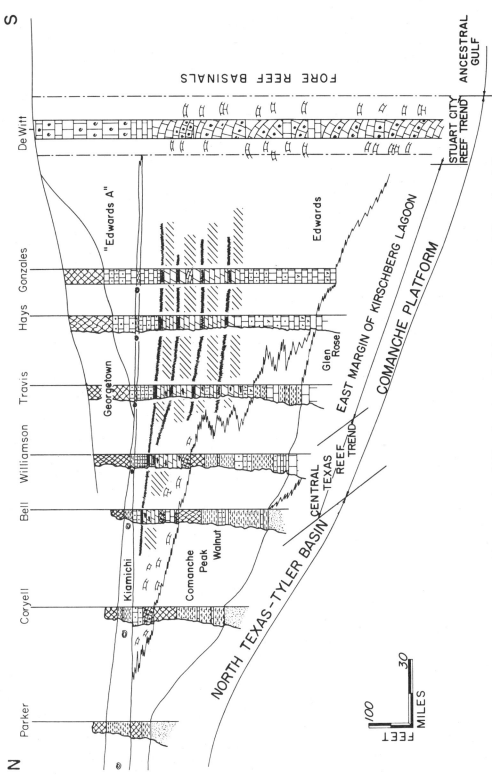

Fig. 3.—Stratigraphic cross section of Edwards and associated formations, east side of Comanche platform. Location of cross section and explanation of symbols shown on Figure 2.

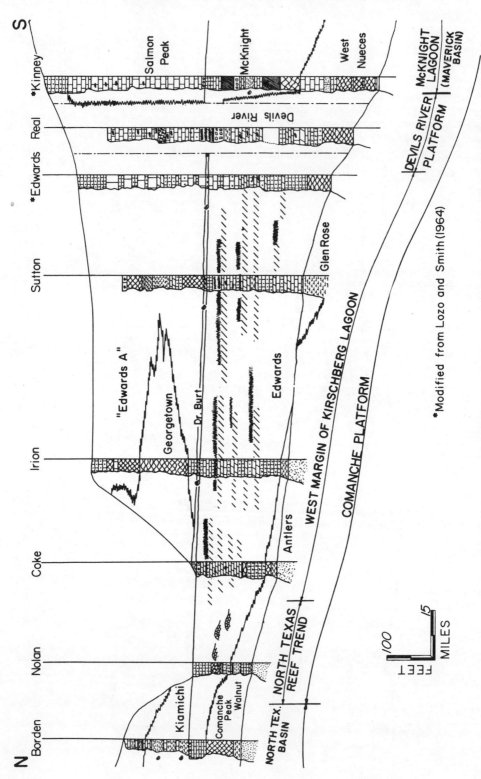

Fig. 4.—Stratigraphic cross section of Edwards and associated formations, west side of Comanche platform. Modified from Lozo and Smith (1964). Location of cross section and explanation of symbols shown on Figure 2.

and Smith, 1964; Moore, 1967; Young, 1966). In the subsurface of South Texas, it has been called "middle Edwards" by Tucker (1962) and "regional dense marker" by oil geologists. Along the northern and northwestern margin of the Comanche platform the Edwards overlies nodular carbonate mudstone of the Comanche Peak Formation and marl and clay of the Walnut Formation; a nodular mudstone (Goodland Formation) is the basinal equivalent of the Edwards. Southwest, toward the Colorado River, these underlying rocks interfinger with and gradually are replaced by the Edwards Formation. South of the Colorado the Edwards generally overlies limestone and marl of the Glen Rose Formation. The southern limit of the Edwards Formation is defined arbitrarily by lateral termination of the overlying oxytropidoceroid beds; rocks equivalent to the Edwards are included with overlying units and form an undifferentiated platform sequence designated the Devils River Formation. This platform sequence (Lozo and Smith, 1964; Smith, 1966, and in press) separates the Kirschberg and McKnight lagoonal facies.

Component facies.—Three primary depositional facies and one diagenetic facies are recognized in the outcrop and subcrop of the Edwards Formation (Fig. 1). Primary facies are (1) rudist biohermal-biostromal facies, (2) platform grainstone facies, and (3) lagoonal facies. A diagenetic dolomitic facies is superimposed on parts of the primary facies.

Rudist biohermal-biostromal facies.—Biohermal and biostromal reefs are well developed at many places in the Edwards Formation, chiefly along the edge of the Comanche platform and partly around platform lagoons (Figs. 1, 5). In outcrop, reefs are especially numerous in north-central Texas (Frost, 1967; Nelson, 1959; Rodda et al., 1966; Young, 1959) along the north edge of the Comanche platform and fringing the northern side of the Kirschberg lagoon. Few bioherms grew along the southern margin of the lagoon. Individual bioherms are relatively small-scale features with average relief of 10–20 ft (3–6 m) but locally as great as 50 ft (15 m). They consist of a core of whole rudist pelecypod shells (in growth position) enclosed in a dense, generally nonalgal, carbonate mudstone matrix. Flank deposits of coarse, angular, poorly sorted shell fragments dip radially from the core at inclinations of up to 35°. Several types of rudists are present in the reef, although the chief building forms are caprinids and radiolitids. Generally, well-bedded carbonate grainstone and local carbonate mudstone constitute the interreef areas. Interreef sedimentary rocks have well-sorted grains, except adjacent to reefs; nodular to locally bedded chert is a characteristic adjunct. Bioherms along the platform edge bordering the Ancestral Gulf of Mexico (commonly designated the Stuart City reef trend) are constructed of corals and stromatoporoids, in addition to a variety of rudists. Biohermal and biostromal reefs also fringed parts of the McKnight lagoon in South Texas and northern Mexico, and locally developed randomly on the Comanche platform. Oölitic shoal units (*e.g.*, Whitestone, Moffat, and Sweetwater lentils) developed locally along the northern margin of the Comanche platform. Some are coincident with present structural lows that may reflect local embayments of the platform, at the heads of which the oölite units developed under conditions of relatively high tidal velocity. The oölite units are comparable with tidal bar belts and marine sand belts in the Bahamas described by Ball (1967). Certain units in the Whitestone Member (Walnut Formation) have been interpreted as "offshore-bar" and "surge-channel" deposits by Moore (1964) and Moore and Martin (1966).

Platform facies.—The platform (or nonbiohermal) facies consists chiefly of bedded, well-

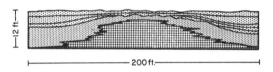

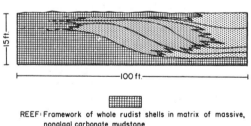

REEF: Framework of whole rudist shells in matrix of massive, nonalgal carbonate mudstone

REEF FLANK: Cross-bedded, poorly sorted, reef detritus and carbonate grainstone

INTERREEF: Flat-bedded, well sorted carbonate grainstone

FIG. 5.—Diagrammatic cross sections, rudist bioherms, Edwards Formation, Central Texas reef trend. Adapted from Nelson (1959).

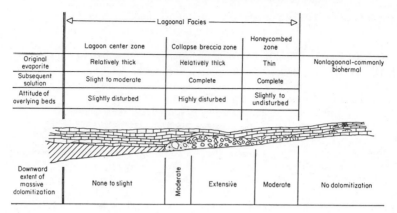

Fig. 6.—Lateral relation of evaporite and evaporite-solution units, Edwards Formation.

sorted, carbonate grainstone similar to interreef deposits, interbedded with carbonate mudstone and rudist carbonate wackestone. Grainstone commonly is cross-bedded and contains abundant miliolid foraminifers; mudstone is thin bedded or massive. Nodular to locally bedded chert is present in much of this facies in outcrop.

Lagoonal facies.—The lagoonal facies, as judged from marginal outcrop sections, consists chiefly of thin-bedded to nodular, carbonate mudstone and thoroughly burrowed carbonate mudstone; carbonate grainstone is less common. Gypsum is preserved locally. Evaporite-solution units composed of brecciated and randomly dipping beds are numerous and indicate original large extent of several evaporite deposits. Four principal evaporite or evaporite-solution units can be mapped in the centrally preserved parts of the Kirschberg lagoon (vicinity of Menard), but along the periphery of the lagoon only one or two evaporite-solution units are distinct. The lagoonal facies expanded through time; successively younger evaporite deposits are more widespread and onlap the surrounding, underlying platform and biohermal facies. Expansion coincided with northward progradation of bioherms fringing the north side of the Kirschberg lagoon. Evaporite deposits comparable in lithology and stratigraphic position with those of the Kirschberg lagoon characterize the McKnight lagoon of South Texas and northern Mexico.

Much of the original evaporite of the Kirschberg lagoon and most of the central part of the lagoonal facies are not preserved Reconstruction of the original lagoon is based on recognition in preserved sections of various evaporite-solution units that mark the original extent of deposition. The present nature of these evaporite-solution units depends chiefly on the original thickness of underlying evaporites, amount of evaporite removed, and subsequent groundwater solution and alteration. Lateral variation in the evaporite-solution units reflects position relative to the center of the original lagoon (Fig. 6). The lagoon center zone is marked by at least partial preservation of the underlying gypsum, and overlying beds are not disturbed extensively; this zone also is marked by the thickest original accumulation of evaporites. Marginal to the lagoon center is the collapse-breccia zone where the original underlying evaporite unit has been removed completely by subsequent solution; this zone is characterized by a lower unit of highly brecciated or conglomeratic rocks succeeded by a sequence with beds dipping in random directions, commonly at relatively steep angles. The brecciated and contorted rocks are especially permeable and therefore are typically altered; secondary coarsely crystalline calcite is common. Collapse-breccia units generally are 5–10 ft (1.5–3 m) thick; overlying strata commonly are undisturbed, which indicates that deposition and subsequent solution removal of evaporites within the Edwards Formation (Kirschberg facies) were intraformational and episodic. Marginal to the collapse-breccia zone and representing the maximum extent of original evaporite deposition is an evaporite-solution unit characterized by a thin, finely honeycombed rock of cuneiform or gash texture resulting from solution of the relatively minor amount of original evaporite. Overlying beds are disturbed only slightly. Evaporite-solution units generally are poorly exposed and difficult to see except in artificial exposures, but in the

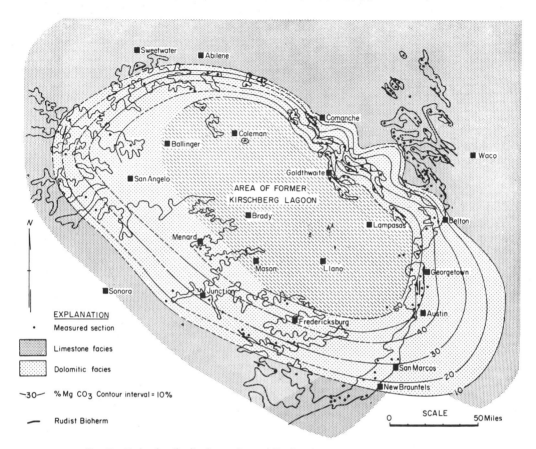

Fig. 7.—Dolomite distribution and generalized outcrop area, Edwards Formation. Contours are percent MgCO$_3$; CI = 10 percent.

preserved parts of the lagoon proper they form distinct patterns on aerial photographs. In outcrops of the marginal facies of the lagoon, only the upper one or two solution units, representing the maximum extent of the lagoon, are distinct.

Dolomite facies.—Regional geologic mapping and chemical analyses of approximately 1,200 samples from 250 localities throughout the outcrop area of the Edwards Formation (Rodda *et al.*, 1966) permit reasonably detailed delineation of the dolomite facies (Fig. 7). Subsurface distribution of dolomite was delineated by Rogers (1967). Dolomite in the Edwards Formation is partly coincident with, but chiefly marginal to, the Kirschberg lagoonal facies. Association of Edwards dolomite and evaporite is similar to that observed on carbonate shelves and platforms throughout the geologic column.

Two main types of dolomite are present in the Edwards Formation, designated stratal dolomite and massive dolomite (Table I). Stratal dolomite is a tightly knit and intergrown mosaic of dolomite crystals. Individual crystals are very small, generally less than 10 μ in diameter; rhombs are poorly developed because of the crowded or intergrown fabric. Porosity is slight and permeability is practically nil. Stratal dolomite is regularly to irregularly laminated or thin bedded and generally is associated with thin-bedded carbonate mudstone. Locally, thin beds of ripple-marked, cross-laminated carbonate grainstone containing carbonate mudstone clasts are interbedded with laminated dolomite; the dolomitic mudstone fills troughs of grainstone ripples (Fig. 8). Fossils are scarce except in associated stromatolitic carbonate mudstone. Mudcracks and rip-up mud clasts are common in stratal dolomite and associated carbonate mudstone. Individual dolomite units generally are less than 2 ft (0.6 m) thick. Evaporite-so-

Table I. Characteristics of Principal Dolomite Types, Edwards Formation, Texas

	Stratal Dolomite	Massive Dolomite
Grain size	Very fine, average 10 μ	Very fine to coarse, 10–150 μ, generally more than 30 μ
Fabric	Tightly knit and intergrown mosaic	Loosely knit mass of euhedral crystals
Porosity	Very low	Moderate to high; pinpoint to vuggy
Crystal shape	Rhombs poorly developed because of crowding and packing	Well-developed rhombs
Occurrence	As specific mudstone beds, selectively in burrow fill and as fossil replacement, and as matrix of carbonate grainstone	Replacement for variety of original rock and textural types other than mudstone
Fossils	Scarce	Common to abundant, as typical of Edwards grainstone facies
$MgCO_3$ content among representative samples	Low to high, irregular	Low to very high, but with strongly bimodal frequency; few intermediate values
Fe_2O_3 content	Relatively high	Relatively low
Thickness of individual dolomite units	Thin, generally less than 2 ft (0.6 m)	Thick, generally greater than 2 ft (0.6 m), commonly 10 ft (3 m) up to 80 ft (24 m)
Bedding	Generally laminated to thin bedded	Variable, that of host limestone
Sedimentary structures	Mudcracks, stromatolites, mud clasts	Variable, that of host limestone
Evaporite or evaporite-solution units	Very thin, less than 1 in. (1.5 cm) to absent, commonly crustlike	Generally predominant, collapse-breccia or honeycombed solution units overlie dolomite unit
Geographic occurrence	Best developed along southern margin of lagoonal facies	Best developed along northern reef margin of lagoonal facies
Time of formation	Prelithification, early diagenetic	Postlithification, early secondary
Mode of formation	Metasomatic replacement of carbonate muds in supratidal zone	Metasomatic replacement of permeable grainstone by seepage refluxion of lagoonal brine

lution units are scarce to common, less than 1 in. (1.5 cm) thick, and crustlike (Fig. 8). Magnesium carbonate content of stratal dolomite and associated carbonate mudstone and dolomitic mudstone ranges irregularly from low to high (Fig. 9). In addition to dolomite mudstone beds, dolomite of similar grain size and fabric locally is present as matrix in carbonate grainstone and as fill of animal burrows (Fig. 8).

Massive dolomite ranges in grain size from fine to coarse, though most crystals are greater than 30 μ in diameter. Fabric is a loosely knit mass of euhedral crystals, resulting in a moderately to highly porous and moderately permeable rock (Table I). Massive dolomite typically replaces fossiliferous carbonate grainstone and contains relict structures of the host limestone. Carbonate mudstone or reef cores of rudist bioherms with matrix of dense carbonate mudstone rarely show massive dolomitization, even where adjacent grainstone is extensively dolomitized, a fact that indicates an initial permeability control for this type of dolomitization.

Massive dolomite sequences are thick, commonly about 10 ft (3 m), although locally up to 80 ft (24 m). Many different relict structures indicate a variety of host limestones in a single massive dolomite sequence. Sequences consisting of several original carbonate grainstone types may be uniformly dolomitized. Dolomite of this type generally underlies prominent evaporite-solution units and is best developed below the lagoonward edge of the honeycombed zone and outer margin of the collapse-breccia zone (Fig. 6). Magnesium carbonate content of massive dolomite and associated limestone or dolomitic limestone is either high or low, with a strong bimodal frequency among several samples (Fig. 9); similar distribution of magnesium carbonate has been reported for dolomite and dolomitic limestone from many areas (e.g., Graf, 1960). Such distribution indicates that, where conditions for massive dolomitization were suitable (e.g., permeable grainstone), dolomitization typically was thorough.

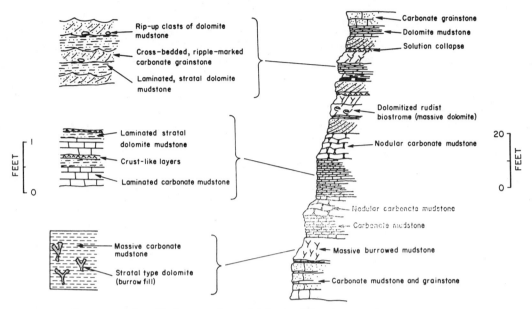

FIG. 8.—Common stratigraphic features of stratal dolomite, Edwards Formation, Mountain Home section, northwestern Kerr County.

EDWARDS DOLOMITE ORIGIN

Both stratal and massive dolomite bodies are judged to be products of metasomatic replacement of calcium carbonate from contact with magnesium-enriched brines. The two dolomite types originated in similar chemical environments, although the physical environments during formation and the time of dolomitization differed. Stratal dolomite is most common along the southern margin of the Kirschberg lagoon; massive dolomite is most common along the northern margin of the lagoon. This distribution was chiefly a function of the configuration and composition of the lagoon, its surrounding areas, and associated facies (Fig. 10).

The northern margin of the lagoon was marked by fringing, rudist, biohermal, and biostromal reefs, separating the lagoon from the shallow, open-marine North Texas-Tyler basin. The northern reef margin was characterized by relatively high and rugged relief and was a depositional environment of moderate to high energy. Reef-flank and interreef deposits are fine to very coarse carbonate grainstone and reef detritus, and relatively small amounts of carbonate mudstone. The southern margin of the lagoon, in contrast, was a zone of very low relief with an extensive intertidal-supratidal area; biohermal growth was sparse. Except during periods of storms, this area was a low-energy environment in which carbonate mud and platform grainstone were deposited.

Stratal dolomite.—The very fine grain size, thin bedding, mudcracks, paucity of fossils and the association of stratal dolomite with thin-bedded carbonate mudstone and stromatolites (Table I) suggest deposition under conditions described for certain modern carbonate supratidal and intertidal environments (Curtis et al., 1963; Deffeyes et al., 1965; Ebanks and Tebbutt, 1966; Illing et al., 1965; Kinsman, 1966; Shearman, 1963; Shinn, 1964; Shinn et al., 1965; Skinner, 1963). These are chiefly low-latitude, carbonate-mud tidal flats adjacent to warm shallow seas, where dolomite forms in soft aragonitic or calcitic mud commonly in association with gypsum or halite. Requisites for dolomitization appear to be sedimentation essentially at sea level in a climate that produces a relatively high net evaporation loss. Under such conditions, evaporation of water trapped at the surface, and salt water brought near the surface by capillary movement, results in concentration of brines in a thin zone at and just below the sediment surface. Calcium content is reduced by precipitation of either calcium carbonate or gypsum; the remaining heavy brine with a high Mg:Ca ratio is capable of inducing metasomatic replacement of carbonate mud on contact. First to form is poorly ordered, calcium-rich

Ca·Mg(CO$_3$) or protodolomite. Subsequently, protodolomite crystals grow larger and are better ordered and ultimately, depending on conditions of pH, salinity, and temperature, convert to dolomite. Crystals are generally 1–10 μ in diameter. Gypsum precipitated in this process may be removed by solution during subsequent flooding of the tidal flat or during periods of rainfall. Specific geochemical features of the process of dolomitization in supratidal sediments are given by Berner (1966), Deffeyes et al. (1965), Degens (1965), Illing et al. (1965), Kinsman (1966), Land (1966), Peterson and von der Borch (1966), Schmalz (1966), and Shinn et al. (1965).

In addition to sedimentary features of Edwards stratal dolomite which are similar to those described for modern carbonate tidal flats, several lines of evidence show that the dolomite was formed before lithification: (1) very fine grain size and intergrown, dense fabric, resulting from replacement in nonsupported soft muds, are in contrast to the porous fabric derived from certain postlithification replacement; (2) nondolomitization of solid shells and protected zones beneath shells is common, though surrounding sediment is thoroughly dolomitized; (3) dolomitized and nondolomitized rocks of identical texture are interbedded; (4) dolomitized units are thin; and (5) occurrence is chiefly as mudstone, a rock type of low permeability. Mudstone generally is not dolomitized by refluxing brines, as noted in the discussion of massive dolomite.

The replacement of thin carbonate grainstone or shelly mudstone by dolomite and the interbedding of the dolomitized grainstone with dolomite mudstone of the stratal type suggest that dolomitization of the grainstone could have occurred within a thin zone below the supratidal surface by downward- and outward-sinking magnesium-enriched brines. Nondolomitized areas under shells suggest that there was downward movement of solutions with the dense shells acting as shields. In such manner, underlying or adjacent intertidal deposits also might be dolomitized by a process involving refluxion, mechanically similar to that described below as massive seepage refluxion. Dolomitization was, however, fundamentally a part of the supratidal environment. Distinction of grainstone dolomitized in this manner from grainstone dolomitized by massive seepage refluxion is based on stratigraphic association (thin-bedded dolomite grainstone interbedded with thin stratal-type dolomite mudstone). That the heavy brine solutions did not sink far and probably were confined to unconsolidated sediments is indicated by the characteristic thinness of stratal dolomite units—generally less than 2 ft (0.6 m).

Dolomite beds similar to Edwards stratal dolomite are known from diverse areas and from practically all parts of the geologic column; some have been interpreted specifically as the product of early diagenesis in the supratidal zone. Dolomites that have features of the stratal type have been reported recently from ancient rocks, including the Ordovician Platteville Formation, Wisconsin (Asquith, 1967); Devonian rocks, Indiana (Bluck, 1965); Permian Phosphoria Formation, Bighorn basin (Campbell, 1962); Triassic Dachstein Formation, Salzburg (Fischer, 1964); Silurian rocks, Pennsylvania (Gwinn and Glack, 1965); Ordovi-

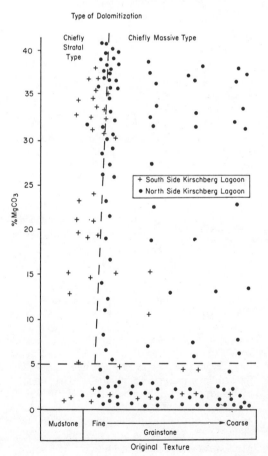

Fig. 9.—Distribution of magnesium carbonate content *versus* relict texture.

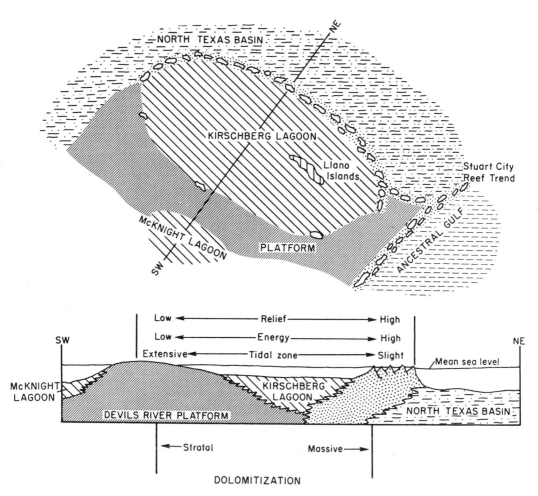

Fig. 10.—Reconstructed physiography and depositional system of Kirschberg lagoon.

cian Newmarket Formation, Maryland (Matter, 1967); Devonian Manlius Formation, New York (LaPorte, 1967); Devonian Sevy Dolomite, Utah and Nevada (Osmond, 1962); Devonian Jeffersonville Limestone, Indiana (Perkins, 1963); Ordovician Stony Mountain and Silurian Interlake Formations, Williston basin (Roehl, 1967); Ordovician Beckmantown Group, Maryland (Sarin, 1962); Mississippian Macumber Formation, Maritime Provinces, Canada (Schenck, 1967); Ordovician Nittany Dolomite, Pennsylvania (Spelman, 1966); and Devonian Martin Formation, Arizona (Teichert, 1965).

Massive dolomite.—In contrast to the very early diagenetic, prelithification formation of stratal dolomite, the formation of massive dolomite was clearly secondary. Abundant relict sedimentary structures and fossils indicate that the host rock was a carbonate grainstone or coarse bioclastic limestone deposited in a medium- to high-energy marine environment, rather than a carbonate mudstone deposited in a low-energy environment. The fabric is loosely knit and grain size is relatively large.

The relation of Edwards massive dolomite to evaporite (or evaporite-solution) units in the adjacent Kirschberg lagoon indicates that formation clearly is attributable to the seepage-refluxion model that was developed by Adams and Rhodes (1960) to explain the origin of Permian shelf-edge dolomite of West Texas. Structural and petrologic features as well as depositional facies of Edwards massive dolomites are similar to shelf- and platform-edge dolomites of the West Texas Permian and to many others.

According to the seepage-refluxion model,

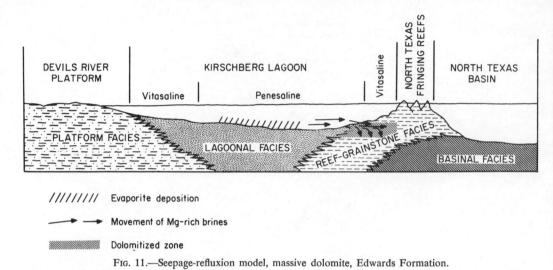

Fig. 11.—Seepage-refluxion model, massive dolomite, Edwards Formation.

highly concentrated brines, particularly those remaining after precipitation of lower rank salts (*e.g.*, gypsum), migrate to depressions of the lagoon floor; overlying lighter waters prevent further concentration. As depressions fill, brines spill over and escape basinward through permeable substrate zones, generally formed by carbonate sand and clastic sediments bordering the fringing reef or other barrier at the lagoon margin; the parts of the lagoon floor covered by dense anhydrite or halite are practically impermeable. Seepage through permeable zones may be slow and not affect lagoonal circulation pattern by rapid loss of heavy brines; hence the supply of magnesium-enriched brines could be maintained. Refluxing brines are denser than the connate water they replace; seepage is downward and outward toward the basin. Extent of dolomitization is limited by changes in composition and character of refluxing brine and host rock. The changes involve chiefly a lowering of the high Mg:Ca ratio as dolomitization proceeds, and alteration of brine temperature, salinity, and alkalinity as brine and connate water interact. Lateral migration of the seepage zone results from transgression of the evaporite lagoon; during regression of the lagoon the previous spillway probably is sealed by a cover of impermeable evaporites.

Application of the seepage-refluxion concept to Edwards massive dolomite suggests the model in Figure 11. Anhydrite was precipitated in the central and north-central part of the Kirschberg lagoon which was barred on the north by a series of fringing rudist reefs (Fig. 10). Heavy, magnesium-enriched brines, formed after precipitation of the anhydrite, spilled from the lower part of the lagoon into the relatively permeable reef-grainstone facies along the northern reef margin. Dolomitization was accomplished by downward and outward migration of refluxing brines through permeable grainstone.

Field relations of Edwards massive dolomite support the concept of origin by seepage refluxion. Although massive dolomite is present along the entire margin of the Kirschberg lagoonal facies, it is developed best on the north side in a belt coincident with the nearly uniformly permeable grainstone-biohermal facies (Fig. 12; *cf.* Fig. 7). Within the dolomitized grainstone facies, reef cores with dense carbonate mudstone matrix are not dolomitized, though surrounding reef-flank and interreef deposits may be altered completely. Locally, nondolomitized grainstone underlies lenses of impermeable mudstone; beyond the mudstone lens, dolomitization extends farther downward. Finally, variation in magnesium content of rocks in this facies is markedly bimodal; values are either high or low but generally not intermediate (Fig. 9). This frequency suggests that if host-rock permeability was sufficient to permit refluxing, it was sufficient for extensive dolomitization; if not, little dolomitization occurred. The bimodal variation contrasts with the irregular variation of stratal dolomite that formed in soft muds.

Massive dolomite bodies of the Edwards Formation characteristically underlie evaporite-solution units. Thickest dolomite units generally underlie the outer part of the col-

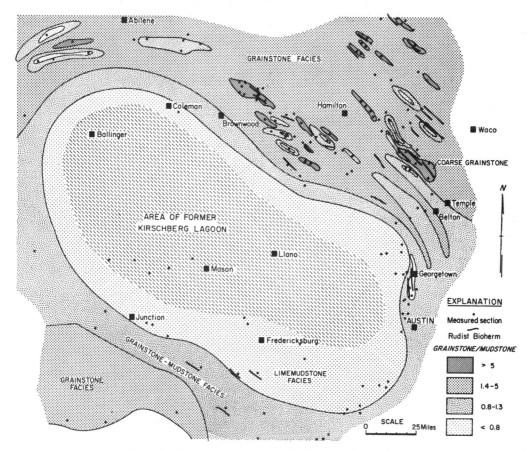

FIG. 12.—Grainstone:mudstone ratio, Edwards Formation.

lapse-breccia zone and inner part of the honeycombed zone (Fig. 6), a feature which suggests that the most extensive dolomitization occurred just marginal to the thicker lagoonal evaporite sequences, approximately at the boundary of the lagoonal and grainstone facies (Fig. 11). The lagoonal margin of the grainstone facies would have been the optimum zone for brine escape. Only moderately thick massive dolomite units, that average 5 ft (1.5 m) in thickness, underlie the lagoonward margin of the collapse-breccia zone (Fig. 6), presumably in consequence of earlier seal by impervious anhydrite as the lagoon expanded. Similarly, only moderately thick zones of massive dolomite underlie the marginal zone of the original lagoon as indicated by thin honeycombed units (Fig. 6). As interpreted, these zones were in the path of spilling brines for a shorter period than the intermediate zone.

Alternation of laterally persistent massive dolomite units and nondolomitized units suggests that distinct episodes of dolomitization occurred. Alternation of disturbed evaporite-solution units and undisturbed strata indicates further that anhydrite deposited during a specific period was removed by solution within a relatively short period of time, before the next period of evaporite deposition.

Possible modern examples of massive dolomitization by seepage refluxion are found in Pliocene-Pleistocene rocks on Bonaire, Netherlands Antilles, reported by Deffeyes et al. (1965), and the rocks of certain Pacific atolls reported by Berner (1965) and Schlanger (1965). Lack of definite modern examples probably is due to scarcity of extensive modern carbonate-evaporite shelves or platforms, and the obvious difficulty in observing the process in a modern environment. By contrast, many ancient dolomite bodies apparently were formed by seepage refluxion. Significant parts of many ancient carbonate shelves and platforms contain dolomite which has petrologic

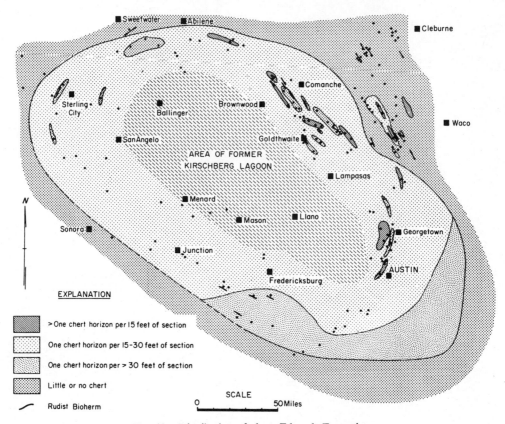

Fig. 13.—Distribution of chert, Edwards Formation.

and structural properties, as well as the association of lagoonal evaporites and shelf-edge dolomite, typical of dolomite herein described as massive and attributed to seepage refluxion. Massive dolomite bodies, some specifically interpreted as the product of seepage refluxion, have been described recently in Permian rocks, West Texas (Adams and Rhodes, 1960); Mississippian Madison Formation, Wyoming and Montana (Andrichuk, 1955); Upper Devonian rocks, Alberta (Andrichuk, 1958); Mississippian Lodgepole Formation, Montana (Cotter, 1966); middle and lower Paleozoic rocks, Teton Mountains (Dixon and Reeves, 1965); Mississippian rocks, Saskatchewan (Edie, 1958); Devonian rocks, Montana (Lewis and Elliott, 1962); Mississippian rocks, Alberta (Murray and Lucia, 1967); Silurian rocks, Michigan (Sharma, 1966); Ordovician Nittany Dolomite, Pennsylvania (Spelman, 1966); and Silurian rocks, Nevada (Winterer and Murphy, 1960).

Fault and fracture dolomite.—Concentration of dolomite along faults, joints, and fractures has been noted widely (Fairbridge, 1957, p. 159, 172). Locally in the Edwards, dolomite is present along small factures or lining cavities associated with both stratal and massive dolomites; much larger grain size (100–200 μ) readily distinguishes this type of dolomite from surrounding finer grained dolomite. It presumably is of late secondary origin and a product of Pleistocene-Holocene weathering. Although widespread, such dolomite is volumetrically insignificant. No regional relation was noted between Edwards dolomite and regional fault or fracture patterns. Pulverulent limestone, another Pleistocene-Holocene surficial weathering product associated with the Edwards Formation, also is widespread but volumetrically insignificant.

CHERT DISTRIBUTION

Chert distribution in the Edwards Formation is coextensive with Edwards dolomite (Fig. 13). Like dolomite, chert is present in a belt marginal to the Kirschberg lagoon, although the chert facies is slightly more extensive. A

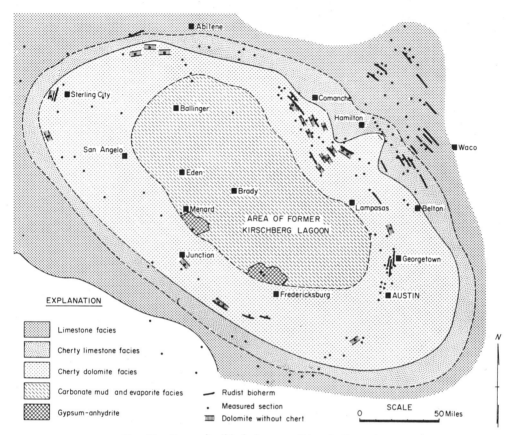

Fig. 14.—Composite lithofacies map, Edwards Formation.

composite lithofacies map of the Edwards shows a lagoon-margin zone of cherty dolomite, grading peripherally to a narrow cherty limestone zone, which grades outward to limestone free of dolomite and chert (Fig. 14). Only about 3 percent of the chert in the Edwards is bedded; the rest is present as nodules from less than 1 in. in diameter to masses as large as 4 by 10 ft (1.2 by 3 m). Of approximately 150 outcrops of dolomite examined, 70 percent contain chert. Biohermal reef-flank deposits generally are not cherty, but within the dolomite facies they are characteristically thoroughly dolomitized. As such they constitute the only dolomite rock unit consistently free of chert. Absence of chert in reef-core and reef-flank deposits apparently is characteristic of many ancient reefs. Most dolomite in the Edwards is the massive type formed by replacement of carbonate grainstone, and a high correlation of chert and dolomite means a high correlation of chert and original carbonate grainstone. However, within the cherty limestone facies less than 20 percent of the chert is present with carbonate mudstone or carbonate wackestone, on the basis of approximately 250 field observations in this facies. This fact indicates a high correlation of chert and carbonate grainstone independent of dolomite. Edwards chert typically contains relict structures and fossils of enclosing limestone which, with other evidence, indicate a secondary origin for most of the chert. Interstitial silica is less than 1 percent in both Edwards limestone and dolomite; there is no significant difference among the several rock types (*i.e.*, dolomite, cherty dolomite, limestone, and cherty limestone).

An explanation of the origin of Edwards chert must be based on consideration of such factors as (1) a secondary origin, (2) association of chert with carbonate grainstone deposited in a medium- to high-energy environment, which suggests a permeability control in formation, (3) regional coextensive distribution of chert and dolomite marginal to a restricted carbonate-evaporite lagoon, which suggests a mode of origin related to that of dolomite, and (4) consistent absence of chert in the reef-core and reef-flank environments within the cherty grainstone facies, which suggests an environmental control of the silica source (*e.g.*, possible environmental restriction of siliceous sponges to the now cherty interreef environment). Investigations of the origin of Edwards chert by the writers are now in progress.

REFERENCES CITED

Adams, J. E., and M. L. Rhodes, 1960, Dolomitization by seepage refluxion: Am. Assoc. Petroleum Geologists Bull., v. 44, p. 1912–1920.

Andrichuk, J. M., 1955, Mississippian Madison Group stratigraphy and sedimentation in Wyoming and southern Montana: Am. Assoc. Petroleum Geologists Bull., v. 39, p. 2170–2210.

———— 1958, Stratigraphy and facies analysis of Upper Devonian reefs in Leduc, Stettler, and Redwater areas, Alberta: Am. Assoc. Petroleum Geologists Bull., v. 42, p. 1–93.

Asquith, G. B., 1967, The marine dolomitization of the Mifflin Member, Platteville Limestone in southwest Wisconsin: Jour. Sed. Petrology, v. 37, p. 311–326.

Ball, M. M., 1967, Carbonate sand bodies of Florida and the Bahamas: Jour. Sed. Petrology, v. 37, p. 556–591.

Barnes, V. E., 1944, Gypsum in the Edwards Limestone of Central Texas: Texas Univ. Pub. 4301, p. 35–46.

Berner, R. A., 1965, Dolomitization of the mid-Pacific atolls: Science, v. 147, p. 1297–1299.

———— 1966, Chemical diagenesis of some modern carbonate sediments (abs.): Am. Assoc. Petroleum Geologists Bull., v. 50, p. 606.

Bluck, B. J., 1965, Sedimentation of Middle Devonian carbonates, southeastern Indiana: Jour. Sed. Petrology, v. 35, p. 656–682.

Campbell, C. V., 1962, Depositional environments of Phosphoria Formation (Permian) in southeastern Bighorn basin, Wyoming: Am. Assoc. Petroleum Geologists Bull., v. 46, p. 478–503.

Cotter, E., 1966, Limestone diagenesis and dolomitization in Mississippian carbonate banks in Montana: Jour. Sed. Petrology, v. 36, p. 765–774.

Curtis, R., *et al.*, 1963, Association of dolomite and anhydrite in recent sediments of the Persian Gulf: Nature, v. 197, p. 679–680.

Deffeyes, K. S., F. J. Lucia, and P. K. Weyl, 1964, Dolomitization—Observations on the island of Bonaire, Netherlands Antilles: Science, v. 143, p. 678–679.

———— and ———— 1965, Dolomitization of recent and Plio-Pleistocene sediments by marine evaporite waters on Bonaire, Netherlands Antilles, *in* Dolomitization and limestone diagenesis: Soc. Econ. Paleontologists and Mineralogists Spec. Pub. 13, p. 71–88.

Degens, E. T., 1965, Geochemistry of sediments—a brief survey: New York, Prentice-Hall, 342 p.

Dixon, J. R., and C. C. Reeves, Jr., 1965, Rhythmic carbonate petrology of some lower and middle Paleozoic rocks, west flank, Teton Mountains, Wyoming: Jour. Sed. Petrology, v. 35, p. 704–721.

Ebanks, W. J., Jr., and G. E. Tebbutt, 1966, Diagenetic modification of recent sediments associated with a limestone island (abs.): Am. Assoc. Petroleum Geologists Bull., v. 50, p. 611–612.

Edie, R. W., 1958, Mississippian sedimentation and oil fields in southeastern Saskatchewan: Am. Assoc. Petroleum Geologists Bull., v. 42, p. 94–126.

Fairbridge, R. W., 1957, The dolomite question, *in* Regional aspects of carbonate deposition—a symposium: Soc. Econ. Paleontologists and Mineralogists Spec. Pub. 5, p. 125–178.

Fischer, A. G., 1964, The Lofer cyclothems of the Alpine Triassic, *in* Symposium on cyclic sedimentation: Kansas Geol. Survey Bull. 169, p. 107–149.

Fisher, W. L., and P. U. Rodda, 1967, Stratigraphy and genesis of dolomite, Edwards Formation (Lower Cretaceous) of Texas, *in* E. E. Angino and R. G. Hardy, eds., Proceedings of Third Forum on Geology of Industrial Minerals: Kansas Geol. Survey Spec. Dist. Pub. 34, p. 52–75.

Frost, J. G., 1967, The Edwards Limestone of Central Texas, *in* Symposium on Comanchean stratigraphy: Permian Basin Sec., Soc. Econ. Paleontologists and Mineralogists Spec. Pub. 67–8, p. 133–156.

Graf, D. L., 1960, Geochemistry of carbonate sediments and sedimentary carbonate rocks, pt. 2, Sedimentary carbonate rocks; Illinois Geol. Survey Circ. 298, 45 p.

Gwinn, V. E., and W. C. Glack, 1965, Penecontemporaneous dolomite, Upper Silurian, Pennsylvania (abs.): Geol. Soc. America Abstracts for 1964 Ann. Mtg., p. 80–81.

Illing, L. V., A. J. Wells, and J. C. M. Taylor, 1965, Penecontemporary dolomite in the Persian Gulf, *in* Dolomitization and limestone diagenesis: Soc. Econ. Paleontologists and Mineralogists Spec. Pub. 13, p. 89–111.

Kepper, J. C., Jr., 1966, Primary dolostone patterns in the Utah-Nevada Middle Cambrian: Jour. Sed. Petrology, v. 36, p. 548–562.

Kinsman, D. J. J., 1966, Supratidal diagenesis of carbonate and noncarbonate sediments in arid regions (abs.): Am. Assoc. Petroleum Geologists Bull., v. 50, p. 620.

Land, L. S., 1966, Diagenetic *versus* post-diagenetic dolomitization (abs.): Am. Assoc. Petroleum Geologists Bull., v. 50, p. 621–622.

LaPorte, L. F., 1964, Supratidal dolomitic horizons within the Manlius Formation (Devonian) of New York, *in* E. G. Purdy and J. Imbrie, Geol. Soc. America Guidebook Field Trip No. 2, Miami Beach Mtg., p. 59–66.

———— 1967, Carbonate deposition near mean sea-level and resultant facies mosaic: Manlius Formation (Lower Devonian) of New York State: Am. Assoc. Petroleum Geologists Bull., v. 51, p. 73–101.

Lewis, P. J., and J. K. Elliott, 1962, Creation of the Devonian dolomites, *in* The Devonian System of Montana and adjacent areas: Billings Geol. Soc. 13th Ann. Fld. Conf., p. 35–41.

Lozo, F. E., and C. I. Smith, 1964, Revision of Comanche Cretaceous stratigraphic nomenclature, southern Edwards Plateau, southwest Texas: Gulf Coast Assoc. Geol. Socs. Trans., v. 14, p. 285–306.

Matter, A., 1967, Tidal flat deposits in the Ordovician of western Maryland: Jour. Sed. Petrology, v. 37, p. 601–609.

Maxwell, R. A., and J. E. Dietrich, 1965, Geologic summary of the Big Bend region: West Texas Geol. Soc. Pub. 65-51 (Field Trip Guidebook), p. 10–33.

—— et al., 1967, Geology of Big Bend National Park, Brewster County, Texas: Texas Univ. Pub. 6711, 320 p.

Moore, C. H., Jr., 1964, Stratigraphy of the Fredericksburg division, south-central Texas: Texas Univ. Bur. Econ. Geology Rept. Inv. 52, 48 p.

—— 1967, Stratigraphy of the Edwards and associated formations, west-central Texas: Gulf Coast Assoc. Geol. Socs. Trans., v. 17, p. 61–75.

—— and K. G. Martin, 1966, Comparison of quartz and carbonate shallow marine sandstones, Fredericksburg Cretaceous, central Texas: Am. Assoc. Petroleum Geologists Bull., v. 50, p. 981–1000.

Murray, R. C., and F. J. Lucia, 1967, Cause and control of dolomite distribution by rock selectivity: Geol. Soc. America Bull., v. 78, p. 21–36.

Nelson, H. F., 1959, Deposition and alteration of the Edwards Limestone, central Texas, in F. E. Lozo, Symposium on Edwards Limestone in central Texas: Texas Univ. Pub. 5905, p. 21–95.

Osmond, J. C., 1962, Stratigraphy of Devonian Sevy Dolomite in Utah and Nevada: Am. Assoc. Petroleum Geologists Bull., v. 46, p. 2033–2056.

Perkins, R. D., 1963, Petrology of the Jeffersonville Limestone (Middle Devonian) of southeastern Indiana: Geol. Soc. America Bull., v. 74, p. 1335–1354.

Peterson, M. N. A., and C. C. von der Borch, 1966, Rates and mechanisms in formation of dolomite (abs.): Am. Assoc. Petroleum Geologists Bull., v. 50, p. 631–632.

Rodda, P. U., et al., 1966, Limestone and dolomite resources, Lower Cretaceous rocks, Texas: Texas Univ. Bur. Econ. Geology Rept. Inv. 56, 286 p.

Roehl, P. O., 1967, Stony Mountain (Ordovician) and Interlake (Silurian) facies analogs of recent low-energy marine and subaerial carbonates, Bahamas: Am. Assoc. Petroleum Geologists Bull., v. 51, p. 1979–2032.

Rogers, J. K., 1967, Comparison of some Gulf Coast Mesozoic carbonate shelves: Gulf Coast Assoc. Geol. Socs. Trans., v. 17, p. 49–60.

Rose, P. R., 1963, Comparison of the type El Abra of Mexico with "Edwards reef trend" of south-central Texas, in Geology of Peregrina Canyon and Sierra de El Abra, Mexico: Corpus Christi Geol. Soc. Ann. Field Trip Guidebook, p. 57–64.

Sarin, D. D., 1962, Cyclic sedimentation of primary dolomite and limestone: Jour. Sed. Petrology, v. 32, p. 451–471.

Schenck, P. E., 1967, The Macumber Formation of the Maritime Provinces, Canada—a Mississippian analogue to recent strand-line carbonates of the Persian Gulf: Jour. Sed. Petrology, v. 37, p. 365–376.

Schlanger, S. O., 1965, Dolomite-evaporite relations on Pacific Islands: Tohoku Univ. Sci. Repts., 2d ser. (Geology), v. 37, p. 15–29.

Schmalz, R. E., 1966, Role of kinetics in early diagenesis of carbonate sediments (abs.): Am. Assoc. Petroleum Geologists Bull., v. 50, p. 633–634.

Sharma, G. D., 1966, Geology of Peters Reef, St. Clair County, Michigan: Am. Assoc. Petroleum Geologists Bull., v. 50, p. 327–350.

Shearman, D. J., 1963, Demonstration of recent anhydrite, gypsum, dolomite, and halite from the coastal faults of the Arab shore of the Persian Gulf: Geol. Soc. London Proc., no. 1607, p. 63–64.

Shinn, E. A., 1964, Recent dolomite, Sugarloaf Key: Geol. Soc. America Guidebook Field Trip No. 1, South Florida carbonate sediments, p. 62–67.

—— R. N. Ginsburg, and R. M. Loyd, 1965, Recent supratidal dolomite from Andros Island, Bahamas, in Dolomitization and limestone diagenesis: Soc. Econ. Paleontologists and Mineralogists Spec. Pub. 13, p. 112–123.

Siegel, F. R., 1961, Factors influencing the precipitation of dolomitic carbonates: Kansas Geol. Survey Bull. 152, pt. 5, p. 127–158.

Skinner, H. C. W., 1963, Precipitation of calcium dolomite and magnesian calcites in the southeast of South Australia: Am. Jour. Science, v. 261, p. 449–472.

Smith, C. I., 1966, Physical stratigraphy and facies analysis, Lower Cretaceous formations, northern Coahuila, Mexico: Unpub. dissert., Dept. Geology, Michigan Univ., 157 p.

—— in press, Physical stratigraphy and facies analysis, Lower Cretaceous formations, northern Coahuila, Mexico: Texas Univ. Bur. Econ. Geology Rept. Inv.

Spelman, A. R., 1966, Stratigraphy of Lower Ordovician Nittany Dolomite in central Pennsylvania: Pennsylvania Geol. Survey Gen. Geol. Rept. G 47, 187 p.

Teichert, C., 1965, Devonian rocks and paleogeography of central Arizona: U. S. Geological Survey Prof. Paper 464, 181 p.

Tucker, D. R., 1962, Subsurface Lower Cretaceous stratigraphy, Central Texas, in Contributions to the geology of South Texas: San Antonio, South Texas Geol. Soc., p. 177–216.

Winter, J. A. F., 1962, Fredericksburg and Washita strata (subsurface Lower Cretaceous), southwest Texas, in Contributions to the geology of South Texas: San Antonio, South Texas Geol. Soc., p. 81–115.

Winterer, E. L., and M. A. Murphy, 1960, Silurian reef complex and associated facies, central Nevada: Jour. Geology, v. 68, p. 117–139.

Young, K. P., 1959, Edwards fossils as depth indicators, in F. E. Lozo, Symposium on Edwards Limestone in central Texas: Texas Univ. Pub. 5905, p. 97–104.

—— 1966, Texas Mojsisovicziinae (Ammonoidea) and the zonation of the Fredericksburg: Geol. Soc. America Mem. 100, 225 p.

Detrital Dolomite in Onondaga Limestone (Middle Devonian) of New York: Its Implications to the "Dolomite Question"[1]

R. C. LINDHOLM[2]
Washington, D. C. 20006

Abstract Dolomite occurs in the matrix of the Onondaga Limestone (Middle Devonian) in New York as scattered grains ranging in size from 4 to 150 μ. Detrital quartz is associated with the dolomite. Study of etched and stained thin sections shows a correlation in grain size between the dolomite and quartz. Limited data show a correlation in grain size among dolomite, quartz, and detrital calcite (silt to fine sand) matrix. In addition, there is a correlation in abundance between dolomite and quartz, where high dolomite values are present with high quartz values.

These data suggest that dolomite in the Onondaga is detrital. Source of the dolomite is uncertain, but reworked penecontemporaneous supratidal sediments and older (e.g., Silurian) dolomite are suggested possibilities. Wind is a likely mechanism for transport of the detritus.

Deposition of detrital dolomite followed by later diagenetic overgrowth on the detrital nuclei is suggested as a mechanism for "dolomitization." This process is compatible with three phenomena observed in dolomitic rocks: (1) association of insoluble detritus with dolomite, (2) presence of dolomite in fine-grained limestone, (3) fine-grained texture of dolomite interpreted as "primary" and coarse-grained texture of dolomite interpreted as "replacement." Two models for the origin of dolomitic rocks are proposed.

INTRODUCTION

This paper is part of a general petrologic study of the Onondaga Limestone (Lindholm, 1967). The Onondaga included in the study crops out for 270 mi between Buffalo and Albany, New York, with the best exposures in quarries. During the summers of 1964 and 1965, 29 outcrops were studied.

STRATIGRAPHY OF ONONDAGA LIMESTONE

Before 1964, the most comprehensive study of Onondaga stratigraphy was that done by

[1] Manuscript received, April 3, 1968; accepted, June 12, 1968.
Read at the SEPM Ancient Carbonates Session, AAPG-SEPM annual meeting, Dallas, Texas, April 14–16, 1969.

[2] Assistant professor of geology, The George Washington University. Field work for this study was made possible by a Sigma Xi Grant-in-Aid of Research and by funds supplied by the New York State Museum and Science Service. The writer is indebted to many persons for suggestions and helpful criticism during the course of work and preparation of the manuscript, especially W. A. Oliver, who suggested the problem, F. J. Pettijohn (supervisor), R. N. Ginsburg, and J. W. Pierce.

Oliver (1954, 1956a,b, 1960). He divided the formation into four members, from oldest to youngest, the Edgecliff, Nedrow, Moorehouse, and Seneca. Members are subdivided into 12 zones (A–L, from oldest to youngest). The subdivisions are based mainly on paleontology, although bedding and gross lithology also are considered.

Lithofacies are distinguished by the proportion of allochems (almost entirely fossil debris) and fine-grained carbonate matrix (plus sparry calcite cement; Lindholm, 1967). Carbonate rock classification follows the scheme proposed by Folk (1959, 1962). Rocks with the highest allochem content are most abundant in eastern (Albany area) and western (Buffalo area) New York, as well as in the lowermost beds of the Onondaga throughout the area (Fig. 1). The Onondaga in central New York (Syracuse) is characterized by less abundant fossil debris.

DOLOMITE

Techniques.—One quarter of each thin section was stained and one quarter was etched to facilitate study of noncalcareous materials. Dolomite and quartz contents were determined by point counting on the etched part of each slide. To check the results, x-ray analyses were made on nine samples. Dolomite percentages were obtained by comparison of x-ray intensities for the 3.30 Å calcite peak and 2.88 Å dolomite peak (Tennant and Berger, 1957). These data agreed with dolomite content determined by point counting.

Grain-size distribution of fine-grained calcite, as well as dolomite and quartz, was determined by point counting stained and etched thin sections under the highest power objective available. The longest dimension of each grain was used in measuring grain size.

Morphology.—Dolomite grains are generally subhedral to euhedral rhombohedrons; anhedral grains constitute less than 20 percent of the total dolomite. A few grains are polycrystalline. Many rhombs contain dark, cloudy interiors; less commonly, the interior contains many opaque inclusions (pyrite?), which range

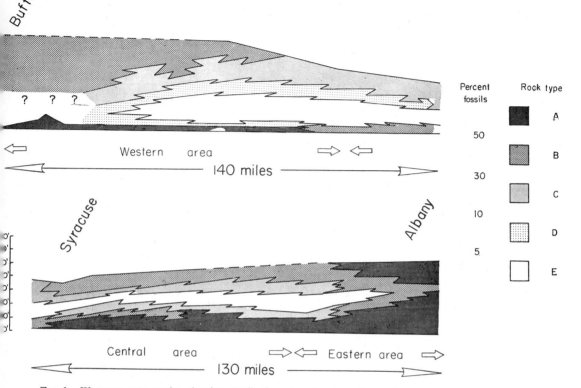

Fig. 1.—West-east cross section showing distribution of carbonate rock type in Onondaga Limestone. Fossil content is indicated as less than 5 percent, 5–10 percent, 10–30 percent, 30–50 percent, and greater than 50 percent.

in size from 1 to 5μ. In most grains the shape of the interior shows no crystal form, suggesting overgrowth on a round to subround core.

Composition.—X-ray analyses of the dolomite from the Onondaga show the principal peak at 2.90 Å. This value indicates that the dolomite has a "non-ideal" crystal structure, and contains between 3 and 6 percent excess $CaCO_3$ (Goldsmith and Graf, 1958).

Distribution.—Dolomite is present in most samples as small grains scattered throughout the limestone matrix. Dolomite replacement of fossils was not observed, although dolomite grains were found in spar-filled cavities of skeletal debris. The cavities are commonly primary features of the fragment or shell, e.g., the cavity between brachiopod valves, or the axial canal in crinoid columnals. Less commonly, the cavities are the result of postmortem boring.

Dolomite is most abundant (5–20 percent of total rock) in limestone where allochems are subordinate to fine-grained calcite matrix. Such rocks generally are present in central New York (Fig. 2). Dolomite and detrital quartz are absent in biosparites.

Grain size and abundance.—There are significant relations between size and abundance of dolomite grains and size and abundance of detrital quartz grains. Maximum size of both dolomite and quartz grains was determined in 70 thin sections (Fig. 3). The data suggest that dolomite and quartz grains present together are approximately the same size (correlation coefficient = +0.76).

Comparable data were obtained from grain-size distribution in two calcisiltites (fine-grained carbonate, 4–62 μ), one coarse grained and the other very fine grained (Fig. 4). All three constituents (calcite, dolomite, and quartz) are much finer in the very fine-grained calcisiltite (Fig. 4B) than in the coarse-grained calcisiltite (Fig. 4A). Correspondence between calcite grain size and grain size of the other two components is significant because the calcite matrix in the Onondaga is considered to be calcareous silt, formed by comminution of skeletal material (Lindholm, 1967, p. 54–59). The samples described in Figure 4 were chosen to show the maximum contrast present in the Onondaga.

In addition to the grain-size relation, there is a correlation in relative abundance between dolomite and detrital quartz (correlation coefficient = +0.66). High dolomite values are present with high quartz values, as shown in Figure 5.

ORIGIN OF DOLOMITE

Scattered dolomite rhombs in other formations are attributed to preferential growth of dolomite in $CaCO_3$ mud (Bergenback and Terriere, 1953; Bluck, 1965; Lucia, 1962; Murray, 1960; Murray and Lucia, 1967). Correlation in abundance between quartz silt and dolomite (Fig. 5), and the association of dolomite, quartz silt, and calcareous silt of approximately the same size (Figs. 3, 4) suggest that dolomite rhombs in the Onondaga are detrital. Detrital dolomite also is known from other ancient (Amsbury, 1962; Bluck, 1965; Sabins, 1962) and recent (Illing et al., 1965; Sugden, 1963) carbonates.

Source of the dolomite in the Onondaga is unknown. It might have been penecontemporaneous supratidal sediments which subsequently were eroded. The supratidal area is the only normal marine environment in which Holocene dolomite is known to form.

The source also might have been older dolomite exposed to erosion during Onondaga deposition. Silurian dolomite underlies the Onondaga in western New York and may have been exposed in land areas adjacent to the Onondaga sea.

The mechanism of transport of detrital dolomite found in the Onondaga is purely speculative. Transport by water currents is one obvious possibility and cannot be ruled out.

Windborne dust is an alternative, both for the dolomite and the quartz. Studies of continental deposits (Arrhenius, 1963; Scheidig, 1934) known to be of eolian origin show that the material is composed largely of grains in the 20–40 μ range. Sediments of that size are blown at least 50 mi into the Pacific Ocean from the deserts of northern Mexico (Bonatti and Arrhenius, 1965). Eolian sediment from the Sahara is present 1,000 mi off the west coast of Africa; at approximately 200 mi, nearly 60 percent of the quartz in some deep-sea deposits has been identified as of desert origin (Barth et al., 1939). Samples collected by ships in the same area demonstrate eolian transport in ex-

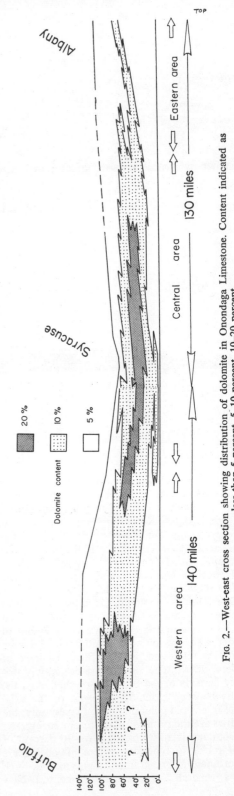

FIG. 2.—West-east cross section showing distribution of dolomite in Onondaga Limestone. Content indicated as less than 5 percent, 5–10 percent, 10–20 percent.

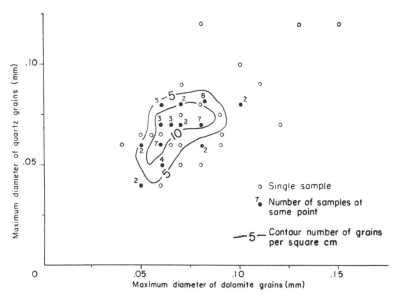

Fig. 3.—Maximum size of quartz grains plotted against maximum size of dolomite grains. Based on 70 thin sections of Onondaga Limestone. Correlation coefficient = +0.76; for N = 70, correlation is significant at the 1-percent level (Dixon and Massey, 1957, p. 468, Table A-30a).

cess of 1,000 mi (Kuenen, 1950, p. 214; Schott, 1942). Airborne sediment derived from the Sahara and containing dolomite silt has been collected at Barbados (Delany et al., 1967). Eolian dust collected on ships in the Red Sea and the Persian Gulf has a grain-size range from 5 to 100 μ, with an average size of 24 μ (Emery, 1956, p. 2367).

Thus, sediment of approximately the same size as the dolomite and quartz in the Onondaga can be transported great distances as wind-blown dust. The argument for this means of transport is strengthened by the association with the Tioga Bentonite, a very widespread unit in the Onondaga which is obviously a wind-transported deposit. In addition, volcanic biotite is present in numerous samples of the fine-grained part of the Onondaga.

IMPLICATIONS TO DOLOMITIZATION PROBLEM

Deposition of detrital[3] dolomite, followed by diagenetic overgrowth on the detrital nuclei, is proposed as a possible mechanism of "dolomiti-

[3] Detrital, in this case, means that the dolomite was eroded and transported. Erosion from older dolomitic rocks is not implied, although it is possible. Penecontemporaneous diagenetic dolomite, subjected to erosion and transportation, is also considered to be detrital.

zation." The process is compatible with several geologic phenomena.

First, large amounts of insoluble residues of detrital origin are associated with abundant dolomite in limestone (Lesley, 1879; Fairbridge, 1957, p. 154–156) and recent carbonate sediments (Taft and Harbaugh, 1964, p. 126–127). This relation would be expected if detrital dolomite was introduced concurrently with detrital quartz into the calcareous sediment.

Second is the association of dolomite with fine-grained limestone and its scarcity in current-deposited carbonate sand (Murray and Lucia, 1967). Detrital dolomite silt would be washed out of current-deposited sand, and concentrated in the $CaCO_3$ mud.

Third, the kinetics needed to initiate crystallization of dolomite are more rigorous than those required to continue growth on previously formed nuclei. Supratidal flats (Deffeyes et al., 1965; Illing et al., 1965; Shinn, 1967; Shinn et al., 1965) and saline lakes (Alderman, 1959; Alderman and Skinner, 1957; Alderman and Von Der Borch, 1960, 1961; Skinner, 1963) are the only modern environments where conditions favor initiation of dolomite crystallization. If supratidal dolomites were reworked into intertidal and subtidal carbonate sediments, the stage would be set for later dolomitization of the entire rock. Later dolomitiza-

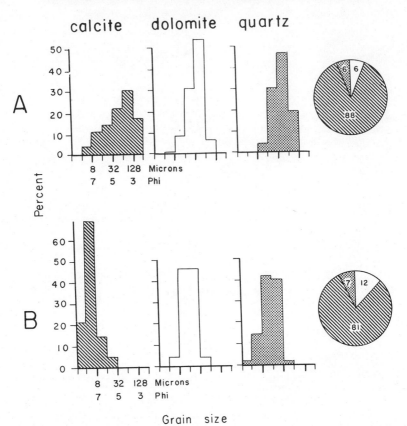

Fig. 4.—Grain-size distribution of calcite, dolomite, and quartz in coarse-grained calcisiltite (A) and very fine-grained dolomite and quartz are present in fine-grained calcisiltite, and coarse-grained dolomite and quartz are present in coarse-grained calcisiltite. Percentage of calcite, dolomite, and quartz is shown by pie diagram.

A. Coarse-grained calcisiltite, Moorehouse Member, coral facies (western area).

	M_z	S_I	S_G
calcite	4.4 ϕ (0.048 mm)	1.44 ϕ	1.22 ϕ
dolomite	4.9 ϕ (0.032 mm)	0.73 ϕ	0.75 ϕ
quartz	4.7 ϕ (0.038 mm)	0.70 ϕ	0.75 ϕ

B. Very fine-grained calcisiltite, Moorehouse Member (central area).

	M_z	S_I	S_G
calcite	7.6 ϕ (0.005 mm)	0.75 ϕ	0.76 ϕ
dolomite	6.0 ϕ (0.016 mm)	0.58 ϕ	0.58 ϕ
quartz	6.2 ϕ (0.014 mm)	0.70 ϕ	0.64 ϕ

tion (at any time) could proceed under conditions much less restrictive than those needed to nucleate dolomite. If, in a limestone containing 4 percent detrital dolomite, the grain size of the dolomite were increased by a factor of 3 (e.g., from 20 to 60μ) by diagenetic growth, the final rock would be 100 percent dolomite.

Ancient dolomite interpreted as being "primary" or detrital is fine grained, as is recent dolomite; ancient dolomite showing clear indications of replacement is coarse grained (Krynine, 1957). The proposed mechanism of dolomitization by initial deposition of detrital dolomite followed by later diagenetic growth, fits these observations. If detrital dolomite (silt size) were not modified by later growth, the evidence of detrital origin would be easily recognized. If, given a primary dolomite content of

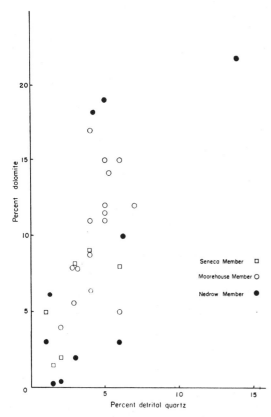

Fig. 5.—Percentage of dolomite plotted against percentage of detrital quartz. Based on point counts of 32 thin sections from Onondaga Limestone, central area. Correlation coefficient = +0.66; for N = 32, correlation is significant at 1 percent level (Dixon and Massey, 1957, p. 468, Table A-30a).

4 percent, dolomite overgrowths enlarged the original grain dimensions by a factor of 3 (volume increased by a factor of 27) and formed a rock composed entirely of dolomite, evidence of replacement would mask any evidence of a "primary" or detrital origin. In the latter example a combination of mechanisms, commonly considered to be mutually exclusive, is operative.

This mechanism of "dolomitization" is presented as one hypothesis that should be considered in determining the origin of a particular dolomite and is not intended to explain all dolomites. Certainly, some are formed by reflux of hypersaline brines through carbonate sediments that probably contained no detrital dolomite (Adams and Rhodes, 1960; Deffeyes et al., 1965). Others may be produced by late diagenetic replacement of dolomite-free (detrital) limestone. Before a detrital origin is proposed for any dolomite, the evidence should be well established. Evidence observed in the Onondaga includes correlation in abundance between dolomite and insoluble detritus and correlation in grain size among dolomite, detrital calcite, and insoluble detritus.

The role of detrital dolomite in carbonate sedimentation and diagenesis warrants attention in future work. Two models might be used in analyzing dolomitic rocks.

The first model involves introduction of detrital dolomite and quartz into calcareous sediments, without later diagenetic increase in dolomite grain size (Fig. 6A). The grain size of the detritus increases toward the source (to the right). If the source of detrital dolomite were penecontemporaneous supratidal flats, the size range would be small, from 1 to 5 μ (based on studies of Holocene dolomite; e.g., Deffeyes et al., 1965; Illing et al., 1965). A much broader range of grain sizes would be available if the source were older dolomite exposed on the land. In addition, the percentage of detritus (dolomite and quartz) in the sediment increases toward the source.[4] Values for size and content of detritus are only estimates, given to suggest order of magnitude. The abscissa (percentage values) indicates distance from the source area and may be in tens or hundreds of miles. Trends, not absolute values, are intended in these models.

The second model (Fig. 6B) involves introduction of detrital dolomite and quartz into calcareous sediment, followed by diagenetic growth of the dolomite, so that the calcareous sediment is totally "replaced" by dolomite. The grain size of the detrital dolomite is kept constant (10μ) to simplify the illustration (in fact it probably would be varied as in the first model). The grain size and amount of detrital quartz increase toward the source (to the right), but the grain size of the dolomite (diagenetic) increases away from the source. It is assumed that diagenetic dolomite crystallizes only on the detrital nuclei, and that no later diagenetic modifications occur. Grain size therefore is controlled by mutual interference of dolomite crystals as the rock nears total replacement, and hence is a function of the spacing of the detrital nuclei (i.e., percentage of nuclei in rock).

[4] Deep-sea samples from the Atlantic show an increase in dolomite toward the North African coast. This dolomite is derived from the desert regions of North Africa and transported westward by the trade winds (Delany et al., 1967).

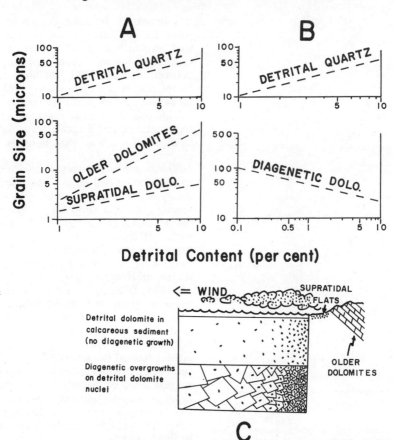

Fig. 6.—Models for dolomitic rocks.
  A. Grain size and abundance of detrital dolomite and quartz in calcareous sediment. Grain size of dolomite would depend on nature of source. If source were supratidal, variation in grain size would be small (1–5 μ); if source were older dolomite, exposed on land, expected variation would be much wider. Assumed source to right.
  B. Grain size and abundance of detrital quartz and dolomite after diagenetic crystallization on detrital nuclei has completely "dolomitized" calcareous sediment (or rock). Grain size of diagenetic dolomite plotted against percentage of detrital dolomite (nuclei for diagenetic dolomite) in sediment. Grain size of detrital dolomite kept constant at 10 μ. Assumed source to right.
  C. Schematic summary of *A* and *B*, showing abundance of detrital dolomite and grain size of "diagenetic dolomite" resulting from crystallization on detrital nuclei.

The models are presented as an aid to stratigraphic interpretation of dolomite and dolomitic limestone. Many variables would affect the models; for example, a source area with no silt-size quartz. Further, the models are entirely conceptual, based on observations of detrital dolomite in the Onondaga Limestone, as well as general characteristics of dolomitic rocks. The ideas warrant attention in future work on dolomites.

REFERENCES CITED

Adams, J. E., and M. L. Rhodes, 1960, Dolomitization by seepage refluxion: Am. Assoc. Petroleum Geologists Bull., v. 44, p. 1912–1920.

Alderman, A. R., 1959, Aspects of carbonate sedimentation: Geol. Soc. Australia Jour., v. 6, p. 1–10.
——— and H. C. W. Skinner, 1957, Dolomite sedimentation in the southeast of South Australia: Am. Jour. Sci., v. 255, p. 561–567.
——— and C. C. Von Der Borch, 1960, Occurrence of hydromagnesite in sediments in South Australia: Nature, v. 188, p. 931.
——— and ——— 1961, Occurrence of magnesite-dolomite sediments in South Australia: Nature, v. 192, p. 861.
Amsbury, D. L., 1962, Detrital dolomite in central Texas: Jour. Sed. Petrology, v. 32, p. 5–14.
Arrhenius, G., 1963, Pelagic sediments, *in* The sea, ideas and observations on progress in the study of the seas—V. 3, The earth beneath the sea: New York, Interscience, p. 655–727.
Barth, T. F. W., C. W. Correns, and P. Eskola, 1939,

Die Entstehung der Gestein: Berlin, Springer, 422 p.
Bergenback, R. E., and R. T. Terriere, 1953, Petrography and petrology of Scurry reef, Scurry County, Texas: Am. Assoc. Petroleum Geologists Bull., v. 37, p. 1014–1029.
Bluck, B. J., 1965, Sedimentation of Middle Devonian carbonates, southeastern Indiana: Jour. Sed. Petrology, v. 35, p. 656–682.
Bonatti, E., and G. Arrhenius, 1965, Eolian sedimentation in the Pacific off northern Mexico: Marine Geology, v. 3, p. 337–348.
Deffeyes, K. S., F. J. Lucia, and P. K. Weyl, 1965, Dolomitization of recent and Plio-Pleistocene sediments by marine evaporite waters on Bonaire, Netherlands Antilles, in Dolomitization and limestone diagenesis—a symposium: Soc. Econ. Paleontologists and Mineralogists Spec. Pub. No. 13, p. 71–88.
Delaney, A. C., et al., 1967, Airborne dust collected at Barbados: Geochim. et Cosmochim. Acta, v. 31, p. 885–909.
Dixon, W. J., and F. J. Massey, 1957, Introduction to statistical analysis: New York, McGraw-Hill, 488 p.
Emery, K. O., 1956, Sediments and water of Persian Gulf: Am. Assoc. Petroleum Geologists Bull., v. 40, p. 2354–2383.
Fairbridge, R. W., 1957, The dolomite question, in R. J. LeBlanc and J. G. Breeding, eds., Regional aspects of carbonate deposition—a symposium: Soc. Econ. Paleontologists and Mineralogists Spec. Pub. No. 5, p. 125–178.
Folk, R. L., 1959, Practical petrographic classification of limestones: Am. Assoc. Petroleum Geologists Bull., v. 43, p. 1–38.
―――― 1962, Spectral subdivision of limestone types, in Classification of carbonate rocks: Am. Assoc. Petroleum Geologists Mem. 1, p. 62–84.
Goldsmith, J. R., and D. L. Graf, 1958, Structural and compositional variations in some natural dolomites: Jour. Geology, v. 66, p. 678–693.
Illing, L. V., A. J. Wells, and J. C. M. Taylor, 1965, Penecontemporary dolomite, Persian Gulf, in Dolomitization and limestone diagenesis: Soc. Econ. Paleontologists and Mineralogists Spec. Pub. No. 13, p. 89–111.
Krynine, P. D., 1957, Dolomites (abs.): Geol. Soc. America Bull., v. 68, p. 1757.
Kuenen, P. H., 1950, Marine geology: New York, Wiley and Sons, 568 p.
Lesley, J. P., 1879, Notes on a series of analyses of dolomitic limestone rocks of Cumberland County, Pa., . . .: 2d Geol. Survey Pennsylvania (1876–78), MM, p. 311–362.
Lindholm, R. C., 1967, Petrology of the Onondaga Limestone (Middle Devonian), New York: Unpub. Ph.D. thesis, Johns Hopkins Univ., 188 p.
Lucia, F. J., 1962, Diagenesis of crinoidal sediment: Jour. Sed. Petrology, v. 32, p. 848–866.
Murray, R. C., 1960, Origin of porosity in carbonate rocks: Jour. Sed. Petrology, v. 30, p. 59–84.
―――― and F. J. Lucia, 1967, Cause and control of dolomite distribution by rock selectivity: Geol. Soc. America Bull., v. 78, p. 21–36.
Oliver, W. A., 1954, Stratigraphy of the Onondaga Limestone (Devonian) in central New York: Geol. Soc. America Bull., v. 65, p. 621–652.
―――― 1956a, Stratigraphy of the Onondaga Limestone in eastern New York: Geol. Soc. America Bull., v. 67, p. 1441–1474.
―――― 1956b, Biostromes and bioherms of the Onondaga Limestone in eastern New York: New York State Mus. and Sci. Service Circ. 45, 23 p.
―――― 1960, Coral faunas in the Onondaga Limestone of New York: U.S. Geol. Survey Prof. Paper 400-B, p. 172–174.
Sabins, F. F., 1962, Grains of detrital, secondary and primary dolomite from Cretaceous strata of the western interior: Geol. Soc. America Bull., v. 73, p. 1183–1196.
Scheidig, A., 1934, Der Löss und seine gastechnischen Eigenschaften: Dresden-Leipzig, Steinkoph, 223 p.
Schott, W., 1942, Geographie des Atlantischen Ozeans: Hamburg, C. Boysen, 438 p.
Shinn, E. A., 1964, Recent dolomite, Sugarloaf Key, in R. N. Ginsburg, comp., South Florida carbonate sediments: Geol. Soc. America Fld. Trip 1 Guidebook, p. 62–67.
―――― R. N. Ginsburg, and R. M. Lloyd, 1965, Recent supratidal dolomite from Andros Island, Bahamas, in Dolomitization and limestone diagenesis—a symposium: Soc. Econ. Paleontologists and Mineralogists Spec. Pub. No. 13, p. 71–88.
Skinner, H. C. W., 1963, Precipitation of calcian dolomites and magnesian calcites in the southeast of South Australia: Am. Jour. Sci., v. 261, p. 449–472.
Sugden, W., 1963, Some aspects of sedimentation in the Persian Gulf: Jour. Sed. Petrology, v. 33, p. 355–364.
Taft, W. H., and J. W. Harbaugh, 1964, Modern carbonate sediments of southern Florida, Bahamas, and Espiritu Santo Island, Baja California: a comparison of their mineralogy and chemistry: Stanford Univ. Pubs. Geol. Sci., v. 8, no. 2, 133 p.
Tennant, C. B., and R. W. Berger, 1957, X-ray determination of dolomite-calcite ratio of a carbonate rock: Am. Mineralogist, v. 42, nos. 1–2, p. 23–29.

Mississippian Dolomites from Lisburne Group, Killik River, Mount Bupto Region, Brooks Range, Alaska[1]

AUGUSTUS K. ARMSTRONG[2]
Menlo Park, California 94025

Abstract Allochthonous Osagian and Meramecian shelf carbonate rocks of the Lisburne Group that crop out from the Killik River west to Mount Bupto are extensively dolomitized and silicified. The carbonates were deposited as open-marine echinoderm-bryozoan wackestone, packstone, and grainstone with a thick sequence of interbedded carbonate mudstone in a shallow, restricted, marine to supratidal environment. Sequence of major diagenetic events in the dolomite is silicification of limestone to form chert nodules, followed by dolomitization of remaining limestone. Petrographic studies of the dolomites indicate that they have (1) dolomite rhombs with cloudy centers and clear rims, (2) voids formed by the partial dissolution of monocrystalline crinoid fragments, and (3) preserved organic skeletal voids. These features indicate dolomitization of limestone by hypersaline brines deficient in CO_2. Field and stratigraphic relations suggest reflux dolomitization by hypersaline brines formed in a supratidal environment.

The dolomite porosity, apparently not related to an ancient erosion surface or recent weathering processes, is thought to have been caused by the dolomitization process. Some pores contain anthroaxolite. Dolomite reservoir porosity may be present in the subsurface of the Brooks Range foothills and the North Slope from the Killik River to the Kiligwa River.

[1] Manuscript received, August 4, 1969; accepted, October 15, 1969. Publication authorized by the Director, U.S. Geological Survey.
[2] U.S. Geological Survey.

The field work, stratigraphic studies, and collections were made in the summers of 1964, when I was employed by Shell Oil Company, and 1968, when I was with the U.S. Geological Survey.

I thank Sigmund Snelson, the 1964 field party chief, and Irvin Tailleur, the 1968 field party chief, for their support of my field activities and their stimulating discussions on the structural and stratigraphic complexities of the region. Also I had the pleasure of spending a day with Perry O. Roehl, Union Research Center, on the Lisburne Ridge section. Field studies and discussions with Dr. Roehl strengthen the environmental interpretation of this report.

The collections of petrographic thin sections and fossils from the 1964 field season were made available to the U.S. Geological Survey by Shell Oil Company. Thin sections were cut by Robert Shely from the stratigraphic samples collected in 1968, and the photographs were made by Kenji Sakamoto, both of the U.S. Geological Survey.

I am grateful to my colleagues who helped in preparation of the manuscript and provided critical review: J. Thomas Dutro, Jr., and George Gryc of the U.S. Geological Survey, F. J. Lucia and Sigmund Snelson, Shell Development Company, and Perry O. Roehl, Union Research Center.

© 1970. The American Association of Petroleum Geologists. All rights reserved.

INTRODUCTION

The location of the study area is shown in Figure 1; Figure 2 shows the outcrop pattern of dolomites in the Lisburne Group of this study. The three measured sections discussed were measured with a Jacob's staff and tape, and lithologic samples of the carbonate rocks and chert were taken at 10-ft intervals.

Thin sections were cut from the samples and examined under the petrographic microscope. Identification of calcite was made by Alizarin-red staining techniques described by Friedman (1959). The carbonate-rock classification used is that of Dunham (1962).

The sedimentary features and structures used in this study to delineate intertidal and supratidal facies have been described in detail by Logan et al. (1964), Shinn et al. (1965), Illing et al. (1965), Roehl (1967), Wilson (1967), and Shinn (1968b). Detiled descriptions of the sedimentary features and how they are formed are not given herein.

The regional geologic relations of the Lisburne Ridge and Mount Bupto sections are shown on the large-scale geologic and outcrop maps of Tailleur et al. (1966). Discussion and description of the regional geology and structure are given by Snelson and Tailleur (1968) and Tailleur and Snelson (1968).

Porosity and petrographic studies on the Mississippian carbonate rocks of the Brooks Range have been reported by Krynine et al. (1950) for the dolomite and by Krynine and Folk (1950) for the limestone of the Lisburne Group, Endicott Mountains.

TECHNIQUES OF INTERPRETING ORIGINAL DEPOSITIONAL TEXTURE

Observations of replacement chert.—When chert replaces calcium carbonate sediments before dolomitization, the sedimentary fabric of the limestone may be preserved in the chert. Murray and Lucia (1967, p. 26) found that in the Mississippian Turner Valley Formation, Alberta, Canada, "the fabric of the original limestone can commonly be recognized more easily in replacement chert nodules than in the host dolomites." The technique of using thin sec-

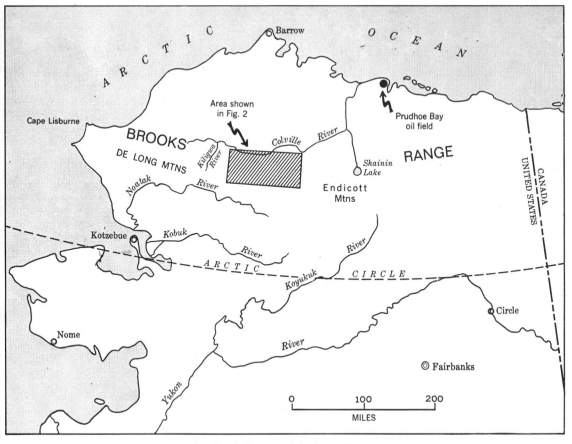

Fig. 1.—Index map of northern Alaska.

tions of chert nodules to determine the texture of the predolomite carbonate has been used in the study of dolomitized Mississippian carbonate rocks in Canada by Illing (1959, p. 43) and in New Mexico by Armstrong (1967, p. 10).

Thin-section studies of the chert indicate the sequence of major diagenetic events which resulted in the formation of the dolomite. The depositional fabric of the limestone is preserved in the chert, and fossil fragments (*Solenopora* algae, bryozoans, echinoderms, endothyrids) and sedimentary structures (*e.g.*, stromatolites and birdseye structures) are clearly discernible. In most samples the preservation of the detrital organic remains and microscopic sedimentary structures in the chert indicates that the chert directly replaced the limestone before dolomitization. Within some of the chert (Fig. 3B) there are floating rhombs of dolomite which range in size from 5 to 200 μ and which have cloudy centers and clear rims. Typically, these rhombs are corroded and have embayments of chert. The chert surrounding the rhombs retains the sedimentary fabric of the limestone. This feature suggests that the dolomitization process may have been initiated in some beds before the formation of chert. Some chert thin sections may show irregular dolomite rhombs (0.5–2 mm), which represent dolomite pseudomorphs of crinoid fragments, and associated calcite rim-cement overgrowths. These rhombs may have been large single crystals of calcite that resisted chertification but later were replaced by a single crystal of dolomite during dolomitization of the limestone. Within these dolomite pseudomorphs the relict outline of the crinoid fragment and the clearer rim cement are discernible.

The sequence of major diagenetic events is (1) deposition of limestone and rim cementation, (2) formation of chert nodules and beds

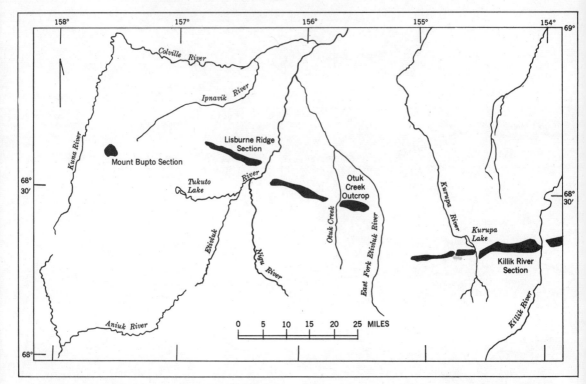

Fig. 2.—Generalized locations of three sections of Lisburne Group discussed herein. Solid black represents outcrop areas of Lisburne dolomite with associated shallow-water and supratidal sedimentary structures. Index map with outcrop pattern is modified from Lathram (1965).

by the replacement of limestone, generally completed before the onset of dolomitization, and (3) dolomitization of the remaining limestone.

Observations of relict fabrics in dolomite.—Lucia (1962, p. 850) stated that the most obvious early diagenetic process in grain-supported crinoidal limestone is the formation of an optically continuous calcite overgrowth on crinoid fragments. Bathurst (1958, p. 24) has shown that the overgrowth can be formed by filling pore space or by replacing carbonate mud surrounding the crinoid fragment. He named the pore-filling overgrowth "rim cement" and the carbonate mud replacement "syntaxial rims." Lucia's (1962, p. 850) studies indicate that more than 90 percent of sparry calcite overgrowth is pore-filling rim cement.

Dolomite (Lucia, 1962, p. 855) tends pseudomorphically to replace, in optical continuity, crinoid fragments and rim cement. A grain-supported encrinite composed of crinoid fragments 0.2 mm and larger, cemented by sparry calcite rim cement, upon dolomitization results in a dolomite composed of 0.2-mm and larger interlocking anhedral crystals.

Lucia (1962, p. 856) and Murray and Lucia (1967, p. 26) have shown that in the dolomitization of a crinoid wackestone and packstone, the crinoid fragments are the last to be dolomitized and either are replaced as a single-crystal dolomite pseudomorph or are dissolved to form molds. The distribution of these molds and pseudomorphs in completely dolomitized rocks provides information concerning the original distribution of the sand-size and larger particles in the original sediment. If the molds and pseudomorphs are not spaced closely enough to form a supporting framework, some material finer than sand size must have supported them. The nature of this finer supporting sediment is difficult to determine, and the sediment is called carbonate mud for lack of more definitive evidence (Murray and Lucia, 1967, p. 26).

Saccharoidal dolomite, composed of randomly oriented, euhedral rhombs (20–100 μ), is interpreted to have been formed by the replacement of carbonate mudstone. This in-

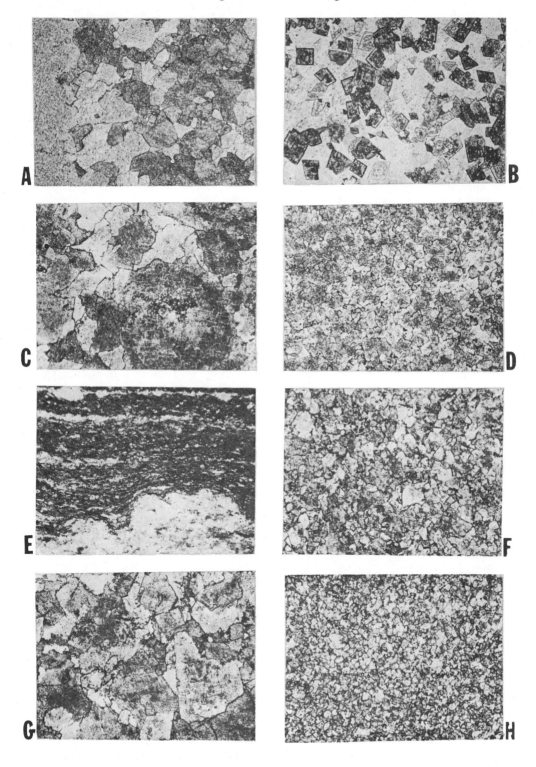

terpretation is supported by thin-section studies of the fabrics preserved in the chert nodules associated with the saccharoidal dolomite.

STRATIGRAPHIC SECTIONS

Mount Bupto section.—The section was measured from the top of Mount Bupto, down the west-central side of the hill into a small valley, S¼, Sec. 4, T11S, R24W (Figs. 4, 5).

The folded and somewhat anticlinally bent beds which form the south flank and the top of Mount Bupto are composed of chert. These massive chert beds are about 100 ft thick and are exposed best along the creek bed on the southwest flank of Mount Bupto. Hand specimens of the chert from the stratigraphically highest beds are light gray to gray and have fine (millimeter) banding, algal mats with antigravitational structures, imbricated chips, and burrowlike structures cutting the laminations. Thin-section studies of the chert show it to be formed by 10–40-μ-size areas of chalcedonic silica, and some samples preserve the moundlike lamination characteristic of stromatolites as defined by Logan et al. (1964, p. 69). The chert beds are believed to be the result of silicification of stromatolitic carbonate mudstone deposited in an intertidal to supratidal environment. Thin sections of the chert from the lower part of the massive cliff-forming unit show well-preserved silicified relict texture of ostracode-bryozoan-echinoderm wackestone indicating that the beds are silicified shallow-marine carbonates.

The section from 110 to 650 ft below the top can be divided into several lithic units. The divisions are based primarily on the percentage of chert and dolomite within segments of the section, as shown graphically in Figure 4 and in the photograph of the outcrop in Figure 6. For simplicity of description, this 540-ft-thick sequence is treated as a single unit.

The dolomite in this part of the section is light gray to gray, has porosity, and is composed of subhedral to anhedral, 20–400-μ dolomite crystals. The dolomite in hand specimens and in thin sections has abundant dolomite pseudomorphs of crinoid fragments (Fig. 3A). Calcite is found as scattered (2–50 μ) corroded crystals within some of the larger dolomite rhombs.

The chert is light gray to dark gray, in small nodules to massive 10-ft-thick beds. In thin sections the chert is composed of 5–100-μ areas of chalcedonic silica, with most crystals in the 10–25-μ size range. Relict carbonate fabric is apparent in most thin sections and indicates that the chert replaced marine echinoderm-bryozoan wackestone and packstone. Some thin sections of the chert show abundant but isolated corroded euhedral rhombs of dolomite (Fig. 3B).

A small dolomite outcrop above the talus slope from 780 to 830 ft stratigraphically below the top of the section is similar to the overlying part of the section except that the chert consists of black nodules 2–3 in. thick and 12–20 in. long (Fig. 3C).

A dolomite outcrop from 1,000 to 1,030 ft below the top is composed of fine-grained dolomite in 1-in.- to 2-ft-thick beds with thin, 1-in.- to 5-ft-long lenses of dark-gray chert.

A talus slope is present from 1,030 to 1,100 ft. From 1,100 to 1,170 ft below the top of the

---

FIG. 3.—Photomicrographs of carbonate rocks ($\times$24, plane polarized light).

A. Dolomite, 250 ft from top of Mount Bupto section, composed of large subhedral dolomite crystals. Large dolomite crystal at left is pseudomorph of crinoid fragment that has smaller externally impinged dolomite crystals.
B. Chert with floating dolomite rhombs, 410 ft from top of Mount Bupto section. Some dolomite rhombs have cloudy centers and clear rims and embayments of chert.
C. Dolomite, 810 ft from top of Mount Bupto section. Dolomite crystals are crinoid fragments and rim-cement pseudomorphs. Note ghost outline of crinoid columnal and its rim cement in bottom center of photograph.
D. Saccharoidal dolomite, 1 ft above base of Lisburne Ridge section.
E. Chert, 1 ft above base of Lisburne Ridge section. Thin section of chert which has algal mat laminated sedimentary structure. Note preservation of laminations and fine details of structure.
F. Dolomite, 580 ft above base of Lisburne Ridge section. Larger crystals are dolomite pseudomorphs of crinoid fragment.
G. Dolomite, 875 ft above base of Lisburne Ridge section. Dolomite pseudomorphs of crinoid fragment and rim cement. Note visible porosity.
H. Saccharoidal dolomite, 1,170 ft above base of Killik River section. At higher magnification, dolomite rhombs show cloudy centers and clear rims.

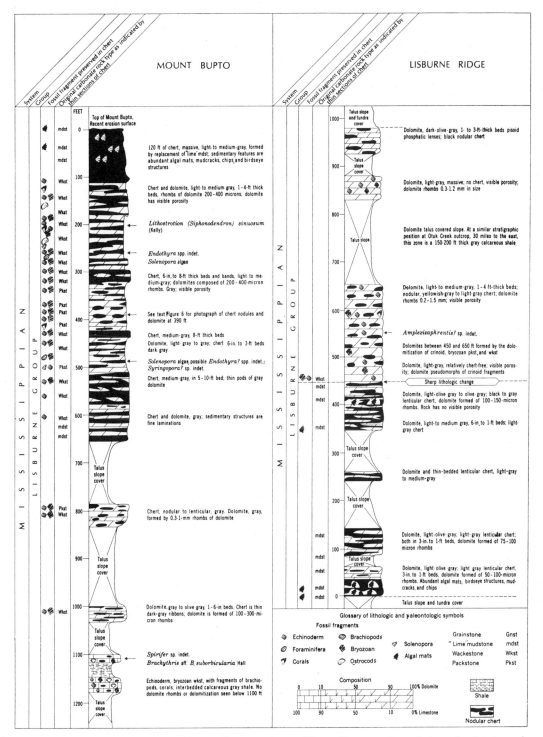

FIG. 4.—Stratigraphic sections of Lisburne Group outcrops. Mount Bupto section; photograph of outcrop is shown in Figure 5; Lisburne Ridge section; photograph of outcrop is shown in Figure 7.

ceous, brown-gray wackestone (limestone) which in thin section shows no evidence of dolomitization. The wackestone beds contain abundant fragments of bryozoans, echinoderms, brachiopods, corals, and ostracods. Below 1,170 ft are tundra cover and rock rubble. The relation of the Lisburne Group shale and its interbedded wackestone to the overlying dolomite is well displayed on the hill on the opposite side of the creek and in the creek bottom from 1,000 to 2,000 ft west of the base of the Mount Bupto measured section.

Lisburne Ridge section.—The Lisburne Ridge section was measured on the east side of one of a series of east-west-aligned hills that form Lisburne Ridge (SE¼, Sec. 32, T9S, R20W). The location of the measured section is shown in Figures 4 and 7.

Lisburne Ridge is composed of the resistant carbonate rocks and chert of the Lisburne Group. These rocks are involved in a complex structure that appears to be part of a rootless, imbricated and folded thrust sheet which overlies strata of Mesozoic age (S. Snelson, oral commun., March 1969; Tailleur *et al.,* 1966). The north side of the hill, on which the section was measured, has a talus apron across the lower part of the section and across the thrust plane.

The outcrop from the bottom of the section to 30 ft is composed of olive-gray dolomite and

FIG. 6.—Dolomite beds and chert nodules 390 ft below top of Mount Bupto section.

light-gray chert. Sedimentary structures include abundant algal mats, lithic chips, birdseye structures, and burrow fillings (Fig. 3E). Thin sections of the dolomite show that it consists of anhedral dolomite crystals which range in size from 50 to 100 μ (Fig. 3D). The environment of deposition is interpreted as supratidal. A significant part of the section, from 30 to 350 ft stratigraphically above the base, is covered by talus (Fig. 4). The limited outcrops within this interval are gray to olive-gray dolomite with light-gray to gray, nodular and lenticular chert. Bedding surfaces are 1–6 in. apart, and the beds have wavy laminations. Other sedimentary structures include poorly defined stromatolite-like features and burrow fillings. The dolomite in thin section is made up of euhedral rhombs

FIG. 5.—Mount Bupto section showing location and footage number of measured section. Photograph was taken at ground level in deep valley on west side of mountain. View is toward east.

FIG. 7.—Photograph of Lisburne Ridge, taken from helicopter, showing location of measured section. View is toward west.

in the 75–100-μ size range (Fig. 3F). The chert consists of interlocking 5–40-μ areas of chalcedonic silica with a few isolated crystals of dolomite. The rocks within this unit are believed to have been formed by the silicification and dolomitization of carbonate mudstone deposited in an intertidal to shallow-marine environment.

The chert and dolomite beds between 400 and 450 ft are 6-in.- to 2-ft-thick, with the chert composing 50–60 percent of the unit. The chert and dolomite are light gray to gray. The dolomite is formed of rhombs which range in size from 100 to 150 μ. Sedimentary structures in the lower 10 ft of the unit include well-developed algal mats and birdseye structures; those in the upper part of the unit are poorly defined and consist of lamination and small-scale cut-and-fill structures.

A sharp lithologic change is present 450 ft above the base. The underlying dolomite heretofore described generally is somber gray, shows no porosity in hand specimens, is composed of dolomite rhombs ranging in size from 100–150 μ, and has abundant chert. The carbonate rocks overlying these beds are light-gray to gray dolomite composed of large rhombs, with visible hand-specimen porosity and a relict fabric formed by dolomite pseudomorphs of crinoid fragments. These rock types characterize the section to a stratigraphic level of 650 ft above the base. Within this interval, the size variation of dolomite crystals is considerable. Dolomitization of the carbonate rocks is virtually complete; only a few corroded calcite crystals (5–30 μ) may be found within larger dolomite rhombs. The relict fabric of the dolomite suggests that these rocks were deposited as echinoderm wackestone and packstone (Fig. 3G).

The south-facing slope at a stratigraphic level of about 650–830 ft above the base of the section is covered by dolomite talus (Figs. 4, 7). The Lisburne Group section approximately 30 mi east at Otuk Creek has, at a similar stratigraphic position above the stromatolitic chert and saccharoidal dolomite beds, a 150–200-ft-thick unit of gray calcareous shale. At Lisburne Ridge this shale unit may be present under the dolomite talus.

Dolomite outcrops from 830 to 880 ft are light gray, massively bedded, and relatively chert free. The dolomite is composed of 0.3- to 1.2-mm anhedral rhombs. Vuggy porosity is well developed. Relict fabric within the dolomite suggests that it was formed by dolomitization of pelletoid, echinoderm packstone and wackestone. The pores in the dolomite between 450 and 880 ft commonly contain organic matter which Harry A. Tourtelot (written commun., April 8, 1969) identified as anthroaxolite.

The youngest beds in the section are exposed from 930 to 980 ft above the base and are composed of dark-gray phosphatic dolomite with nodular dark-gray chert in beds 1–3 ft thick. Some beds have zones of concentrated phosphate pisoids. The dolomite consists of rhombs ranging in size from 100 to 200 μ. These beds appear to be in normal contact with the underlying gray coarse-grained dolomite. The top of the phosphatic dolomite strata is hidden by talus slopes and tundra cover.

Killik River section.—The Killik River section was measured on the south side of a round hill, on the west bank of the Killik River, at lat. 68°21′ N, long. 154°05′ W (Figs. 8, 9).

The contact of the Lisburne Group carbonate rocks with the underlying Mississippian Kayak Shale is obscured by a talus slope but is believed to be 300–500 ft stratigraphically below the base of the section. The top of the section is a fault surface.

The carbonate rocks in the lower 400 ft are characterized by wackestone, packstone, and scarce grainstone, composed of fairly to poorly sorted fragments of ostracods, bryozoans, and echinoderms. Some beds are partly dolomitized, and others, such as the beds between 180 and 200 ft, are extensively dolomitized. The chert at the 100–300-ft level is nodular. The shape of the nodules suggests that chert has selectively replaced the carbonate mud which filled burrows. The chert nodules have contours reminiscent of Shinn's (1968a, Pl. 109) casts of shrimp burrows taken from recent carbonate sediments of Florida and the Bahamas.

The wackestone and packstone beds in the lower 400 ft of the section contain carbonate mud, numerous large bryozoan fronds, and articulated echinoderm fragments, all of which, according to Murray and Lucia (1967, p. 34), are indicative and typical of noncurrent-deposited sediments.

The rock types and sedimentary structures between 430 and 540 ft above the base are perplexing. The massively bedded rock is gray to dark-gray siliceous dolomite and chert. Thin sections show the chert to be a silica replacement of carbonate mudstone and ostracod-echinoderm wackestone. Enclosed within the siliceous beds are flat-topped pods and lens-shaped bodies of coarse-grained, porous dolomite 2–12 ft deep and 5–60 ft wide. Hand-specimen and thin-section studies of the dolomite show that it contains abundant dolomite pseudomorphs of crinoid fragments. The dolomite appears to be a replacement of echinoderm packstone and grainstone.

My interpretation of these beds, based on their sedimentary structures and petrographic studies of the chert and dolomite, is that the coarse-grained dolomite was current-laid, grain-supported encrinite deposited in tidal channels cut into noncurrent deposits of carbonate mudstone and wackestone.

The carbonate rocks from 540 to 830 ft are generally almost completely dolomitized echinoderm wackestone and packstone. The beds from 745 to 755 ft are less dolomitized. These dolomitic echinoderm packstone strata contain a limited fauna of relatively well-preserved colonial rugose corals.

From 910 to 1,110 ft above the base are thin-bedded light-gray chert and gray dolomite beds. The dolomite is composed of euhedral, randomly oriented 50–125-μ rhombs. The sedimentary features in these beds are algal mats, mudcracks and chips, and burrows. Thin-section studies of the chert indicate that the original rock was carbonate mudstone. Beds within this interval are interpreted to have been deposited in a restricted-shallow marine to supratidal environment.

The dolomite and chert from 1,110 to 1,270 ft form a homogeneous unit that is characterized by thin-bedded gray to olive-gray dolomite and lenticular to nodular dark-gray chert; the latter forms about half of the unit. The dolomite is composed of euhedral, randomly oriented rhombs in the 40–100-μ size range (Fig. 3H). Thin sections of the chert indicate that the original rock ranged from a spicule carbonate mudstone to echinoderm-bryozoan-ostracod wackestone. Some beds and thin sections suggest algal mat structures. Environment of deposition for this unit was probably restricted-shallow-marine to intertidal.

The echinoderm-bryozoan carbonate wackestone and packstone from 1,370 to 1,400 ft above the base are in sharp contact with the underlying gray dolomite. These limestones, in

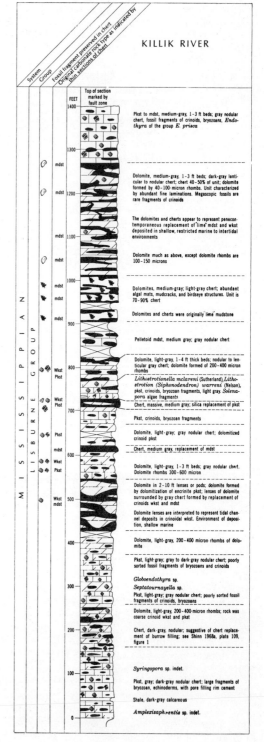

FIG. 8.—Stratigraphic section of Killik River section; photograph showing location of outcrop is shown in Figure 9.

FIG. 9.—Photograph of Killik River section showing location and footage numbers of measured section. View from helicopter west is over Killik River.

beds 1–3 ft thick, are gray, with 10–20 percent of the unit composed of gray nodular chert. Thin-section studies indicate that dolomitization of these carbonate rocks is uncommon and, where present, is restricted to a few dolomite rhombs replacing the carbonate mud particles between the fossil fragments. These beds are interpreted to represent relatively open-marine sedimentation.

A fault surface and structural complications at 1,400 ft above the base delineate the top of the section.

AGE AND REGIONAL CORRELATION

Within the three extensively dolomitized and silicified sections described in this study, few fossils can be identified at the generic or specific level. The oldest fauna found was collected 1,100–1,170 ft below the base of the Mount Bupto section. Fragments of brachiopods include *Spirifer* sp. indet., *Brachythyris* aff. *B. suborbicularis* (Hall), and *Camarotoechia* aff. *C. tuta* (Miller). Rugose corals are represented by *Amplexizaphrentis* sp. indet. This fauna suggests a late Osagian (Early Mississippian) or Meramecian (Late Mississippian) age).

I found, within the dolomite strata of the Mount Bupto section from 150 to 400 ft below the top, numerous poorly preserved dolomitized cerioid and fasciculate colonial corals; generic identification is impossible. Extensive examination of the thick chert nodules within this section did result in finding identifiable colonial corals 200 ft below the top of the section. The specimens are silicified *Lithostrotion (Siphonodendron) sinuosum* (Kelly). Thin-section studies of stratigraphically adjacent chert reveal a small fauna of large endothyrids. The coral species and endothyrids indicate a possibly Meramecian age for this part of the section and a possible correlation with the lower part of Bowsher and Dutro's (1957) Alapah Limestone at Shainin Lake, Endicott Mountains.

Megafossils or microfossils indicative of a biostratigraphic age were not found in the Lisburne Ridge section from 0 to 450 ft above its base. Fossils collected from the dolomite beds 450–650 ft above the base are dolomitized, and positive generic identification is impossible. Within these beds specimens of dolomitized cerioid and fasciculate colonial corals and solitary corals were collected. The latter appear to be *Amplexizaphrentis*? sp. indet. Also collected were external molds of brachiopod fragments, *Spirifer*? sp. indet., *Punctospirifer*? sp. indet., and productids. My impression is that the poorly preserved fossils are Late Mississippian. No recognizable fossils were found at stratigraphic levels above 650 ft.

The Killik River section is the most fossiliferous of the three sections studied. At 320 ft above the base, a poorly preserved microfauna contains *Globoendothyra* sp. indet. and *Septatournayella* sp. indet. This microfauna appears to be of Meramecian age. From 745 to 755 ft the coral fauna consists of *Syringopora* sp. indet., *Lithostrotionella mclareni* (Sutherland), *Lithostrotion (Siphonodendron) warreni* Nelson, and large solitary corals, possibly *Vesiculophyllum*? sp. indet. This coral fauna is of Meramecian age and possibly is equivalent to the lower part of the Alapah Limestone of Bowsher and Dutro (1957) at Shainin Lake. At 1,350 ft above the base of the section, a poorly preserved assemblage resembling *Endothyra* of the group *E. prisca* Rauzer-Chernoussova and Reitlinger was found. It is similar to, if not conspecific with, a fauna found in the upper part of the Alapah Limestone at Shainin Lake. This microfauna is of late Meramecian or early Chesteran age.

Figure 10 presents a tentative correlation of the three measured sections. The interpretation is based on the meager fossil data and the dis-

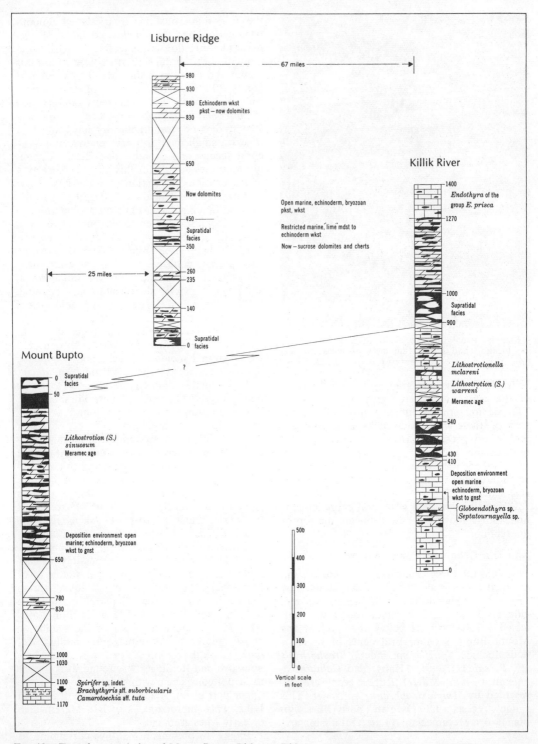

FIG. 10.—Tentative correlation of Mount Bupto, Lisburne Ridge, and Killik River sections of Lisburne Group. For explanation of lithologic symbols, see Figure 4.

tribution of carbonate facies. The concept of a major regional, shallow-water, restricted marine to supratidal facies development and a subsequent open-marine transgression with carbonate deposition is used in correlation of the sections.

More extensive fossil collections and detailed stratigraphic studies should be made at selected localities within the outcrop band shown in Figure 2. Detailed work is necessary to the understanding of the stratigraphic facies within these carbonate rocks and their correlation and facies relations with the Wachsmuth and Alapah Formations of the Endicott Mountains and Sable and Dutro's (1961) Mississippian Kogruk and Tupik Formations of the DeLong Mountains.

DOLOMITE DEVELOPMENT AND FACIES

Friedman and Sanders (1967, p. 334–338) believe all dolomites, whether syngenetic, diagenetic, or epigenetic, form from hypersaline brines. Deffeyes et al. (1965, p. 77–78) have demonstrated that modern hypersaline brines formed on supratidal flats contain 100 times as much magnesium ion as total CO_2 content and are, therefore, very carbonate poor. These brines can migrate downward (reflux dolomitization) into underlying limestone (p. 80–84).

The dolomite of this report probably was formed from hypersaline brines deficient in CO_2. The reasons for this interpretation are (1) the widespread presence of dolomite rhombs with cloudy centers and clear rims, (2) the preservation of intercrystalline porosity and the development of voids by the partial dissolution of monocrystalline crinoid fragments, and (3) the preservation of organic voids, as exemplified by the colonial coral.

The Killik River section shows a possible relation between supratidal sediments and dolomitization. The rocks between 900 and 1,250 ft above the base were deposited in environments that oscillated between shallow, restricted marine and supratidal. These rocks are dolomitized, and the beds beneath them to a level 100 ft above the base have been affected by varied degrees of dolomitization. The beds which originally contained the greatest percentage of carbonate mud are most extensively dolomitized, whereas beds that were relatively mud free and were composed of crinoid fragments with pore-filling rim cement are the least affected by dolomitization.

The youngest beds of the Lisburne Group in the Mount Bupto section contain sedimentary structures and a relict fabric which to me indicate that they were carbonate mudstone deposited in a supratidal environment. Lisburne Group carbonate rocks beneath these beds are dolomitized. The fossiliferous wackestone within the shale beneath the dolomite is not affected by dolomitization. The wackestone beds are separated from the dolomite by more than 50 ft of shale, which could have acted as a seal against downward-moving hypersaline brines, preventing dolomitization of the wackestone in the shale. Lucia (1968, p. 852) found that hypersaline brines which are dolomitizing modern limestones on Bonaire Island are stopped in their downward migration by shale layers, and the underlying limestone is not affected by dolomitization.

The Lisburne Ridge section is dolomite and chert. It contains, from 0 to 450 ft above its base, sedimentary structures and fabrics characteristic of shallow, restricted marine to supratidal environments. The carbonate rocks from 450 to 880 ft were originally crinoid-bryozoan-coral wackestone and packstone which are now porous dolomite. At the Lisburne Ridge and Otuk Creek outcrops, sedimentary structures or beds which indicate the development of a supratidal environment above the porous dolomite were not found. Possibly there are, or were, supratidal beds above the dolomitized crinoidal wackestone and packstone; they may be covered by talus slopes, obscured by structural complexities, or removed by Triassic or earlier erosion.

The formation of dolomite from calcite crinoidal-bryozoan wackestone by dolomitizing hypersaline brines derived from the supratidal environment does not explain the sequence of events in the silicified rocks which contain the stromatolites, lithic chips, and birdseye structures. These massively bedded cherts are believed to have been penecontemporary dolomites, formed in supratidal environments, similar to recent dolomites described by Shinn et al. (1965) from the Bahamas and by Illing et al. (1965) from the Persian Gulf.

F. J. Lucia (written commun., March 1969) stated that most penecontemporary dolomites are composed of fine-grained dolomite rhombs 20 μ or smaller. Supratidal dolomites are so finely crystalline that the sedimentary structures are well preserved, but chertification of these dolomites would retain well-preserved sedimentary textures.

Peterson and Von der Borch (1965, p. 1501–1503) recorded from the Coorong Lagoon of south Australia the deposition of inorganic chert and dolomite in an environment associated with carbonate edgewise conglomerate and slump structures. The means by which those carbonate rocks are being silicified may explain some of the silicification seen in the Lisburne Group chert that has supratidal sedimentary features.

POROSITY DEVELOPMENT

Within the Lisburne Group dolomite, porosity is vuggy and intercrystalline. The presence of dead oil in the present-day porosity suggests the development of porosity before exposure and weathering. The origin of porosity in dolomite has been studied extensively by Murray (1960), Weyl (1960), Lucia (1962), and Murray and Lucia (1967). These writers suggest that it is formed by either (1) leaching of the calcite from a partly dolomitized rock after dolomitization or (2) the dolomitization process itself. Murray (1960) and Lucia (1962) have shown that there is a strong relation between porosity development and percentage of dolomitization.

Calcite dissolution.—The carbonate strata on Lisburne Ridge and Mount Bupto are virtually completely dolomitized. Staining of thin sections with Alizarin-red solution reveals, in only a few thin sections, scattered small corroded rhombs (5–50μ) of calcite within large dolomite rhombs. The latter are interpreted to be dolomite pseudomorphs of crinoid fragments with residual calcite. The vugs within the dolomite are believed to have formed during the dolomitization in which no CO_2 was added to the system (Murray's 1960 local-source dolomitization).

Dolomite rhombs first replaced the carbonate mud between the fossil fragments. As dolomitization proceeded, larger fossil fragments of bryozoans and ostracods were dolomitized. This sequence of events is shown clearly in the incompletely dolomitized limestone in the lower part of the Killik River section. The monocrystalline crinoid ossicles resisted replacement and were the last of the calcite fossil fragments to be dolomitized. If the assumption is made that all the carbonate used in the growth of the dolomite is derived locally from the rocks being dolomitized, then the available CO_2 is insufficient for the growth of dolomite euhedra, and calcite dissolution must accompany growth of dolomite, thus forming and redistributing porosity (Murray, 1960, p. 73).

Because monocrystalline crinoid ossicles are the most resistant to dolomitization and are the last rock-forming element to be dolomitized, they are most subject to partial or complete dissolution to supply carbonate to adjacent growing dolomite crystals.

Thin sections and polished slabs from the Lisburne Ridge and Mount Bupto sections show that most of the vuggy porosity in the dolomite is the result of partial dissolution of crinoid fragments. The excellent porosity from 830 to 880 ft in the Lisburne Ridge section appears to have been formed in this way.

Intercrystalline porosity.—The carbonate strata in the lower 450 ft of the Lisburne Ridge section are saccharoidal dolomite. Thin-section studies of the chert associated with them indicate that the depositional fabric ranged from stromatolitic carbonate mudstone to crinoidal wackestone (Fig. 3D). The dolomite crystal size ranges from 40 to 150 μ. The rhombs are euhedral to subhedral and in thin section have cloudy centers and clear rims. The crystals have in part interlocking boundaries with adjacent rhombs and low visible intercrystalline porosity (Fig. 3D, F).

The Killik River section from 1,000 to 1,250 ft above its base contains a saccharoidal dolomite that is similar lithologically and petrographically to the saccharoidal dolomite found in the lower 450 ft of the Lisburne Ridge section (Fig. 10). Thin-section studies show that individual and isolated rhombs have been fractured or broken, a feature that suggests compaction of the dolomite during diagenesis. The meager voids between the rhombs are not filled with dolomite or calcite. As explained by Murray (1960, p. 72–74), the voids between rhombs suggests a local source for the CO_2 during dolomitization of the carbonate mud. The many dolomite rhombs with cloudy centers and clear rims further support a local source for the carbonate in these dolomites (Murray, 1964, p. 400–403, Figs. 7–9).

Preservation of organic voids.—The open-marine carbonate shelf facies of the Lisburne Group in the Brooks Range generally contains a relatively diversified coral fauna. Within the study area, dolomite that was originally open-marine limestone contains dolomitized and poorly preserved corals. The corallite walls are preserved by 100–200-μ dolomite rhombs, but the internal structures either are preserved very poorly or have been destroyed. The internal cavities of the corallites are still cavities, and dolomite rhombs project into the voids.

The tendency of dolomite to grow by replacement of existing carbonate and to avoid the growth of new crystals in preexisting void space has been interpreted (Murray, 1960, p. 73; 1964, p. 391) as an indication that the dolomite grew by utilizing a local source of carbonate ions. Weyl (1960) has shown that, if dolomitization is the result of waters relatively low in total CO_2 with respect to Ca and Mg, local utilization of carbonate in volume-for-volume replacement occurs. Growth of a replacement dolomite rhomb is accompanied by dissolution of calcium carbonate beyond the volume of the rhomb.

This kind of void preservation within corallites collected from the Lisburne Group dolomites strongly suggests dolomitization by Ca- and Mg-rich fluids poor in CO_2, with the dolomites formed from a local source carbonate.

References Cited

Armstrong, A. K., 1967, Biostratigraphy and carbonate facies of the Mississippian Arroyo Peñasco Formation, north-central New Mexico: New Mexico Bur. Mines and Mineral Resources Mem. 20, 79 p.

Bathurst, R. G. G., 1958, Diagenetic fabrics in some British Dinantian limestones: Liverpool and Manchester Geol. Jour., v. 2, pt. 1, p. 11–36.

Bowsher, A. L., and J. T. Dutro, Jr., 1957, The Paleozoic section in the Shainin Lake area, central Brooks Range, Alaska: U.S. Geol. Survey Prof. Paper 303-A, 39 p.

Deffeyes, K. S., F. J Lucia, and P. K. Weyl, 1965, Dolomitization of Recent and Plio-Pleistocene sediments by marine evaporite waters on Bonaire, Netherlands Antilles, in L. C. Pray and R. C. Murray, eds., Dolomitization and limestone diagenesis, a symposium: Soc. Econ. Paleontologists and Mineralogists Spec. Pub. 13, p. 71–88.

Dunham, R. J., 1962, Classification of carbonate rocks according to depositional texture, in Classification of carbonate rocks—a symposium: Am. Assoc. Petroleum Geologists Mem. 1, p. 108–121.

Friedman, G. M., 1959, Identification of carbonate minerals by staining methods: Jour. Sed. Petrology, v. 29, p. 87–97.

——— and J. E. Sanders, 1967, Origin and occurrence of dolostones, Chap. 7 in G. V. Chilingar, H. J. Bissell, and R. W. Fairbridge, eds., Carbonate rocks—origin, occurrence, and classification, 9A al Developments in sedimentology: New York, Elsevier Publishing Co., p. 267–348.

Illing, L. V., 1959, Deposition and diagenesis of some upper Paleozoic carbonate sediments in western Canada: 5th World Petroleum Cong. Proc., New York, sec. 1, p. 23–52.

——— A. J. Wells, and J. C. M. Taylor, 1965, Penecontemporary dolomite in the Persian Gulf, in L. C. Pray and R. C. Murray, eds., Dolomitization and limestone diagenesis, a symposium: Soc. Econ. Paleontologists and Mineralogists Spec. Pub. 13, p. 89–111.

Krynine, P. D., and R. L. Folk, 1950, Petrology of the Lisburne Limestone: U.S. Geol. Survey Open File Rept., Geol. Inv. Naval Petroleum Reserve No. 4, Alaska, Spec. Rept. 22, 25 p.

——— ——— and M. A. Rosenfeld, 1950, Porosity and petrography of Lisburne Limestone samples from the Kanayut, Nanushuk and Itkillik Lakes area; with a discussion of the distribution of porous zones in the Lisburne Limestone by A. L. Bowsher: U.S. Geol. Survey Open File Rept., Geol. Inv. Naval Petroleum Reserve No. 4, Spec. Rept. 17, 18 p.

Lathram, E. H., 1965, Preliminary geologic map of northern Alaska: U.S. Geol. Survey Open File Map, May 3, 1965, scale 1:1,000,000.

Logan, B. W., R. Rezak and R. N. Ginsburg, 1964, Classification and environmental significance of algal stromatolites: Jour. Geology, v. 72, no. 1, p. 68–83.

Lucia, F. J., 1962, Diagenesis of a crinoidal sediment: Jour. Sed. Petrology, v. 32, p. 848–865.

——— 1968, Recent sediments and diagenesis of south Bonaire, Netherlands Antilles: Jour. Sed. Petrology, v. 38, no. 3, p. 845–858.

Murray, R. C., 1960, Origin of porosity in carbonate rocks: Jour. Sed. Petrology, v. 30, no. 1, p. 59–84.

——— 1964, Preservation of primary structures and fabrics in dolomite, in J. Imbrie and N. D. Newell, eds., Approaches to paleoecology: New York, John Wiley & Sons, p. 388–403.

——— and F. J. Lucia, 1967, Cause and control of dolomite distribution by rock selectivity: Geol. Soc. America Bull., v. 78, p. 21–36.

Peterson, M. N. A., and C. C. von der Borch, 1965, Chert: modern inorganic deposition in a carbonate-precipitating locality: Science, v. 149, p. 1501–1503.

Roehl, P. O., 1967, Stony Mountain (Ordovician) and Interlake (Silurian) facies analogs of Recent low-energy marine and subaerial carbonates, Bahamas: Am. Assoc. Petroleum Geologists Bull., v. 51, no. 10, p. 1979–2032.

Sable, E. G., and J. T. Dutro, Jr., 1961, New Devonian and Mississippian formations in De Long Mountains, northern Alaska: Am. Assoc. Petroleum Geologists Bull., v. 45, no. 5, p. 585–593.

Shinn, E. A., 1968a, Burrowing in recent lime sediments of Florida and the Bahamas: Jour. Paleontology, v. 42, no. 4, p. 879–894.

——— 1968b, Practical significance of birdseye structures in carbonate rocks: Jour. Sed. Petrology, v. 38, no. 1, p. 215–223.

——— R. N. Ginsburg, and R. M. Lloyd, 1965, Recent supratidal dolomites from Andros Island, Bahamas, in L. C. Pray and R. C. Murray, eds., Dolomitization and limestone diagenesis, a symposium: Soc. Econ. Paleontologists and Mineralogists Spec. Pub. 13, p. 112–123.

Snelson, Sigmund, and I. L. Tailleur, 1968, Large-scale thrusting and migrating Cretaceous foredeeps in western Brooks Range and adjacent regions of northwestern Alaska (abs.): Am. Assoc. Petroleum Geologists Bull., v. 52, no. 3, p. 567.

Tailleur, I. L., and Sigmund Snelson, 1966, Large-scale flat thrusts in the Brooks Range orogen, northern Alaska (abs.): Geol. Soc. America Prog. Ann. Mtg., p. 217; 1968: Geol. Soc. America Spec. Paper 101, p. 217.

——— B. H. Kent, and H. N. Reiser, 1966, Outcrop geologic maps of the Nuka-Etivuluk region, northern Alaska: U.S. Geol. Survey Open File Maps, scale 1:63,360.

Weyl, P. K., 1960, Porosity through dolomitization: Conservation of mass requirements: Jour. Sed. Petrology, v. 30, no. 1, p. 85–90.

Wilson, J. L., 1967, Carbonate evaporite cycles in lower Duperow Formation of Williston basin: Bull. Canadian Petroleum Geology, v. 15, no. 3, p. 230–312.